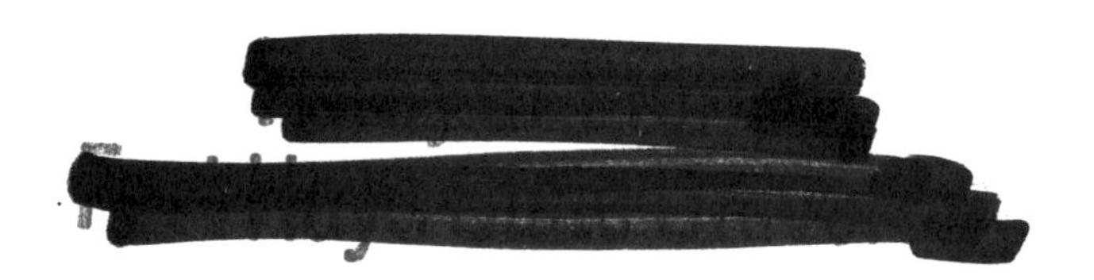

Fermionic Expressions for Minimal Model Virasoro Characters

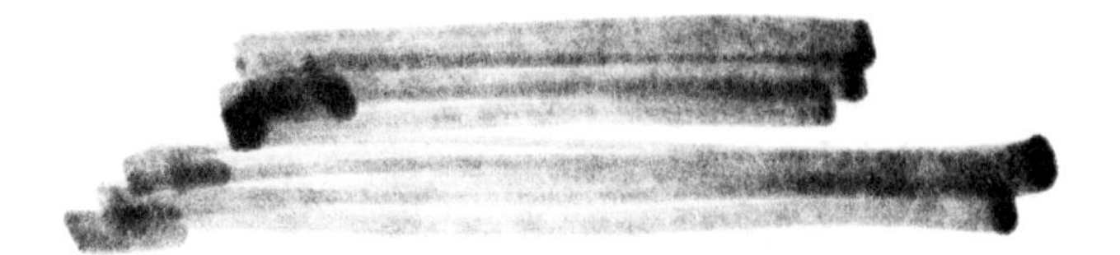

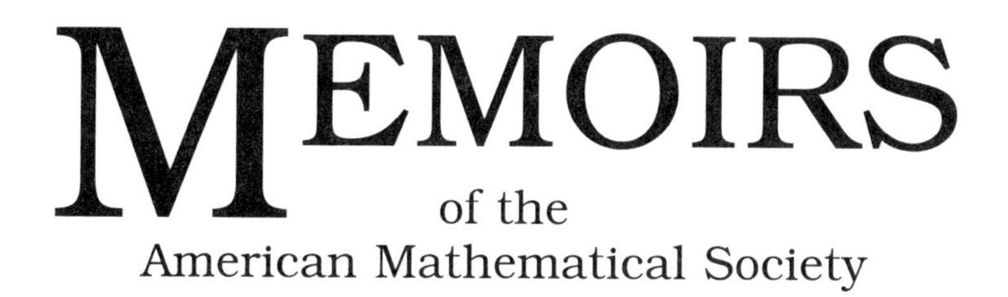

Number 827

Fermionic Expressions for Minimal Model Virasoro Characters

Trevor A. Welsh

May 2005 • Volume 175 • Number 827 (third of 4 numbers) • ISSN 0065-9266

American Mathematical Society
Providence, Rhode Island

2000 *Mathematics Subject Classification.*
Primary 82B23; Secondary 05A15, 05A19, 17B68, 81T40.

Library of Congress Cataloging-in-Publication Data

Welsh, Trevor A. (Trevor Alan), 1963–
Fermionic expressions for minimal model Virasoro characters / Trevor A. Welsh.
p. cm. — (Memoirs of the American Mathematical Society, ISSN 0065-9266 ; no. 827)
"Volume 175, number 827 (third of 4 numbers)."
Includes bibliographical references.
ISBN 0-8218-3656-0 (alk. paper)
1. Combinatorial enumeration problems. 2. Lie algebras. I. Title. II. Series.

QA3.A57 no. 827
[QA164.8]
510 s—dc22
[511′.62] 2005041982

Memoirs of the American Mathematical Society

This journal is devoted entirely to research in pure and applied mathematics.

Subscription information. The 2005 subscription begins with volume 173 and consists of six mailings, each containing one or more numbers. Subscription prices for 2005 are $606 list, $485 institutional member. A late charge of 10% of the subscription price will be imposed on orders received from nonmembers after January 1 of the subscription year. Subscribers outside the United States and India must pay a postage surcharge of $31; subscribers in India must pay a postage surcharge of $43. Expedited delivery to destinations in North America $35; elsewhere $130. Each number may be ordered separately; *please specify number* when ordering an individual number. For prices and titles of recently released numbers, see the New Publications sections of the *Notices of the American Mathematical Society.*

Back number information. For back issues see the *AMS Catalog of Publications.*

Subscriptions and orders should be addressed to the American Mathematical Society, P. O. Box 845904, Boston, MA 02284-5904, USA. *All orders must be accompanied by payment.* Other correspondence should be addressed to 201 Charles Street, Providence, RI 02904-2294, USA.

Memoirs of the American Mathematical Society is published bimonthly (each volume consisting usually of more than one number) by the American Mathematical Society at 201 Charles Street, Providence, RI 02904-2294, USA. Periodicals postage paid at Providence, RI. Postmaster: Send address changes to Memoirs, American Mathematical Society, 201 Charles Street, Providence, RI 02904-2294, USA.

This publication is indexed in *Science Citation Index*®, *SciSearch*®, *Research Alert*®, *CompuMath Citation Index*®, *Current Contents*®/*Physical, Chemical & Earth Sciences.*
Printed in the United States of America.

♾ The paper used in this book is acid-free and falls within the guidelines established to ensure permanence and durability.
Visit the AMS home page at `http://www.ams.org/`

10 9 8 7 6 5 4 3 2 1 10 09 08 07 06 05

Contents

Abstract

Fermionic expressions for all minimal model Virasoro characters $\chi^{p,p'}_{r,s}$ are stated and proved. Each such expression is a sum of terms of *fundamental fermionic form* type. In most cases, all these terms are written down using certain trees which are constructed for s and r from the Takahashi lengths and truncated Takahashi lengths associated with the continued fraction of p'/p. In the remaining cases, in addition to such terms, the fermionic expression for $\chi^{p,p'}_{r,s}$ contains a different character $\chi^{\hat{p},\hat{p}'}_{\hat{r},\hat{s}}$, and is thus recursive in nature.

Bosonic-fermionic q-series identities for all characters $\chi^{p,p'}_{r,s}$ result from equating these fermionic expressions with known bosonic expressions. In the cases for which $p = 2r$, $p = 3r$, $p' = 2s$ or $p' = 3s$, Rogers-Ramanujan type identities result from equating these fermionic expressions with known product expressions for $\chi^{p,p'}_{r,s}$.

The fermionic expressions are proved by first obtaining fermionic expressions for the generating functions $\chi^{p,p'}_{a,b,c}(L)$ of length L Forrester-Baxter paths, using various combinatorial transforms. In the $L \to \infty$ limit, the fermionic expressions for $\chi^{p,p'}_{r,s}$ emerge after mapping between the trees that are constructed for b and r from the Takahashi and truncated Takahashi lengths respectively.

Received by the editor March 12, 2003, and in revised form December 23, 2003.

2000 *Mathematics Subject Classification.* Primary 82B23; Secondary 05A15, 05A19, 17B68, 81T40.

Key words and phrases. Fermionic expressions, Virasoro characters, Rogers-Ramanujan identities, path generating functions.

Research supported by the Australian Research Council (ARC).

1. Prologue

1.1. Introduction

The rich mathematical structure of the two-dimensional conformal field theories of Belavin, Polyakov, and Zamolodchikov [**9, 18**] is afforded by the presence of infinite dimensional Lie algebras such as the Virasoro algebra in their symmetries. Indeed, the presence of these algebras ensures the solvability of the theories. In the conformal field theories known as the minimal models, which are denoted $\mathcal{M}^{p,p'}$ where $1 < p < p'$ with p and p' coprime, the spectra are expressible in terms of the Virasoro characters $\hat{\chi}^{p,p'}_{r,s}$ where $1 \leq r < p$ and $1 \leq s < p'$. The irreducible highest weight module corresponding to the character $\hat{\chi}^{p,p'}_{r,s}$ has central charge $c^{p,p'}$ and conformal dimension $\Delta^{p,p'}_{r,s}$ given by:

$$c^{p,p'} = 1 - \frac{6(p'-p)^2}{pp'}, \qquad \Delta^{p,p'}_{r,s} = \frac{(p'r-ps)^2-(p'-p)^2}{4pp'}. \tag{1.1}$$

In [**19, 36, 20**], it was shown that $\hat{\chi}^{p,p'}_{r,s} = q^{\Delta^{p,p'}_{r,s}} \chi^{p,p'}_{r,s}$, where the (normalised) character $\chi^{p,p'}_{r,s}$ is given by:

$$\chi^{p,p'}_{r,s} = \frac{1}{(q)_\infty} \sum_{\lambda=-\infty}^{\infty} \left(q^{\lambda^2 pp' + \lambda(p'r-ps)} - q^{(\lambda p + r)(\lambda p' + s)} \right), \tag{1.2}$$

and as usual, $(q)_\infty = \prod_{i=1}^\infty (1-q^i)$. Note that $\chi^{p,p'}_{r,s} = \chi^{p,p'}_{p-r,p'-s}$.

An expression such as (1.2) is known as bosonic because it arises naturally via the construction of a Fock space using bosonic generators. Submodules are factored out from the Fock space using an inclusion-exclusion procedure, thereby leading to an expression involving the difference between two constant-sign expressions.

However, there exist other expressions for $\chi^{p,p'}_{r,s}$ that provide an intrinsic physical interpretation of the states of the module in terms of quasiparticles. These expressions are known as fermionic expressions because the quasiparticles therein are forbidden to occupy identical states. Two of the simplest such expressions arise when $p' = 5$ and $p = 2$:

$$\chi^{2,5}_{1,2} = \sum_{n=0}^{\infty} \frac{q^{n^2}}{(q)_n}, \qquad \chi^{2,5}_{1,1} = \sum_{n=0}^{\infty} \frac{q^{n(n+1)}}{(q)_n}. \tag{1.3}$$

Here, $(q)_0 = 1$ and $(q)_n = \prod_{i=1}^n (1-q^i)$ for $n > 0$. Equating expressions of this fermionic type with the corresponding instances of (1.2) yields what are known as bosonic-fermionic q-series identities. In this paper, we give fermionic expressions for all $\chi^{p,p'}_{r,s}$. In doing so, we obtain a bosonic-fermionic identity for each character $\chi^{p,p'}_{r,s}$.

From both a mathematical and physical point of view, further interest attaches to the fermionic expressions for $\chi^{p,p'}_{r,s}$ because in certain cases, another expression is available for these characters. This expression is of a product form and is obtained from (1.2) by means of Jacobi's triple product identity [**27**, eq. (II.28)] or Watson's quintuple product identity [**27**, ex. 5.6]. The former applies in the cases where $p = 2r$ or $p' = 2s$, giving:

$$\chi^{2r,p'}_{r,s} = \prod_{\substack{n=1 \\ n\not\equiv 0,\pm rs \,(\mathrm{mod}\, rp')}}^{\infty} \frac{1}{1-q^n}, \qquad \chi^{p,2s}_{r,s} = \prod_{\substack{n=1 \\ n\not\equiv 0,\pm rs \,(\mathrm{mod}\, sp)}}^{\infty} \frac{1}{1-q^n}. \tag{1.4}$$

The latter applies in the cases where $p = 3r$ or $p' = 3s$, giving:

$$\chi^{3r,p'}_{r,s} = \prod_{\substack{n=1 \\ n\not\equiv 0,\pm rs \,(\mathrm{mod}\, 2rp') \\ n\not\equiv \pm 2r(p'-s) \,(\mathrm{mod}\, 4rp')}}^{\infty} \frac{1}{1-q^n}, \qquad \chi^{p,3s}_{r,s} = \prod_{\substack{n=1 \\ n\not\equiv 0,\pm rs \,(\mathrm{mod}\, 2sp) \\ n\not\equiv \pm 2s(p-r) \,(\mathrm{mod}\, 4sp)}}^{\infty} \frac{1}{1-q^n}. \tag{1.5}$$

It may be shown (see [**15, 14**]) that apart from these and those resulting from identifying $\chi^{3r,p'}_{2r,p'-s} = \chi^{3r,p'}_{r,s}$ and $\chi^{p,3s}_{p-r,2s} = \chi^{p,3s}_{r,s}$ in (1.5), there exist no other expressions for $\chi^{p,p'}_{r,s}$ as products of terms $(1-q^n)^{-1}$.

In the case where $r = 1$ and $p' = 5$, the first expression in (1.4) yields via (1.3), the celebrated Rogers-Ramanujan identities [**37, 38**]:

$$\sum_{n=0}^{\infty} \frac{q^{n^2}}{(q)_n} = \prod_{j=1}^{\infty} \frac{1}{(1-q^{5j-4})(1-q^{5j-1})}, \tag{1.6}$$

$$\sum_{n=0}^{\infty} \frac{q^{n(n+1)}}{(q)_n} = \prod_{j=1}^{\infty} \frac{1}{(1-q^{5j-3})(1-q^{5j-2})}. \tag{1.7}$$

Via (1.4) or (1.5), the fermionic expressions for $\chi^{p,p'}_{r,s}$ given in this paper thus lead to generalisations of the Rogers-Ramanujan identities in the cases where $p = 2r$, $p = 3r$, $p' = 2s$ or $p' = 3s$.

In fact, prior to the advent of conformal field theory, the search for generalisations of the Rogers-Ramanujan identities led to many expressions that are now recognised as fermionic expressions for certain $\chi^{p,p'}_{r,s}$. Slater's compendium [**41**] contains, amongst other things, fermionic expressions for all characters when $(p,p') = (3,4)$, when $(p,p') = (3,5)$ and, of course, when $(p,p') = (2,5)$. The multisum identities of Andrews and Gordon [**2, 28**] deal with all cases when $(p,p') = (2, 2k+1)$ for $k > 2$. In [**3**], Andrews showed how the Bailey chain may be used to yield further identities, and in particular, produced fermionic expressions for the characters $\chi^{3,3k+2}_{1,k}$, $\chi^{3,3k+2}_{1,k+1}$, $\chi^{3,3k+2}_{1,2k+1}$ and $\chi^{3,3k+2}_{1,2k+2}$ for $k \geq 1$. The Bailey chain together with the Bailey lattice was further exploited in [**23**] to yield more infinite sequences of characters (see [**23**, p1652] for these sequences).

Another route to the Virasoro characters is provided by two-dimensional statistical models or one-dimensional quantum spin chains that exhibit critical behaviour [**8, 42**]. In this paper it suffices for us to concentrate on the restricted solid-on-solid (RSOS) statistical models of Forrester-Baxter [**26, 6**] in regime III, because all Virasoro characters $\chi^{p,p'}_{r,s}$ arise in the calculation of the one-point functions of such models in the thermodynamic limit.

The RSOS models of [**26**] are parametrised by integers p', p, a, b and c, where $1 \leq p < p'$ with p and p' coprime, $1 \leq a, b, c < p'$ and $c = b \pm 1$. (The special cases where $p = 1$ or $p = p' - 1$ were first dealt with in [**6**]: the $p = 1$ case may be associated with regime II of these models.) Here, the use of the corner transfer matrix method [**8**] naturally leads to expressions for the one-point functions in terms of generating functions of certain lattice paths. Such a path of length L is a sequence $h_0, h_1, h_2, \ldots, h_L$, of integers such that:

(1) $1 \leq h_i \leq p' - 1$ for $0 \leq i \leq L$,
(2) $h_{i+1} = h_i \pm 1$ for $0 \leq i < L$,
(3) $h_0 = a, h_L = b$.

The set of all such paths is denoted $\mathcal{P}^{p,p'}_{a,b,c}(L)$.

Each element of $\mathcal{P}^{p,p'}_{a,b,c}(L)$ is readily depicted on a two-dimensional $L \times (p' - 2)$ grid by connecting the points (i, h_i) and $(i + 1, h_{i+1})$ for $0 \leq i < L$. A typical element of $\mathcal{P}^{3,8}_{2,4,3}(14)$ is shown in Fig. 1.1.

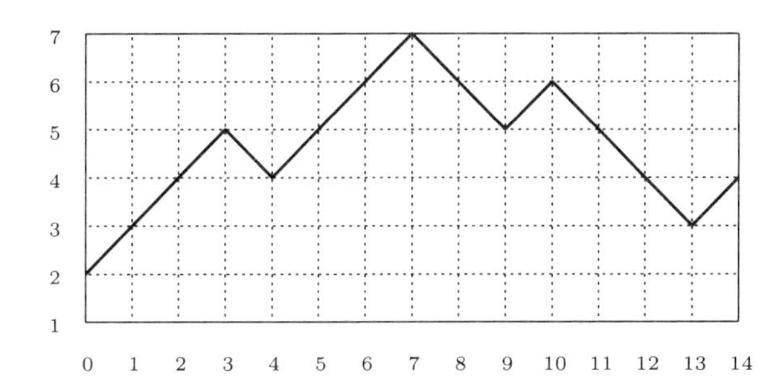

FIGURE 1.1. Typical path.

Each path $h \in \mathcal{P}^{p,p'}_{a,b,c}(L)$ is assigned a weight[1] $wt(h) \in \mathbb{Z}_{\geq 0}$, whereupon the generating function $\chi^{p,p'}_{a,b,c}(L)$ is defined by:

$$\chi^{p,p'}_{a,b,c}(L) = \sum_{h \in \mathcal{P}^{p,p'}_{a,b,c}(L)} q^{wt(h)}. \tag{1.8}$$

By setting up recurrence relations for $\chi^{p,p'}_{a,b,c}(L)$ (see Appendix B), it may be verified that:

$$\begin{aligned} \chi^{p,p'}_{a,b,c}(L) = & \sum_{\lambda=-\infty}^{\infty} q^{\lambda^2 pp' + \lambda(p'r - pa)} \begin{bmatrix} L \\ \frac{L+a-b}{2} - p'\lambda \end{bmatrix}_q \\ & - \sum_{\lambda=-\infty}^{\infty} q^{(\lambda p + r)(\lambda p' + a)} \begin{bmatrix} L \\ \frac{L+a-b}{2} - p'\lambda - a \end{bmatrix}_q, \end{aligned} \tag{1.9}$$

where

$$r = \left\lfloor \frac{pc}{p'} \right\rfloor + \frac{b - c + 1}{2}, \tag{1.10}$$

[1]This weighting function is described in Section 2.3. It makes use of the values of p and c. The weighting function is not required for this introduction.

and, as usual, the Gaussian polynomial $\begin{bmatrix} A \\ B \end{bmatrix}_q$ is defined to be:

$$\begin{bmatrix} A \\ B \end{bmatrix}_q = \begin{cases} \dfrac{(q)_A}{(q)_{A-B}(q)_B} & \text{if } 0 \le B \le A; \\ 0 & \text{otherwise.} \end{cases} \tag{1.11}$$

So as to be able to take the $L \to \infty$ limit in (1.9), we assume that $|q| < 1$. Then, if $r = 0$ or $r = p$,

$$\lim_{L\to\infty} \chi^{p,p'}_{a,b,c}(L) = 0. \tag{1.12}$$

Otherwise, if $0 < r < p$,

$$\lim_{L\to\infty} \chi^{p,p'}_{a,b,c}(L) = \chi^{p,p'}_{r,a}. \tag{1.13}$$

In view of this, we refer to $\chi^{p,p'}_{a,b,c}(L)$ as a finitized character.

Since the expression (1.9) is of a similar form to (1.2) and the former yields the latter in the $L \to \infty$ limit, we refer to (1.9) as a bosonic expression for $\chi^{p,p'}_{a,b,c}(L)$. In this paper we first derive fermionic expressions for $\chi^{p,p'}_{a,b,c}(L)$. The fermionic expressions for $\chi^{p,p'}_{r,a}$ then arise in the $L \to \infty$ limit. For example, for L even, we have (c.f. [**33**, p43]):

$$\chi^{2,5}_{2,2,3}(L) = \sum_{n=0}^{\infty} q^{n^2} \begin{bmatrix} L-n \\ n \end{bmatrix}_q, \qquad \chi^{2,5}_{1,3,2}(L) = \sum_{n=0}^{\infty} q^{n(n+1)} \begin{bmatrix} L-1-n \\ n \end{bmatrix}_q. \tag{1.14}$$

Using (1.13) and (1.10), in the $L \to \infty$ limit, these yield the fermionic expressions for the characters $\chi^{2,5}_{1,2}$ and $\chi^{2,5}_{1,1}$ given in (1.3).

Note that en route to the bosonic-fermionic q-series identities for the characters $\chi^{p,p'}_{r,s}$, we will have proved bosonic-fermionic polynomial identities for the finitized characters $\chi^{p,p'}_{a,b,c}(L)$.

From a physical point of view, a systematic study of fermionic expressions for $\chi^{p,p'}_{r,s}$ was first undertaken in [**30**]. An excellent overview of this and related work is given in [**31**]. In [**30**], fermionic expressions were conjectured for all the unitary characters $\chi^{p,p+1}_{r,s}$, as well as for characters $\chi^{p,kp+1}_{1,k}$ where $k \ge 2$, and characters $\chi^{p,p+2}_{(p-1)/2,(p+1)/2}$ for p odd. In fact, the expressions for $\chi^{p,p+1}_{r,s}$ and $\chi^{p,p+1}_{p-r,p+1-s}$ in [**30**] are different, but since these two characters are equal, two different fermionic expressions for each character $\chi^{p,p+1}_{r,s}$ are thus provided. A further two expressions for each character $\chi^{p,p+1}_{r,s}$ were conjectured in [**34**], and proofs of these and the expressions of [**30**] were given in the cases of $p = 3$ and $p = 4$. A proof of the expressions of [**30**] in the special case where $s = 1$ was given in [**10**] using the technique of "telescopic expansion". A similar technique was used in [**39**] to give a proof of all four expressions for each $\chi^{p,p+1}_{r,s}$. In [**43, 44**], an analysis of the lattice paths described above, provided a combinatorial proof of two of the fermionic expressions for $\chi^{p,p+1}_{r,s}$. In [**24**], a combinatorial proof along the lines of that used in the current paper, was given for all four fermionic expressions for $\chi^{p,p+1}_{r,s}$. A proof for the $\chi^{p,p+1}_{1,1}$ case using Young tableaux and rigged-configurations was presented in [**17**].

Fermionic expressions for $\chi^{p,p'}_{r,s}$ for all p and p', and certain r and s, were stated in [**11**]. To specify the restrictions on r and s, let $\mathcal{T}$ be the set of Takahashi lengths and $\tilde{\mathcal{T}}$ the set of truncated Takahashi lengths associated with the continued fraction of p'/p (these values are defined in Section 1.5). Then either both $p' - s \in \mathcal{T}$ and $r \in \tilde{\mathcal{T}}$, or $s \in \mathcal{T}$ and $r \in \tilde{\mathcal{T}}$ and either $s \geq s_0$ or $r \geq r_0$, where r_0 and s_0 are the smallest positive integers such that $|p'r_0 - ps_0| = 1$. From the point of view of path generating functions, the reason for this strange restriction is expounded in [**25**].

The expressions of [**11**] were proved and vastly extended in [**12**] to give fermionic expressions for all $\chi^{p,p'}_{r,s}$ where $s \in \mathcal{T}$ and r is unrestricted except for $1 \leq r < p$. In the most general cases, these expressions are positive sums of *fundamental fermionic forms* whereas in all previous expressions a single fundamental fermionic form was involved. The expressions of [**12**] are obtained through first using telescopic expansion to derive fermionic expressions for the finitizations $\chi^{p,p'}_{a,b,c}(L)$: these finitized expressions are linear combinations, not necessarily positive sums, of fundamental fermionic forms. An interesting feature of the resulting fermionic expressions is that they make use of a modified definition of the Gaussian polynomial [**12**, eqn. (3.16)]. As explained in [**25**], the modified definition of the Gaussian polynomial serves to account for terms that correspond to a character $\chi^{\hat{p},\hat{p}'}_{\hat{r},\hat{s}}$ where $\hat{p}$ and $\hat{p}'$ are smaller than p and p' respectively. In this article, we provide and prove fermionic expressions for all $\chi^{p,p'}_{r,s}$ and $\chi^{p,p'}_{a,b,c}(L)$. Here, the modified Gaussian polynomial does not correctly account for the extra terms that sometimes arise. Thus we revert to the original definition (1.11) of the Gaussian polynomial, and deal with the extra terms, which again correspond to a character $\chi^{\hat{p},\hat{p}'}_{\hat{r},\hat{s}}$, by writing a recursive expression for $\chi^{p,p'}_{r,s}$: a sum of a number of fundamental fermionic forms plus $\chi^{\hat{p},\hat{p}'}_{\hat{r},\hat{s}}$. Since $\hat{p}' < p'$, this recursive expression terminates, and in fact, the degree of recursion is small compared to p'.

As described later, obtaining the fundamental fermionic forms that occur in the fermionic expression for $\chi^{p,p'}_{r,s}$ requires the construction of two trees: the first depends on s and the Takahashi lengths for p'/p, and the second depends on r and the truncated Takahashi lengths for p'/p. The first of these tree constructions is similar to the recursion underlying the "branched chain of vectors" of [**12**, p340]. Obtaining the fundamental fermionic forms that occur in the fermionic expression for $\chi^{p,p'}_{a,b,c}(L)$ also requires the construction of two trees — one for a and one for b, with each constructed using the Takahashi lengths for p'/p. In fact, in our proof, these expressions are derived first, and the fermionic expressions for $\chi^{p,p'}_{r,s}$ emerge after taking the $L \to \infty$ limit.

1.2. Structure of this paper

In the remainder of this prologue, we describe our fermionic expressions for $\chi^{p,p'}_{r,s}$ and $\chi^{p,p'}_{a,b,c}(L)$ in detail. Central to these expressions is the continued fraction of p'/p, for which the notation is fixed in Section 1.3. The fermionic expressions for the Virasoro characters $\chi^{p,p'}_{r,s}$ are stated in Section 1.4 as a positive sum over *fundamental fermionic forms* $F(\boldsymbol{u}^L, \boldsymbol{u}^R)$ where $\boldsymbol{u}^L$ and $\boldsymbol{u}^R$ are certain vectors. The $F(\boldsymbol{u}^L, \boldsymbol{u}^R)$ are manifestly positive definite. The construction of the vectors $\boldsymbol{u}^L$ and $\boldsymbol{u}^R$, and the notation required to define $F(\boldsymbol{u}^L, \boldsymbol{u}^R)$ are described and discussed in the subsequent Sections 1.5 through 1.12.

The fermionic expressions for $\chi^{p,p'}_{r,s}$ and $\chi^{p,p'}_{a,b,c}(L)$ have a recursive component in that they (sometimes) involve a term $\chi^{\hat{p},\hat{p}'}_{\hat{r},\hat{s}}$ or $\chi^{\hat{p},\hat{p}'}_{\hat{a},\hat{b},\hat{c}}(L)$ respectively, where $\hat{p}' < p'$ and $\hat{p} < p$. These terms are specified in Section 1.13, where a limit to the degree of recursion is also given.

The fermionic expression for a particular $\chi^{p,p'}_{a,b,c}(L)$ is given in Section 1.14 or 1.15 depending on the value of b. In either case, $\chi^{p,p'}_{a,b,c}(L)$ is expressed in terms of fundamental fermionic forms $F(\boldsymbol{u}^L, \boldsymbol{u}^R, L)$ which are also manifestly positive definite. In the case in Section 1.14, the expression for $\chi^{p,p'}_{a,b,c}(L)$ is a positive sum of such terms. However, as described in Section 1.15, in the other case we use linear combinations of such terms with both positive and negative coefficients.

Complete examples of the constructions that are described in this section are provided in Appendix A. Maple programs that construct (and evaluate) the fermionic expressions for any case are available from the author.

The proof of these expressions occupies Sections 2 through 9. In Section 10, we discuss various interesting aspects of these expressions and their proof.

Appendices C, D and E give some technical details that are required in the proof. Appendix B provides a proof of the bosonic expression (1.9).

1.3. Continued fractions

The continued fraction of p'/p lies at the heart of the constructions of this paper. Let p and p' be coprime integers with $1 \le p < p'$. If

$$\frac{p'}{p} = c_0 + \cfrac{1}{c_1 + \cfrac{1}{\ddots \; c_{n-2} + \cfrac{1}{c_{n-1} + \cfrac{1}{c_n}}}}$$

with $c_i \ge 1$ for $0 \le i < n$, and $c_n \ge 2$, then $[c_0, c_1, c_2, \dots, c_n]$ is said to be the *continued fraction* for p'/p.

We refer to n as the *height* of p'/p. We set $t = c_0 + c_1 + \cdots + c_n - 2$ and refer to it as the *rank* of p'/p.

We also define:[2]

$$t_k = -1 + \sum_{i=0}^{k-1} c_i, \tag{1.15}$$

for $0 \le k \le n+1$. Then $t_{n+1} = t + 1$ and $t_n \le t - 1$. If the non-negative integer $j \le t_{n+1}$ satisfies $t_k < j \le t_{k+1}$, we say that j is *in zone* k. We then define $\zeta(j) = k$. Note that there are $n+1$ zones and that for $0 \le k \le n$, zone k contains c_k integers.

1.4. Fermionic character expressions

Here, we present fermionic expressions for all Virasoro characters $\chi^{p,p'}_{r,s}$ where $1 < p < p'$ with p and p' coprime, $1 \le r < p$ and $1 \le s < p'$.

---

[2] The t_{n+1} defined here differs from that defined in [**11, 12**], and that defined in [**21**].

Using notation that will be defined subsequently, when $p' > 3$ the expression for $\chi^{p,p'}_{r,s}$ takes the form:

$$\chi^{p,p'}_{r,s} = \sum_{\substack{\boldsymbol{u}^L\in\mathcal{U}(s)\\ \boldsymbol{u}^R\in\widetilde{\mathcal{U}}(r)}} F(\boldsymbol{u}^L,\boldsymbol{u}^R) + \begin{cases} \chi^{\hat p,\hat p'}_{\hat r,\hat s} & \text{if } \eta(s)=\tilde\eta(r) \text{ and } \hat s\neq 0\neq \hat r;\\ 0 & \text{otherwise,}\end{cases} \tag{1.16}$$

where $\mathcal{U}(s)$ and $\widetilde{\mathcal{U}}(r)$ are sets of t-dimensional vectors that are described below. For t-dimensional vectors $\boldsymbol{u}^L$ and $\boldsymbol{u}^R$, the *fundamental fermionic form* $F(\boldsymbol{u}^L,\boldsymbol{u}^R)$ is defined by:

$$\begin{aligned} F(\boldsymbol{u}^L,\boldsymbol{u}^R) = &\sum_{\substack{n_1,n_2,\ldots,n_{t_1}\in\mathbb{Z}_{\ge0}\\ \boldsymbol{m}\equiv\overline{\boldsymbol{Q}}(\boldsymbol{u}^L+\boldsymbol{u}^R)}} q^{\tilde{\boldsymbol{n}}^T\boldsymbol{B}\tilde{\boldsymbol{n}}+\frac14\boldsymbol{m}^T\overline{\boldsymbol{C}}\boldsymbol{m}-\frac12(\overline{\boldsymbol{u}}^L_\flat+\overline{\boldsymbol{u}}^R_\sharp)\cdot\boldsymbol{m}+\frac14\gamma(\mathcal{X}^L,\mathcal{X}^R)}\\ &\times\frac{1}{(q)_{m_{t_1+1}}}\prod_{j=1}^{t_1}\frac{1}{(q)_{n_j}}\prod_{j=t_1+2}^{t-1}\begin{bmatrix} m_j-\frac12(\overline{\boldsymbol{C}}^*\boldsymbol{m}-\overline{\boldsymbol{u}}^L-\overline{\boldsymbol{u}}^R)_j\\ m_j\end{bmatrix}_q, \end{aligned} \tag{1.17}$$

where the sum is over non-negative integers $n_1,n_2,\ldots,n_{t_1}$, and $(t-t_1-1)$-dimensional vectors[3] $\boldsymbol{m}=(m_{t_1+1},m_{t_1+2},\ldots,m_{t-1})$ each of whose integer components is congruent, modulo 2, to the corresponding component of the vector $\overline{\boldsymbol{Q}}(\boldsymbol{u}^L+\boldsymbol{u}^R)$ which we define below. The t_1-dimensional vector $\tilde{\boldsymbol{n}}$ is defined by:

$$\tilde{\boldsymbol{n}} = (n_1-\tfrac12u_1, n_2-\tfrac12u_2,\ldots,n_{t_1-1}-\tfrac12u_{t_1-1}, n_{t_1}-\tfrac12u_{t_1}+\tfrac12m_{t_1+1}), \tag{1.18}$$

having set $(u_1,u_2,\ldots,u_t)=\boldsymbol{u}^L+\boldsymbol{u}^R$. We also set $\mathcal{X}^L=\mathcal{X}(s,\boldsymbol{u}^L)$ and $\mathcal{X}^R=\tilde{\mathcal{X}}(r,\boldsymbol{u}^R)$, where this and all the other notation occurring in the expressions (1.16) and (1.17) will be defined in the subsequent sections. In the particular cases for which $p=2$, we find that $t=t_1+1$ and yet m_t is undefined. We obtain the correct expressions by setting $m_t=0$.

The exceptional case of $(p,p')=(2,3)$ is dealt with by:

$$\chi^{2,3}_{1,1}=\chi^{2,3}_{1,2}=1, \tag{1.19}$$

which follows from (1.2) and Jacobi's triple product identity, or simply from the first case of (1.4). The expression (1.19) is sometimes required when iterating (1.16).

We note that in the cases where $p'=p+1$, certain choices mean that up to four fermionic expressions for $\chi^{p,p+1}_{r,s}$ arise from (1.16). These are the expressions of [**34**]. One choice is as explained in footnote 5 in Section 1.7, and an analogous choice is available in Section 1.8. In a similar way, two expressions can arise when $p'=kp+1$ for $k\ge2$ and $p\ge4$. In cases other than these, only one fermionic expression for $\chi^{p,p'}_{r,s}$ arises from (1.16).

We also note that when $r=1$, the expression obtained from (1.16) does not always appear to coincide with a known expression for the same character [**11, 12, 21**]. However, the discrepancy appears only in the (t_1+1)th component of $\boldsymbol{u}^R$. It is easily seen (see Lemma 9.7) that this discrepancy has no effect on the summands of (1.17). In these cases, the known expressions may be reproduced by setting $\tilde\sigma_1=t_1$ in the description of Section 1.8 instead of the stated $\tilde\sigma_1=t_1+1$.

[3]All vectors in this paper will be column vectors: for typographical convenience, their components will be expressed in row vector form.

1.5. The Takahashi and string lengths

For p and p' coprime with $1 \le p < p'$, we now use the notation of Section 1.3 to define the corresponding set $\{\kappa_i\}_{i=0}^{t}$ of *Takahashi lengths*, the multiset $\{\tilde{\kappa}_i\}_{i=0}^{t}$ of *truncated Takahashi lengths*, and the multiset $\{l_i\}_{i=0}^{t}$ of *string lengths*. These definitions are based on those of [**42, 12**]. First define y_k and z_k for $-1 \le k \le n+1$ by:

$$\begin{aligned} y_{-1} &= 0; & z_{-1} &= 1; \\ y_0 &= 1; & z_0 &= 0; \\ y_k &= c_{k-1} y_{k-1} + y_{k-2}; & z_k &= c_{k-1} z_{k-1} + z_{k-2}, \quad (1 \le k \le n+1). \end{aligned}$$

In particular, $y_{n+1} = p'$ and $z_{n+1} = p$. Now for $t_k < j \le t_{k+1}$ and $0 \le k \le n$, set:

$$\begin{aligned} \kappa_j &= y_{k-1} + (j - t_k) y_k; \\ \tilde{\kappa}_j &= z_{k-1} + (j - t_k) z_k; \\ l_j &= y_{k-1} + (j - t_k - 1) y_k. \end{aligned}$$

Note that $\kappa_j = l_{j+1}$ unless $j = t_k$ for some k, in which case $\kappa_{t_k} = y_k$ and $l_{t_k+1} = y_{k-1}$. It may be shown (see Lemma E.8) that if j is in zone k then $\kappa_j / \tilde{\kappa}_j$ has continued fraction $[c_0, c_1, \ldots, c_{k-1}, j - t_k]$. In particular, y_k / z_k has continued fraction $[c_0, c_1, \ldots, c_{k-1}]$ for $0 \le k \le n+1$. We define $\mathcal{T} = \{\kappa_i\}_{i=0}^{t-1}$ and $\mathcal{T}' = \{p' - \kappa_i\}_{i=0}^{t-1}$ (we do not include κ_t in the former, nor $p' - \kappa_t$ in the latter). Apart from the cases in which $p = 1$ or $p = p' - 1$, we find that $\mathcal{T} \cap \mathcal{T}' = \emptyset$. We also define $\tilde{\mathcal{T}} = \{\tilde{\kappa}_i\}_{i=t_1+1}^{t-1}$ and $\tilde{\mathcal{T}}' = \{p - \tilde{\kappa}_i\}_{i=t_1+1}^{t-1}$. (We see that these latter two sets are the sets $\mathcal{T}$ and $\mathcal{T}'$ that would be obtained for the continued fraction $[c_1, c_2, \ldots, c_n]$.)

For example, in the case $p' = 223$, $p = 69$, which yields the continued fraction $[3, 4, 3, 5]$, so that $n = 3$, $(t_1, t_2, t_3, t_4) = (2, 6, 9, 14)$ and $t = 13$, we obtain:

$$\begin{aligned} \{y_k\}_{k=-1}^{4} &= \{0, 1, 3, 13, 42, 223\}, \\ \{z_k\}_{k=-1}^{4} &= \{1, 0, 1, 4, 13, 69\}, \\ \{\kappa_j\}_{j=0}^{13} &= \{1, 2, 3, 4, 7, 10, 13, 16, 29, 42, 55, 97, 139, 181\}, \\ \{\tilde{\kappa}_j\}_{j=0}^{13} &= \{1, 1, 1, 1, 2, 3, 4, 5, 9, 13, 17, 30, 43, 56\}, \\ \{l_j\}_{j=1}^{13} &= \{1, 2, 1, 4, 7, 10, 3, 16, 29, 13, 55, 97, 139\}. \end{aligned}$$

Then $\mathcal{T} = \{1, 2, 3, 4, 7, 10, 13, 16, 29, 42, 55, 97, 139\}$ and $\tilde{\mathcal{T}} = \{1, 2, 3, 4, 5, 9, 13, 17, 30, 43\}$.

1.6. Takahashi trees

This and the following two sections describe the sets $\mathcal{U}(s)$ and $\widetilde{\mathcal{U}}(r)$ that occur in (1.16).

Given a with $1 \le a < p'$, we use the Takahashi lengths to produce a *Takahashi tree* for a. This is a binary tree of positive integers for which each node except the root node, is labelled $a_{i_1 i_2 \cdots i_k}$ for some $k \ge 1$, with $i_j \in \{0, 1\}$ for $1 \le j \le k$.[4] The root node of this tree is unlabelled. Each node is either a *branch-node*, a *through-node* or a *leaf-node*. These have 2, 1, or 0 children respectively. Each child

[4]This notation will be abused by letting $a_{i_1 i_2 \cdots i_k}$ denote both the particular node of the tree, and the value at that node: the interpretation will be clear from the context.

of each branch-node is either a branch-node or a through-node, but the child of each through-node is always a leaf-node. Naturally, each child of the node labelled $a_{i_1 i_2 \cdots i_k}$ is labelled $a_{i_1 i_2 \cdots i_k 0}$ or $a_{i_1 i_2 \cdots i_k 1}$. In fact, our construction has $a_{i_1 i_2 \cdots i_k 0} < a$ and $a_{i_1 i_2 \cdots i_k 1} > a$ for non-leaf-nodes and $a_{i_1 i_2 \cdots i_k} = a$ for leaf-nodes.

We obtain the Takahashi tree for a as follows. In the case that $a \in \mathcal{T} \cup \mathcal{T}'$, the Takahashi tree comprises a single leaf-node in addition to the root node. This leaf-node is $a_0 = a$, and the root node is designated a through-node. Otherwise the root node is a branch-node: a_0 is set to be the largest element of $\mathcal{T} \cup \mathcal{T}'$ smaller than a, and a_1 the smallest element of $\mathcal{T} \cup \mathcal{T}'$ larger than a. We now generate the tree recursively. If $a_{i_1 i_2 \cdots i_k} \neq a$, and $|a_{i_1 i_2 \cdots i_k} - a| \in \mathcal{T}$ then we designate $a_{i_1 i_2 \cdots i_k}$ a through-node and define $a_{i_1 i_2 \cdots i_k 0} = a$ to be a leaf-node. Otherwise, when $|a_{i_1 i_2 \cdots i_k} - a| \notin \mathcal{T}$, we make $a_{i_1 i_2 \cdots i_k}$ a branch-node. We take $\kappa_x \in \mathcal{T}$ to be the largest element of $\mathcal{T}$ smaller than $|a_{i_1 i_2 \cdots i_k} - a|$. Then, for $i_{k+1} \in \{0, 1\}$, set $a_{i_1 i_2 \cdots i_k i_{k+1}} = a_{i_1 i_2 \cdots i_k} + (-1)^{i_k} \kappa_{x+|i_k - i_{k+1}|}$. This ensures that $a_{i_1 i_2 \cdots i_k 0} < a$ and $a_{i_1 i_2 \cdots i_k 1} > a$. It is easy to see that the tree is finite. In fact, each leaf-node occurs no deeper than $n+1$ levels below the root node. Thus, there are at most 2^n leaf-nodes. A typical Takahashi tree is shown in Fig. 1.2.

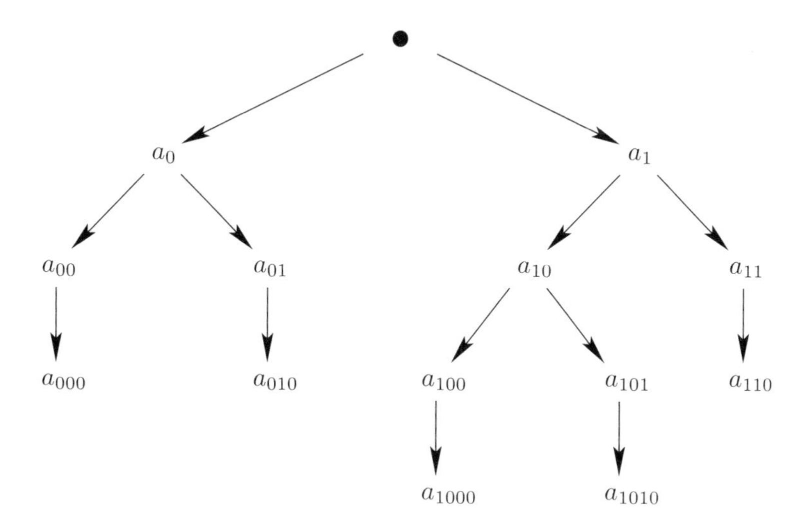

FIGURE 1.2. Typical Takahashi tree.

In fact, precisely this Takahashi tree arises in the case for which $p' = 223$, $p = 69$ and $a = 66$: the nodes take on the values $a_0 = 55$, $a_1 = 84$, $a_{00} = 65$, $a_{01} = 68$, $a_{10} = 55$, $a_{11} = 68$, $a_{100} = 65$, $a_{101} = 68$ and $a_{000} = a_{010} = a_{1000} = a_{1010} = a_{110} = 66$.

Note that the above construction ensures that $|a_{i_1 i_2 \cdots i_{k-1}} - a_{i_1 i_2 \cdots i_k}| \in \mathcal{T}$ for each node $a_{i_1 i_2 \cdots i_k}$ that is present in the Takahashi tree for a. In addition, it is readily seen that if $a_{i_1 i_2 \cdots i_k}$ is a branch-node, then $a_{i_1 i_2 \cdots i_k 1} - a_{i_1 i_2 \cdots i_k 0} \in \mathcal{T}$.

For each leaf-node $a_{i_1 i_2 \cdots i_k 0}$, we also set $a_{i_1 i_2 \cdots i_k 1} = a_{i_1 i_2 \cdots i_k 0} = a$ for later convenience.

1.7. Takahashi tree vectors

Using the Takahashi tree for a, we now define a set $\mathcal{U}(a)$ of t-dimensional vectors. The cardinality of this set is equal to the number of leaf-nodes of the Takahashi tree. First, for $0 \leq j \leq t+1$, define the vector $\boldsymbol{e}_j = (\delta_{1j}, \delta_{2j}, \ldots, \delta_{tj})$ where, as usual, the Kronecker delta is defined by $\delta_{ij} = 1$ if $i = j$ and $\delta_{ij} = 0$ otherwise. Then, for $0 \leq i < j \leq t+1$, define:

$$\boldsymbol{u}_{i,j} = \boldsymbol{e}_i - \boldsymbol{e}_j - \sum_{k:i \leq t_k < j} \boldsymbol{e}_{t_k}. \tag{1.20}$$

Given a leaf-node $a_{i_1 i_2 \cdots i_d}$, we obtain the corresponding vector $\boldsymbol{u} \in \mathcal{U}(a)$ as follows. We first define a *run* $\mathcal{X} = \{\tau_j, \sigma_j, \Delta_j\}_{j=1}^d$. Set $\bar{\imath} = 1 - i$ for $i \in \{0, 1\}$. If $a_{\bar{\imath}_1} \in \mathcal{T}$, set $\Delta_1 = -1$ and define σ_1 such that $\kappa_{\sigma_1} = a_{\bar{\imath}_1}$; otherwise if $a_{\bar{\imath}_1} \in \mathcal{T}'$, set $\Delta_1 = 1$ and define σ_1 such that $\kappa_{\sigma_1} = p' - a_{\bar{\imath}_1}$.[5] For $2 \leq j \leq d$, set $\Delta_j = -(-1)^{i_{j-1}}$ and define σ_j such that $\kappa_{\sigma_j} = |a_{i_1 i_2 \cdots i_{j-1}} - a_{i_1 i_2 \cdots i_{j-1} \bar{\imath}_j}|$. Define $\tau_1 = t+1$. For $2 \leq j \leq d$, define τ_j such that $\kappa_{\tau_j} = a_{i_1 i_2 \cdots i_{j-2} 1} - a_{i_1 i_2 \cdots i_{j-2} 0}$. Finally define:

$$\boldsymbol{u}(\mathcal{X}) = \sum_{m=1}^{d} \boldsymbol{u}_{\sigma_m, \tau_m} + \begin{cases} 0 & \text{if } \Delta_1 = -1; \\ \boldsymbol{e}_t & \text{if } \Delta_1 = 1. \end{cases} \tag{1.21}$$

The set $\mathcal{U}(a)$ comprises all vectors $\boldsymbol{u}(\mathcal{X})$ obtained in this way from the leaf-nodes of the Takahashi tree for a.

To illustrate this construction, again consider the case $p' = 223$, $p = 69$ and $a = 66$. In the case of the leaf-node a_{1010}, we have $|a_{101} - a_{1011}| = 2 = \kappa_1$, $|a_{10} - a_{100}| = 10 = \kappa_5$, $|a_1 - a_{11}| = 16 = \kappa_7$ and $a_0 = 55 = \kappa_{10}$, and hence $\sigma_4 = 1$, $\sigma_3 = 5$, $\sigma_2 = 7$, and $\sigma_1 = 10$. Also, we have $a_{101} - a_{100} = 3 = \kappa_2$, $a_{11} - a_{10} = 13 = \kappa_6$ and $a_1 - a_0 = 29 = \kappa_8$, and thus $\tau_4 = 2$, $\tau_3 = 6$, $\tau_2 = 8$ and $\tau_1 = 14$. Since $a_0 \in \mathcal{T}$, we have $\Delta_1 = -1$ leading to the run $\mathcal{X}_1 = \mathcal{X} = \{\{14, 8, 6, 2\}, \{10, 7, 5, 1\}, \{-1, 1, -1, 1\}\}$, from which using (1.21), we obtain the vector $\boldsymbol{u}(\mathcal{X}_1) = (1, -1, 0, 0, 1, -1, 1, -1, 0, 1, 0, 0, 0)$. In the case of the leaf-node a_{000}, we obtain the run $\mathcal{X}_2 = \mathcal{X} = \{\{14, 8, 2\}, \{12, 6, 0\}, \{1, -1, 1\}\}$, which yields the vector $\boldsymbol{u}(\mathcal{X}_2) = (0, -1, 0, 0, 0, 0, 0, -1, 0, 0, 0, 1, 1)$. After performing similar calculations for the leaf-nodes a_{010}, a_{1000} and a_{110}, we obtain the set:

$$\begin{aligned} \mathcal{U}(66) = \{&(1, -1, 0, 0, 1, -1, 1, -1, 0, 1, 0, 0, 0), \\ &(0, -1, 0, 0, 0, 0, 0, -1, 0, 0, 0, 1, 1), \\ &(1, -1, 0, 0, 1, -1, 0, -1, 0, 0, 0, 1, 1), \\ &(0, -1, 0, 0, 0, 0, 1, -1, 0, 1, 0, 0, 0), \\ &(1, -1, 0, 0, 0, -1, 0, 0, 0, 1, 0, 0, 0)\}. \end{aligned}$$

Given a, the run $\mathcal{X} = \{\tau_j, \sigma_j, \Delta_j\}_{j=1}^d$ may be recovered from the vector $\boldsymbol{u} \in \mathcal{U}(a)$, and thus we define $\mathcal{X}(a, \boldsymbol{u}) = \mathcal{X}$. We also define $\Delta(\boldsymbol{u}) = \Delta_d$. For example, in the case of the vector $\boldsymbol{u} = (1, -1, 0, 0, 1, -1, 1, -1, 0, 1, 0, 0, 0)$ that arises in the above example, we have $\mathcal{X}(66, \boldsymbol{u}) = \mathcal{X}_1$ and $\Delta(\boldsymbol{u}) = 1$.

---

[5] In the cases where $p = 1$ or $p = p' - 1$, we have $\mathcal{T} \cup \mathcal{T}' = \{2, 3, \cdots, p' - 2\}$. If $1 < a_{\bar{\imath}_1} < p' - 1$, we may proceed by considering either $a_{\bar{\imath}_1} \in \mathcal{T}$ or $a_{\bar{\imath}_1} \in \mathcal{T}'$. The two alternatives lead to different equally valid expressions.

1.8. Truncated Takahashi tree

For $1 \leq r < p$, we now mirror the above constructions of the Takahashi tree, the vectors $\boldsymbol{u}(\mathcal{X})$ and the set $\mathcal{U}(a)$ using the truncated Takahashi lengths in place of the Takahashi lengths.

The truncated Takahashi tree for r is obtained by using the prescription given in Section 1.6 for $a = r$, after replacing each occurrence of $\mathcal{T}$ and $\mathcal{T}'$ with $\tilde{\mathcal{T}}$ and $\tilde{\mathcal{T}}'$ respectively, and using truncated Takahashi lengths $\tilde{\kappa}_j$ in place of Takahashi lengths κ_j.

For instance, in the case $p' = 223$ and $p = 69$, where $\tilde{\mathcal{T}} = \{1, 2, 3, 4, 5, 9, 13, 17, 30, 43\}$ and $\tilde{\mathcal{T}}' = \{68, 67, 66, 65, 64, 60, 56, 52, 39, 26\}$, the truncated Takahashi tree for $r = 37$ is given in Fig. 1.3.

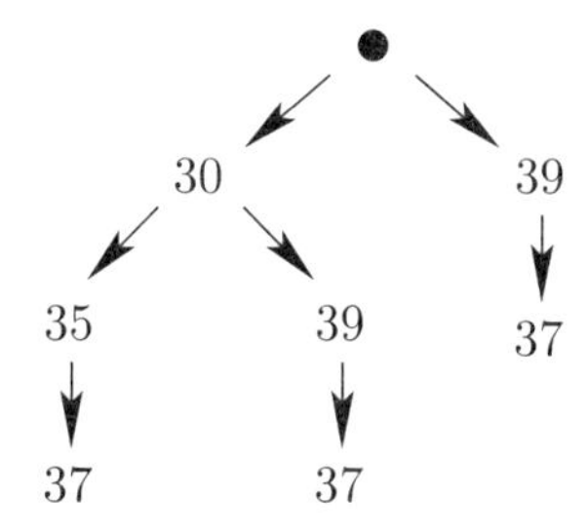

FIGURE 1.3. Truncated Takahashi tree for $r = 37$, $p' = 223$, $p = 69$.

We now use the truncated Takahashi tree for r to produce a set $\widetilde{\mathcal{U}}(r)$ of t-dimensional vectors, the cardinality of this set being equal to the number of leaf-nodes of the tree.

For each leaf-node $r_{i_1 i_2 \cdots i_d}$ of the truncated Takahashi tree for r, we obtain the corresponding vector $\boldsymbol{u} \in \widetilde{\mathcal{U}}(r)$ in a similar way to the prescription given in Section 1.7 as follows. We first define the run $\tilde{\mathcal{X}} = \{\tilde{\tau}_j, \tilde{\sigma}_j, \tilde{\Delta}_j\}_{j=1}^{d}$. If $r_{\bar{\imath}_1} \in \tilde{\mathcal{T}}$, set $\tilde{\Delta}_1 = -1$ and define $\tilde{\sigma}_1$ such that $\tilde{\kappa}_{\tilde{\sigma}_1} = r_{\bar{\imath}_1}$; otherwise if $r_{\bar{\imath}_1} \in \tilde{\mathcal{T}}'$, set $\tilde{\Delta}_1 = 1$ and define $\tilde{\sigma}_1$ such that $\tilde{\kappa}_{\tilde{\sigma}_1} = p - r_{\bar{\imath}_1}$. For $2 \leq j \leq d$, set $\tilde{\Delta}_j = -(-1)^{i_{j-1}}$ and define $\tilde{\sigma}_j$ such that $\tilde{\kappa}_{\tilde{\sigma}_j} = |r_{i_1 i_2 \cdots i_{j-1}} - r_{i_1 i_2 \cdots i_{j-1} \bar{\imath}_j}|$. Define $\tilde{\tau}_1 = t + 1$. For $2 \leq j \leq d$, define $\tilde{\tau}_j$ such that $\tilde{\kappa}_{\tilde{\tau}_j} = r_{i_1 i_2 \cdots i_{j-2} 1} - r_{i_1 i_2 \cdots i_{j-2} 0}$. Finally we obtain the vector $\boldsymbol{u} = \boldsymbol{u}(\tilde{\mathcal{X}})$ using the analogue of (1.21). Since the number of leaf-nodes of the truncated Takahashi tree for r is at most 2^{n-1}, we then have $|\widetilde{\mathcal{U}}(r)| \leq 2^{n-1}$.

To illustrate this construction, again consider the case $p' = 223$, $p = 69$ and $r = 37$. In the case of the leaf-node r_{000}, we have $|r_{00} - r_{001}| = 2 = \tilde{\kappa}_4$, $|r_0 - r_{01}| = 9 = \tilde{\kappa}_8$, and $r_1 = 39 = p - \tilde{\kappa}_{11}$, and hence $\tilde{\sigma}_3 = 4$, $\tilde{\sigma}_2 = 8$, and $\tilde{\sigma}_1 = 11$. Also, we have $r_{00} - r_{01} = 4 = \tilde{\kappa}_6$ and $r_0 - r_1 = 9 = \tilde{\kappa}_8$, and thus $\tilde{\tau}_3 = 6$, $\tilde{\tau}_2 = 8$ and $\tilde{\tau}_1 = 14$. Then, after noting that $r_1 \in \tilde{\mathcal{T}}'$, we obtain the run $\tilde{\mathcal{X}} = \{\{14, 8, 6\}, \{11, 8, 4\}, \{1, -1, -1\}\}$. The use of (1.21) then yields $\boldsymbol{u}(\tilde{\mathcal{X}}) = (0, 0, 0, 1, 0, -1, 0, 0, 0, 0, 1, 0, 1)$. After performing similar calculations for the leaf-nodes r_{010} and r_{10}, we obtain the set:

$$\begin{aligned}\widetilde{\mathcal{U}}(37) = \{&(0,0,0,1,0,-1,0,0,0,0,1,0,1),\\ &(0,0,0,1,0,-1,1,-1,0,0,1,0,1),\\ &(0,0,0,1,0,-1,0,-1,0,0,1,0,0)\}.\end{aligned}$$

1.9. The linear term

With $t_0, t_1, t_2, \dots, t_{n+1}$ and t as defined in Section 1.3, for each t-dimensional vector $\boldsymbol{u} = (u_1, u_2, \dots, u_t)$, the $(t-1)$-dimensional vector $\boldsymbol{u}^\flat = (u_1^\flat, u_2^\flat, \dots, u_{t-1}^\flat)$ is defined by:

$$u_j^\flat = \begin{cases} 0 & \text{if } t_k < j \le t_{k+1},\ k \equiv 0 \,(\mathrm{mod}\, 2); \\ u_j & \text{if } t_k < j \le t_{k+1},\ k \not\equiv 0 \,(\mathrm{mod}\, 2), \end{cases} \tag{1.22}$$

and the $(t-1)$-dimensional vector $\boldsymbol{u}^\sharp = (u_1^\sharp, u_2^\sharp, \dots, u_{t-1}^\sharp)$ by:

$$u_j^\sharp = \begin{cases} u_j & \text{if } t_k < j \le t_{k+1},\ k \equiv 0 \,(\mathrm{mod}\, 2); \\ 0 & \text{if } t_k < j \le t_{k+1},\ k \not\equiv 0 \,(\mathrm{mod}\, 2). \end{cases} \tag{1.23}$$

Then, of course, $(\boldsymbol{u})_j = (\boldsymbol{u}^\flat + \boldsymbol{u}^\sharp)_j$ for $1 \le j < t$.

We also define $\overline{\boldsymbol{u}} = (u_{t_1+2}, u_{t_1+3}, \dots, u_t)$, $\overline{\boldsymbol{u}}^\flat = (u_{t_1+1}^\flat, u_{t_1+2}^\flat, \dots, u_{t-1}^\flat)$, and $\overline{\boldsymbol{u}}^\sharp = (u_{t_1+1}^\sharp, u_{t_1+2}^\sharp, \dots, u_{t-1}^\sharp)$.

For convenience, we sometimes write $\boldsymbol{u}_\flat$, $\boldsymbol{u}_\sharp$, $\overline{\boldsymbol{u}}_\flat$ and $\overline{\boldsymbol{u}}_\sharp$ for $\boldsymbol{u}^\flat$, $\boldsymbol{u}^\sharp$, $\overline{\boldsymbol{u}}^\flat$ and $\overline{\boldsymbol{u}}^\sharp$ respectively.

1.10. The constant term

Here we obtain the constant terms $\gamma(\mathcal{X}^L, \mathcal{X}^R)$ that appear in (1.17).

First, given the run $\mathcal{X} = \{\tau_j, \sigma_j, \Delta_j\}_{j=1}^d$, define the t-dimensional vector $\boldsymbol{\Delta}(\mathcal{X})$ by:[6]

$$\boldsymbol{\Delta}(\mathcal{X}) = \sum_{m=1}^{d} \Delta_m \boldsymbol{u}_{\sigma_m, \tau_m} - \begin{cases} 0 & \text{if } \Delta_1 = -1; \\ \boldsymbol{e}_t & \text{if } \Delta_1 = 1. \end{cases} \tag{1.24}$$

Note the similarity between (1.24) and (1.21). To illustrate this definition, consider the run $\mathcal{X}_1 = \{\{14,8,6,2\}, \{10,7,5,1\}, \{-1,1,-1,1\}\}$ obtained in Section 1.7 above in the case where $p' = 223$, $p = 69$ and $a = 66$. Here, (1.24) yields:

$$\boldsymbol{\Delta}(\mathcal{X}_1) = (1,-1,0,0,-1,1,1,-1,0,-1,0,0,0).$$

Using (1.24), the runs $\mathcal{X}^L = \{\tau_j^L, \sigma_j^L, \Delta_j^L\}_{j=1}^{d^L}$ and $\mathcal{X}^R = \{\tau_j^R, \sigma_j^R, \Delta_j^R\}_{j=1}^{d^R}$ give rise to vectors $\boldsymbol{\Delta}^L = \boldsymbol{\Delta}(\mathcal{X}^L)$ and $\boldsymbol{\Delta}^R = \boldsymbol{\Delta}(\mathcal{X}^R)$. From these, we iteratively generate sequences $(\beta_t, \beta_{t-1}, \dots, \beta_0)$, $(\alpha_t, \alpha_{t-1}, \dots, \alpha_0)$, and $(\gamma_t, \gamma_{t-1}, \dots, \gamma_0)$ as follows. Let $\alpha_t = \beta_t = \gamma_t = 0$. Now, for $j \in \{t, t-1, \dots, 1\}$, obtain α_{j-1}, β_{j-1}, and γ_{j-1} from α_j, β_j, and γ_j in the following three stages. Firstly, obtain:

$$(\beta'_{j-1}, \gamma'_{j-1}) = (\beta_j + (\boldsymbol{\Delta}^L)_j - (\boldsymbol{\Delta}^R)_j, \gamma_j + 2\alpha_j(\boldsymbol{\Delta}^R)_j). \tag{1.25}$$

Then obtain:

$$(\alpha''_{j-1}, \gamma''_{j-1}) = (\alpha_j + \beta'_{j-1}, \gamma'_{j-1} - (\beta'_{j-1})^2). \tag{1.26}$$

Finally set:

$$(\alpha_{j-1}, \beta_{j-1}, \gamma_{j-1}) = \begin{cases} (\alpha''_{j-1}, \alpha_j, -(\alpha''_{j-1})^2 - \gamma''_{j-1}) & \\ & \text{if } j = t_k + 1 \text{ with } 1 \le k \le n; \\ (\alpha''_{j-1}, \beta'_{j-1}, \gamma''_{j-1}) & \text{otherwise.} \end{cases} \tag{1.27}$$

Having thus obtained γ_0, we set $\gamma(\mathcal{X}^L, \mathcal{X}^R) = \gamma_0$ whenever $\sigma_{d^R}^R > 0$. This definition suffices for the purposes of writing down the fermionic expressions for

[6]The values $\{\Delta_j\}_{j=1}^d$ should not be confused with the components of $\boldsymbol{\Delta}(\mathcal{X})$.

$\chi^{p,p'}_{r,s}$ because, with $\tilde{\mathcal{T}} = \{\tilde{\kappa}_i\}_{i=t_1+1}^{t-1}$, the construction of Section 1.8 ensures that $\sigma^R_{d^R} > 0$.

However, for later purposes, values of $\gamma(\mathcal{X}^L, \mathcal{X}^R)$ are required when $\sigma^R_{d^R} = 0$. We specify this and also $\gamma'(\mathcal{X}^L, \mathcal{X}^R)$ as follows. First obtain γ_0 as above. Then, with $\sigma = \sigma^R_{d^R}$ and $\Delta = \Delta^R_{d^R}$:

if $p' > 2p$, $\sigma = 0$, $\Delta = 1$, and either $\left\lfloor \frac{(b+1)p}{p'} \right\rfloor = \left\lfloor \frac{bp}{p'} \right\rfloor$ or $b = p' - 1$, then set

$$\gamma = \gamma_0 - 2(a-b), \qquad \gamma' = \gamma - 2L;$$

if $p' > 2p$, $\sigma = 0$, $\Delta = -1$, and either $\left\lfloor \frac{(b-1)p}{p'} \right\rfloor = \left\lfloor \frac{bp}{p'} \right\rfloor$ or $b = 1$, then set

$$\gamma = \gamma_0 + 2(a-b), \qquad \gamma' = \gamma - 2L;$$

if $p' < 2p$, $\sigma = 0$, $\Delta = 1$, and either $\left\lfloor \frac{(b+1)p}{p'} \right\rfloor \neq \left\lfloor \frac{bp}{p'} \right\rfloor$ or $b = p' - 1$, then set

$$\gamma = \gamma_0 + 2(a-b), \qquad \gamma' = \gamma + 2L;$$

if $p' < 2p$, $\sigma = 0$, $\Delta = -1$, and either $\left\lfloor \frac{(b-1)p}{p'} \right\rfloor \neq \left\lfloor \frac{bp}{p'} \right\rfloor$ or $b = 1$, then set

$$\gamma = \gamma_0 - 2(a-b), \qquad \gamma' = \gamma + 2L.$$

Otherwise, we set $\gamma = \gamma' = \gamma_0$. We then define $\gamma(\mathcal{X}^L, \mathcal{X}^R) = \gamma$ and $\gamma'(\mathcal{X}^L, \mathcal{X}^R) = \gamma'$.

Note that although the constant term $\gamma'(\mathcal{X}^L, \mathcal{X}^R)$ is independent of the components of $\boldsymbol{m}$, in certain cases it does have a dependence on L.

1.11. The quadratic term

In this section, we define the $(t - t_1 - 1) \times (t - t_1 - 1)$ matrices $\overline{\boldsymbol{C}}$, $\overline{\boldsymbol{C}}^*$, and the $t_1 \times t_1$ matrix $\boldsymbol{B}$ that occur in (1.17).

First, with $t_0, t_1, t_2, \ldots, t_{n+1}$ and t as defined in Section 1.3, define C_{ji} for $0 \le i, j \le t$ by, when the indices are in this range,

$$(1.28) \qquad \begin{array}{llll} C_{j,j-1} = -1, & C_{j,j} = 1, & C_{j,j+1} = 1, & \text{if } j = t_k, \quad k = 1, 2, \ldots, n; \\ C_{j,j-1} = -1, & C_{j,j} = 2, & C_{j,j+1} = -1, & 0 \le j < t \text{ otherwise,} \end{array}$$

and set $C_{ji} = 0$ for $|i - j| > 1$. Also define $B_{ji} = \min\{i, j\}$ for $1 \le i, j \le t_1$. Then define the matrices:

$$\overline{\boldsymbol{C}} = \{C_{ji}\}_{\substack{t_1+1 \le j < t, \\ t_1+1 \le i < t}} \qquad \overline{\boldsymbol{C}}^* = \{C_{ji}\}_{\substack{t_1+2 \le j \le t, \\ t_1+1 \le i < t}} \qquad \boldsymbol{B} = \{B_{ji}\}_{\substack{1 \le j \le t_1. \\ 1 \le i \le t_1}}$$

Note that $\overline{\boldsymbol{C}}$ is tri-diagonal, and $\overline{\boldsymbol{C}}^*$ is upper-diagonal. The matrices $\overline{\boldsymbol{C}}$ and $\boldsymbol{B}^{-1}$ may be viewed as minor generalisations of Cartan matrices of type A.

To illustrate these constructions, consider the case $p = 9$ and $p' = 67$, where the continued fraction of p'/p is $[7, 2, 4]$ and $t_1 = 6$, $t_2 = 8$, $t_3 = 12$ and $t = 11$. Here, we obtain:

$$\overline{\boldsymbol{C}} = \begin{pmatrix} 2 & -1 & . & . \\ -1 & 1 & 1 & . \\ . & -1 & 2 & -1 \\ . & . & -1 & 2 \end{pmatrix}, \qquad \overline{\boldsymbol{C}}^* = \begin{pmatrix} -1 & 1 & 1 & \\ . & -1 & 2 & -1 \\ . & . & -1 & 2 \\ . & . & . & -1 \end{pmatrix}, \qquad \boldsymbol{B} = \begin{pmatrix} 1 & 1 & 1 & 1 & 1 & 1 \\ 1 & 2 & 2 & 2 & 2 & 2 \\ 1 & 2 & 3 & 3 & 3 & 3 \\ 1 & 2 & 3 & 4 & 4 & 4 \\ 1 & 2 & 3 & 4 & 5 & 5 \\ 1 & 2 & 3 & 4 & 5 & 6 \end{pmatrix}.$$

The $t \times t$ matrices $\boldsymbol{C}$ and $\boldsymbol{C}^*$ are conveniently defined here as well. With C_{ji} as in (1.28) above, set:

$$\boldsymbol{C} = \{C_{ji}\}_{\substack{0 \le j < t,\\ 0 \le i < t}} \qquad \boldsymbol{C}^* = \{C_{ji}\}_{\substack{1 \le j \le t.\\ 0 \le i < t}}$$

Then $\boldsymbol{C}$ is tri-diagonal, $\boldsymbol{C}^*$ is upper-diagonal, and $\overline{\boldsymbol{C}}$ and $\overline{\boldsymbol{C}}^*$ are the lower right $(t - t_1 - 1) \times (t - t_1 - 1)$ submatrices of $\boldsymbol{C}$ and $\boldsymbol{C}^*$ respectively.

To illustrate these constructions, consider the example of $p = 9$ and $p' = 31$, where the continued fraction of p'/p is $[3, 2, 4]$ and $t_1 = 2$, $t_2 = 4$, $t_3 = 8$ and $t = 7$. Here:

$$\boldsymbol{C} = \begin{pmatrix} 2 & -1 & . & . & . & . & . \\ -1 & 2 & -1 & . & . & . & . \\ . & -1 & 1 & 1 & . & . & . \\ . & . & -1 & 2 & -1 & . & . \\ . & . & . & -1 & 1 & 1 & . \\ . & . & . & . & -1 & 2 & -1 \\ . & . & . & . & . & -1 & 2 \end{pmatrix}, \qquad \boldsymbol{C}^* = \begin{pmatrix} -1 & 2 & -1 & . & . & . & . \\ . & -1 & 1 & 1 & . & . & . \\ . & . & -1 & 2 & -1 & . & . \\ . & . & . & -1 & 1 & 1 & . \\ . & . & . & . & -1 & 2 & -1 \\ . & . & . & . & . & -1 & 2 \\ . & . & . & . & . & . & -1 \end{pmatrix}.$$

1.12. The mn-system and the parity vector

The $\boldsymbol{mn}$-system is a set of t linear equations that depends on the vector $\boldsymbol{u} = (u_1, u_2, \ldots, u_t) = \boldsymbol{u}^L + \boldsymbol{u}^R$, and which defines an interdependence between two t-dimensional vectors $\boldsymbol{n} = (n_1, n_2, \ldots, n_t)$ and $\hat{\boldsymbol{m}} = (m_0, m_1, \ldots, m_{t-1})$. The equations are given by, for $1 \le j \le t$:

$$(1.29) \qquad m_{j-1} - m_{j+1} = m_j + 2n_j - u_j \qquad \text{if } j = t_k, \quad k = 1, 2, \ldots, n;$$

$$(1.30) \qquad m_{j-1} + m_{j+1} = 2m_j + 2n_j - u_j \qquad \text{otherwise,}$$

where we set $m_t = m_{t+1} = 0$.

Then (1.29) and (1.30) imply that for $\hat{\boldsymbol{m}}$ and $\boldsymbol{n}$ satisfying the $\boldsymbol{mn}$-system,

$$(1.31) \qquad -\boldsymbol{C}^* \hat{\boldsymbol{m}} = 2\boldsymbol{n} - \boldsymbol{u}.$$

Since $\boldsymbol{C}^*$ is upper-triangular, its inverse is readily obtained to yield:

$$(1.32) \qquad \hat{\boldsymbol{m}} = (\boldsymbol{C}^*)^{-1}(-2\boldsymbol{n} + \boldsymbol{u}).$$

We now define $Q_i \in \{0, 1\}$ for $0 \le i < t$, by:

$$(1.33) \qquad (Q_0, Q_1, Q_2, \ldots, Q_{t-1}) \equiv (\boldsymbol{C}^*)^{-1}\boldsymbol{u},$$

and define the $(t-1)$-dimensional *parity vector* $\boldsymbol{Q}(\boldsymbol{u}) = (Q_1, Q_2, \ldots, Q_{t-1})$ and the $(t - t_1 - 1)$-dimensional *parity vector* $\overline{\boldsymbol{Q}}(\boldsymbol{u}) = (Q_{t_1+1}, Q_{t_1+2}, \ldots, Q_{t-1})$. It is readily seen that we also have $\overline{\boldsymbol{Q}}(\boldsymbol{u}) \equiv (\overline{\boldsymbol{C}}^*)^{-1}\overline{\boldsymbol{u}}$.

To accord with the labelling of its constituents, we label the components of the $(t - t_1 - 1)$-dimensional vector $(\overline{\boldsymbol{C}}^* \boldsymbol{m} - \overline{\boldsymbol{u}}^L - \overline{\boldsymbol{u}}^R)$ by $t_1 + 2, \ldots, t$. The restriction $\boldsymbol{m} \equiv \overline{\boldsymbol{Q}}(\boldsymbol{u}^L + \boldsymbol{u}^R)$ in expression (1.17) ensures that the quantity $\frac{1}{2}(\overline{\boldsymbol{C}}^* \boldsymbol{m} - \overline{\boldsymbol{u}}^L - \overline{\boldsymbol{u}}^R)_j$ is an integer for $t_1 + 2 \le j \le t$.

1.13. The extra term

Define $\{\xi_\ell\}_{\ell=0}^{2c_n-1}$ and $\{\tilde{\xi}_\ell\}_{\ell=0}^{2c_n-1}$ according to $\xi_0 = \tilde{\xi}_0 = 0$, $\xi_{2c_n-1} = p'$, $\tilde{\xi}_{2c_n-1} = p$, and

$$\begin{aligned} \xi_{2k-1} &= k y_n; & \tilde{\xi}_{2k-1} &= k z_n; \\ \xi_{2k} &= k y_n + y_{n-1}; & \tilde{\xi}_{2k} &= k z_n + z_{n-1}, \end{aligned}$$

for $1 \leq k < c_n$. For example, when $p' = 223$, $p = 69$, we obtain:

$$\begin{aligned} \{\xi_\ell\}_{\ell=0}^{9} &= \{0, 42, 55, 84, 97, 126, 139, 168, 181, 223\}; \\ \{\tilde{\xi}_\ell\}_{\ell=0}^{9} &= \{0, 13, 17, 26, 30, 39, 43, 52, 56, 69\}. \end{aligned}$$

For $1 \leq s < p'$, define $\eta(s) = \ell$ where $\xi_\ell \leq s < \xi_{\ell+1}$. Similarly, for $1 \leq r < p$, define $\tilde{\eta}(r) = \ell$ where $\tilde{\xi}_\ell \leq r < \tilde{\xi}_{\ell+1}$. The values of $\hat{p}'$, $\hat{p}$, $\hat{r}$ and $\hat{s}$ that occur in (1.16) are then defined by:

$$\begin{aligned} \hat{p}' &= \xi_{\eta(s)+1} - \xi_{\eta(s)}, \\ \hat{p} &= \tilde{\xi}_{\eta(s)+1} - \tilde{\xi}_{\eta(s)}, \\ \hat{r} &= r - \tilde{\xi}_{\eta(s)}, \\ \hat{s} &= s - \xi_{\eta(s)}. \end{aligned} \tag{1.34}$$

In fact, it may be shown (see Appendix E) that if $\eta(s) = \tilde{\eta}(r)$ and $\hat{r} \neq 0 \neq \hat{s}$ so that the extra term actually appears in (1.16), then the continued fraction of $\hat{p}'/\hat{p}$ is given by:

$$\begin{array}{ll} [c_0, c_1, \ldots, c_{n-2}] & \text{if } \eta(s) \text{ is odd;} \\ [c_0, c_1, \ldots, c_{n-1}] & \text{if } \eta(s) \in \{0, 2c_n - 2\}; \\ [c_0, c_1, \ldots, c_{n-1} - 1] & \text{if } \eta(s) \text{ is even, } 0 < \eta(s) < 2c_n - 2 \text{ and } c_{n-1} > 1; \\ [c_0, c_1, \ldots, c_{n-3}] & \text{if } \eta(s) \text{ is even, } 0 < \eta(s) < 2c_n - 2 \text{ and } c_{n-1} = 1. \end{array} \tag{1.35}$$

(Here, if a continued fraction $[c_0, \ldots, c_{n'-1}, c_{n'}]$ with $c_{n'} = 1$ arises, then it is to be equated with the continued fraction $[c_0, \ldots, c_{n'-1} + 1]$.)

We then see that the height of the continued fraction of $\hat{p}'/\hat{p}$ is at least one less than that of p'/p. Since, we cannot have $\hat{p} = 1$, the recursive process implied by (1.16) terminates after at most n steps, where n is the height of p'/p.

1.14. Finitized fermionic expressions

In this and the following sections, we provide fermionic expressions for the finitized characters $\chi^{p,p'}_{a,b,c}(L)$ where $1 \leq p < p'$ with p and p' coprime, $1 \leq a, b < p'$ and $c = b \pm 1$ and $L \equiv a - b \pmod 2$. In fact, in the proof that occupies the bulk of this paper, these expressions for $\chi^{p,p'}_{a,b,c}(L)$ are proved first, and the expressions (1.16) for $\chi^{p,p'}_{r,s}$ derived therefrom.

The expressions for $\chi^{p,p'}_{a,b,c}(L)$ fall naturally into two categories. The first deals with those cases for which:[7]

$$2 \leq b \leq p' - 2 \quad \text{and} \quad \left\lfloor \frac{(b+1)p}{p'} \right\rfloor = \left\lfloor \frac{(b-1)p}{p'} \right\rfloor + 1. \tag{1.36}$$

Here we may take either $c = b \pm 1$ because for these values of b, $\chi^{p,p'}_{a,b,b-1}(L) = \chi^{p,p'}_{a,b,b+1}(L)$.

In these cases, we have the identity:

[7]Later in this paper, we refer to values of b which satisfy (1.36) as *interfacial*.

(1.37)

$$\chi^{p,p'}_{a,b,c}(L) = \sum_{\substack{\boldsymbol{u}^L\in\mathcal{U}(a)\\ \boldsymbol{u}^R\in\mathcal{U}(b)}} F(\boldsymbol{u}^L,\boldsymbol{u}^R,L) + \begin{cases} \chi^{\hat{p},\hat{p}'}_{\hat{a},\hat{b},\hat{c}}(L) & \text{if } \eta(a)=\eta(b) \text{ and } \hat{a}\neq 0\neq\hat{b};\\ 0 & \text{otherwise.}\end{cases}$$

where the sets $\mathcal{U}(a)$ and $\mathcal{U}(b)$ are defined in Section 1.7, $F(\boldsymbol{u}^L,\boldsymbol{u}^R,L)$ is defined below, and using the definitions of Section 1.13, we set:

$$\begin{aligned} \hat{p}' &= \xi_{\eta(a)+1}-\xi_{\eta(a)},\\ \hat{p} &= \tilde{\xi}_{\eta(a)+1}-\tilde{\xi}_{\eta(a)},\\ \hat{a} &= a-\xi_{\eta(a)},\\ \hat{b} &= b-\xi_{\eta(a)}, \end{aligned} \tag{1.38}$$

and

$$\hat{c} = \begin{cases} 2 & \text{if } c=\xi_{\eta(a)}>0 \text{ and } \left\lfloor\frac{(b+1)p}{p'}\right\rfloor = \left\lfloor\frac{(b-1)p}{p'}\right\rfloor+1;\\ \hat{p}'-2 & \text{if } c=\xi_{\eta(a)+1}<p' \text{ and } \left\lfloor\frac{(b+1)p}{p'}\right\rfloor = \left\lfloor\frac{(b-1)p}{p'}\right\rfloor+1;\\ c-\xi_{\eta(a)} & \text{otherwise.}\end{cases} \tag{1.39}$$

When $\eta(a)=\eta(b)$ and $\hat{a}\neq 0\neq\hat{b}$ (so that the extra term actually appears in (1.37)), the continued fraction of $\hat{p}'/\hat{p}$ is again as specified in Section 1.13, after setting $s=a$.

For the finitized characters $\chi^{p,p'}_{a,b,c}(L)$, the fundamental fermionic form is defined by:

(1.40) $F(\boldsymbol{u}^L,\boldsymbol{u}^R,L) =$

$$\sum_{\boldsymbol{m}\equiv\boldsymbol{Q}(\boldsymbol{u}^L+\boldsymbol{u}^R)} q^{\frac14\hat{\boldsymbol{m}}^T\boldsymbol{C}\hat{\boldsymbol{m}}-\frac14L^2-\frac12(\boldsymbol{u}^L_\flat+\boldsymbol{u}^R_\sharp)\cdot\boldsymbol{m}+\frac14\gamma'(\mathcal{X}^L,\mathcal{X}^R)} \prod_{j=1}^{t-1} \begin{bmatrix} m_j-\frac12(\boldsymbol{C}^*\hat{\boldsymbol{m}}-\boldsymbol{u}^L-\boldsymbol{u}^R)_j\\ m_j\end{bmatrix}_q,$$

where the sum is over all $(t-1)$-dimensional vectors $\boldsymbol{m}=(m_1,m_2,\ldots,m_{t-1})$, each of whose integer components is congruent, modulo 2, to the corresponding component of the vector $\boldsymbol{Q}(\boldsymbol{u}^L+\boldsymbol{u}^R)$, and where we set $\hat{\boldsymbol{m}}=(L,m_1,m_2,\ldots,m_{t-1})$. We also set $\mathcal{X}^L=\mathcal{X}(a,\boldsymbol{u}^L)$ and $\mathcal{X}^R=\mathcal{X}(b,\boldsymbol{u}^R)$, as defined in Section 1.7. When $t=1$, $F(\boldsymbol{u}^L,\boldsymbol{u}^R,L)$ is still defined by (1.40), after interpreting the (empty) product as 1 and omitting the summation.

As will be seen later, the first component of (1.32) gives:

$$m_0=\sum_{i=1}^{t} l_i\,(2n_i-u_i)\,, \tag{1.41}$$

where $l_1,l_2,\ldots,l_t$ are the string lengths defined earlier. For the practical purposes of evaluating the expression (1.40), we identify m_0 with L and write (1.41) in the form:

$$\sum_{i=1}^{t} l_in_i=\frac12\left(L+\sum_{i=1}^{t} l_iu_i\right). \tag{1.42}$$

The summands in (1.40) then correspond to solutions of this partition problem with $n_i \in \mathbb{Z}_{\geq 0}$ for $1 \leq i \leq t$, with $\{m_j\}_{j=1}^{t-1}$ obtained using (1.32), or equivalently, using (1.29) and (1.30).

Note that when using (1.37), the expression (1.38) can sometimes yield $\hat{b} \in \{1, \hat{p}-1\}$. In the subsequent iteration, (1.36) is thus not satisfied, and the fermionic expression for $\chi^{\hat{p},\hat{p}'}_{\hat{a},\hat{b},\hat{c}}(L)$ must be obtained using the expressions of the following Section 1.15. Also note that (1.39) might yield $\hat{c} \in \{0, \hat{p}'\}$. Although this is no impediment to using the expressions of Section 1.15, the bosonic expression (1.9) does not immediately apply for $c \in \{0, p'\}$. In Section 2.3 we show how (1.9) may be extended to deal with $c \in \{0, p'\}$.

1.15. Fermionic-like expressions

In this section, we provide *fermionic-like* expressions for $\chi^{p,p'}_{a,b,c}(L)$ in the cases for which b does not satisfy (1.36). In contrast to the case where (1.36) holds, the value of c here does affect the resulting fermionic expression. Indeed here, $\chi^{p,p'}_{a,b,b-1}(L) \neq \chi^{p,p'}_{a,b,b+1}(L)$.

The statement of these expressions involves sets $\mathcal{U}^{\pm}(b)$ where, for $\Delta \in \{\pm 1\}$, we define:

$$\mathcal{U}^{\Delta}(b) = \{\boldsymbol{u} \in \mathcal{U}(b) : \Delta(\boldsymbol{u}) = \Delta\}.$$

Therefore, we have the disjoint union $\mathcal{U}(b) = \mathcal{U}^{+}(b) \cup \mathcal{U}^{-}(b)$.

For $p' > 2$, we then have the following identity:[8]

$$\chi^{p,p'}_{a,b,c}(L) = \sum_{\substack{\boldsymbol{u}^L \in \mathcal{U}(a) \\ \boldsymbol{u}^R \in \mathcal{U}^{c-b}(b)}} F(\boldsymbol{u}^L, \boldsymbol{u}^R, L) + \sum_{\substack{\boldsymbol{u}^L \in \mathcal{U}(a) \\ \boldsymbol{u}^R \in \mathcal{U}^{b-c}(b)}} \widetilde{F}(\boldsymbol{u}^L, \boldsymbol{u}^R, L) + \begin{cases} \chi^{\hat{p},\hat{p}'}_{\hat{a},\hat{b},\hat{c}}(L) & \text{if } \eta(a) = \eta(b) \text{ and } \hat{a} \neq 0 \neq \hat{b}; \\ 0 & \text{otherwise,} \end{cases} \tag{1.43}$$

where $F(\boldsymbol{u}^L, \boldsymbol{u}^R, L)$ is as defined in (1.40), $\tilde{F}(\boldsymbol{u}^L, \boldsymbol{u}^R, L)$ is defined below, and the other notation is as in Section 1.14.

Let $\boldsymbol{u} \in \mathcal{U}(b)$ and let $\mathcal{X} = \mathcal{X}(b, \boldsymbol{u})$. If $\mathcal{X} = \{\tau_j, \sigma_j, \Delta_j\}_{j=1}^{d}$, we now define runs $\mathcal{X}^{+} = \{\tau_j, \sigma_j^{+}, \Delta_j\}_{j=1}^{d}$ and $\mathcal{X}^{++} = \{\tau_j, \sigma_j^{++}, \Delta_j\}_{j=1}^{d}$, by setting $\sigma_j^{++} = \sigma_j^{+} = \sigma_j$ for $1 \leq j < d$, and setting $\sigma_d^{++} = \sigma_d + 2$ and $\sigma_d^{+} = \sigma_d + 1$. Using (1.21), we then define $\boldsymbol{u}^{+} = \boldsymbol{u}(\mathcal{X}^{+})$ and $\boldsymbol{u}^{++} = \boldsymbol{u}(\mathcal{X}^{++})$.

We now define the fermionic-like terms $\widetilde{F}(\boldsymbol{u}^L, \boldsymbol{u}, L)$ which appear in (1.43) in terms of $F(\boldsymbol{u}^L, \boldsymbol{u}^R, L')$ given by (1.40), with $\boldsymbol{u}^R$ being set to either of $\boldsymbol{u}$, $\boldsymbol{u}^{+}$ or $\boldsymbol{u}^{++}$. In these three cases, the $\mathcal{X}^R$ that appears in (1.40) should be set to $\mathcal{X}$, $\mathcal{X}^{+}$ and $\mathcal{X}^{++}$ respectively, and $\mathcal{X}^L = \mathcal{X}(a, \boldsymbol{u}^L)$.

Set $\Delta = \Delta_d$ and set:

$$\tau = \begin{cases} \tau_d & \text{if } d > 1; \\ t_n & \text{if } d = 1 \text{ and } \sigma_1 < t_n; \\ t & \text{if } d = 1 \text{ and } \sigma_1 \geq t_n. \end{cases} \tag{1.44}$$

[8]In some very particular cases (1.43) reduces to (1.37). To describe these, let $t_* = t_1$ if $p' > 2p$, and let $t_* = t_2$ if $p' < 2p$. If $1 \leq b < t_*$ then $\mathcal{U}^{-}(b) = \mathcal{U}(b)$ and $\mathcal{U}^{+}(b) = \emptyset$. If $p' - t_* < b \leq p' - 1$ then $\mathcal{U}^{+}(b) = \mathcal{U}(b)$ and $\mathcal{U}^{-}(b) = \emptyset$. Thus, if $1 \leq b < t_*$ and $c = b - 1$, or $p' - t_* < b \leq p' - 1$ and $c = b + 1$ then the second summation in (1.43) is zero, thus yielding (1.37).

Then if $p' > 2p$, define:

$$\widetilde{F}(\boldsymbol{u}^L, \boldsymbol{u}, L) = \begin{cases} q^{\frac{1}{2}(L+\Delta(a-b))} F(\boldsymbol{u}^L, \boldsymbol{u}, L) & \text{if } \sigma_d = 0; \\ q^{-\frac{1}{2}(L-\Delta(a-b))} \big(F(\boldsymbol{u}^L, \boldsymbol{u}^+, L+1) - F(\boldsymbol{u}^L, \boldsymbol{u}^{++}, L)\big) & \\ & \text{if } 0 < \sigma_d < \tau - 1; \\ q^{-\frac{1}{2}(L-\Delta(a-b))} \big(F(\boldsymbol{u}^L, \boldsymbol{u}, L) + (q^L - 1) F(\boldsymbol{u}^L, \boldsymbol{u}^+, L-1)\big) & \\ & \text{if } 0 < \sigma_d = \tau - 1, \end{cases} \tag{1.45}$$

and if $p' < 2p$, define:

$$\widetilde{F}(\boldsymbol{u}^L, \boldsymbol{u}, L) = \begin{cases} q^{-\frac{1}{2}(L+\Delta(a-b))} F(\boldsymbol{u}^L, \boldsymbol{u}, L) & \text{if } \sigma_d = 0; \\ F(\boldsymbol{u}^L, \boldsymbol{u}^+, L+1) - q^{\frac{1}{2}(L+2+\Delta(a-b))} F(\boldsymbol{u}^L, \boldsymbol{u}^{++}, L) & \\ & \text{if } 0 < \sigma_d < \tau - 1; \\ q^{\frac{1}{2}(L-\Delta(a-b))} F(\boldsymbol{u}^L, \boldsymbol{u}, L) + (1 - q^L) F(\boldsymbol{u}^L, \boldsymbol{u}^+, L-1) & \\ & \text{if } 0 < \sigma_d = \tau - 1. \end{cases} \tag{1.46}$$

(c.f. [**12**, Eqs. (3.35) and (12.9)].) Note that the right side of (1.43) depends on the value of c whereas, as will become clear later in this paper, the right side of (1.37) is independent of the value of c.

The exceptional case of $(p, p') = (1, 2)$ is dealt with by:

$$\chi^{1,2}_{1,1,0}(L) = \chi^{1,2}_{1,1,2}(L) = \delta_{L,0}, \tag{1.47}$$

where $\delta_{i,j}$ is the Kronecker delta. This expression (1.47) is sometimes required when iterating (1.37) and (1.43).

Equating expressions (1.37) and (1.43) with the corresponding instances of (1.9), yields polynomial identities. These may be viewed as finitizations of the q-series identities for the characters $\chi^{p,p'}_{r,s}$ obtained by equating (1.16) with (1.2), because these latter identities result on taking the $L \to \infty$ limit of the above polynomial identities. In the case of b satisfying (1.36), the polynomial identities are of genuine bosonic-fermionic type because all terms in (1.37) are manifestly positive. This is not the case for those identities obtained by equating (1.43) with (1.9) when b does not satisfy (1.36), because the fermionic side consists of linear combinations of fundamental fermionic forms, not positive sums. Nonetheless, in the $L \to \infty$ limit, genuine bosonic-fermionic q-series identities result.

2. Path combinatorics

2.1. Outline of proof

In this and the following sections, we prove the fermionic expressions stated in Section 1. This proof makes use of the path picture for the (finitized) Virasoro characters $\chi^{p,p'}_{a,b,c}(L)$ and is combinatorial in the sense that we obtain relationships between different $\chi^{p,p'}_{a,b,c}(L)$ (or in fact, similar generating functions) by manipulating the paths. By means of these relationships, we are able to express $\chi^{p,p'}_{a,b,c}(L)$ in terms of 'simpler' such generating functions. Iterating this process eventually leads to a trivial generating function and thence to a fermionic expression for $\chi^{p,p'}_{a,b,c}(L)$. Taking the $L \to \infty$ limit then yields fermionic expressions for $\chi^{p,p'}_{r,s}$. The techniques that we use refine and extend those that were developed in [**21, 24, 25**]. The account given here is self-contained.

In Section 2.2, we endow the path picture of Section 1.1 with further structure by shading certain regions. This shaded path picture is referred to as the (p,p')-model. The weighting function that we then define for the paths $h \in \mathcal{P}^{p,p'}_{a,b,c}(L)$, although apparently unrelated to that of Forrester and Baxter [**26**], turns out to be a renormalisation thereof ([**21**] describes how it arises via a bijection between the Forrester-Baxter paths and the partitions with hook-difference constraints that appear in [**5**]). Through this weighting function, each vertex of the path h (the ith vertex is the shape of the path at (i, h_i)) is naturally designated as either scoring or non-scoring, with the latter contributing 0 to $wt(h)$. We actually proceed using a slightly different set of paths $\mathcal{P}^{p,p'}_{a,b,e,f}(L)$ which have assigned pre- and post-segments specified by $e, f \in \{0, 1\}$. For these paths, we use a weighting function $\tilde{wt}(h)$ that differs subtly from that defined before. This permits a greater range of consistent combinatorial manipulations. Thus, we carry out the bulk of our analysis under this weighting function, only reverting to the original weighting $wt(h)$ to obtain fermionic expressions for $\chi^{p,p'}_{a,b,c}(L)$ later in the paper.

After further refining the set of paths to $\mathcal{P}^{p,p'}_{a,b,e,f}(L,m)$, each element of which contains exactly m non-scoring vertices, we proceed in Sections 3 and 4 to define transforms that map paths between different models. The $\mathcal{B}$-transform of Section 3 shows how for specific a', b', the elements of $\mathcal{P}^{p,p'+p}_{a',b',e,f}(L',L)$ may be obtained combinatorially from those of $\mathcal{P}^{p,p'}_{a,b,e,f}(L,m)$ for various m. This transform was inspired by [**13, 1**]. It has three stages. The first stage is known as the $\mathcal{B}_1$-transform and enlarges the features of a path, so that the resultant path resides in a larger model. The second stage, referred to as a $\mathcal{B}_2(k)$-transform, lengthens a path by appending k pairs of segments to the path. Each of these pairs is known as a particle. The third stage, the $\mathcal{B}_3(\lambda)$-transform, deforms the path in a particular way. This process may be viewed as the particles *moving* through the path.

The $\mathcal{D}$-transform of Section 4 notes that the elements of $\mathcal{P}^{p'-p,p'}_{a,b,1-e,1-f}(L)$ may be obtained from those of $\mathcal{P}^{p,p'}_{a,b,e,f}(L)$ in a combinatorially trivial way. In fact, it is more convenient to use the $\mathcal{D}$-transform combined with the $\mathcal{B}$-transform. The combined $\mathcal{BD}$-transform is discussed in Section 4.2.

Up to this point, the development deviates only marginally from that of [**25**]. However here, we require something more general than the transformations of generating functions that were specified in Corollaries 3.14 and 4.6 of [**25**]. This is achieved in Section 5, where first, for vectors $\boldsymbol{\mu}$, $\boldsymbol{\nu}$, $\boldsymbol{\mu}^*$ and $\boldsymbol{\nu}^*$, we define $\mathcal{P}^{p,p'}_{a,b,e,f}(L,m)\left\{ {\boldsymbol{\mu}\,;\boldsymbol{\nu} \atop \boldsymbol{\mu}^*;\boldsymbol{\nu}^*} \right\}$ to be the subset of $\mathcal{P}^{p,p'}_{a,b,e,f}(L,m)$ whose elements attain certain heights specified by $\boldsymbol{\mu}$, $\boldsymbol{\nu}$, $\boldsymbol{\mu}^*$ and $\boldsymbol{\nu}^*$, in a certain order. The bijections that the $\mathcal{B}$- and $\mathcal{D}$-transforms yield between different such sets are specified in Lemmas 5.7 and 5.9. The relationships between the corresponding generating functions $\tilde{\chi}^{p,p'}_{a,b,e,f}(L,m)\left\{ {\boldsymbol{\mu}\,;\boldsymbol{\nu} \atop \boldsymbol{\mu}^*;\boldsymbol{\nu}^*} \right\}$ are given in Corollaries 5.8 and 5.10.

In Section 6, we consider extending or truncating the paths on either the right or the left. This yields further relationships between different $\tilde{\chi}^{p,p'}_{a,b,e,f}(L,m)\left\{ {\boldsymbol{\mu}\,;\boldsymbol{\nu} \atop \boldsymbol{\mu}^*;\boldsymbol{\nu}^*} \right\}$. These relationships are given in Lemmas 6.2, 6.4, 6.6 and 6.8.

We bring together all these results in Section 7. In Theorem 7.1 we state that $\tilde{\chi}^{p,p'}_{a,b,e,f}(L)\left\{ {\boldsymbol{\mu}\,;\boldsymbol{\nu} \atop \boldsymbol{\mu}^*;\boldsymbol{\nu}^*} \right\}$ is equal (possibly up to a factor) to the fundamental fermionic form $F(\boldsymbol{u}^L,\boldsymbol{u}^R,L)$ where $\boldsymbol{u}^L$ and $\boldsymbol{u}^R$ are certain vectors that are related to $\boldsymbol{\mu}$, $\boldsymbol{\nu}$, $\boldsymbol{\mu}^*$ and $\boldsymbol{\nu}^*$. Sections 7.1, 7.2 and 7.3 are dedicated to its proof. The centrepiece of the proof is Lemma 7.3. The proof of this lemma is quite long and intricate, requiring an induction argument to show many things simultaneously.

In Section 7.4, this result is transferred back to a similar one concerning the original weighting function. Thereby, in Theorem 7.16, we find $F(\boldsymbol{u}^L,\boldsymbol{u}^R,L)$ to be the generating function for a subset of $\mathcal{P}^{p,p'}_{a,b,c}(L)$ that attains certain heights and does so in a certain order. However, the value of c here sometimes depends on $\boldsymbol{u}^R$. The purpose of Section 7.5 is to obtain the generating function for precisely the same set of paths, but with the value of c switched. This generating function turns out to be the *fermion-like* term $\widetilde{F}(\boldsymbol{u}^L,\boldsymbol{u}^R,L)$.

Section 7.6 describes the $\boldsymbol{mn}$-system which aids the evaluation of $F(\boldsymbol{u}^L,\boldsymbol{u}^R,L)$ or $\widetilde{F}(\boldsymbol{u}^L,\boldsymbol{u}^R,L)$.

As indicated above, the fermionic forms $F(\boldsymbol{u}^L,\boldsymbol{u}^R,L)$ and $\widetilde{F}(\boldsymbol{u}^L,\boldsymbol{u}^R,L)$ are the generating functions for certain subsets of $\mathcal{P}^{p,p'}_{a,b,c}(L)$. In Section 8, we sum these fermionic forms over all $\boldsymbol{u}^L \in \mathcal{U}(a)$ and $\boldsymbol{u}^R \in \mathcal{U}(b)$, where $\mathcal{U}(a)$ and $\mathcal{U}(b)$ are obtained from the Takahashi trees for a and b respectively as described in Section 1.7. As shown in Theorems 8.10 and 8.12, for many cases, this gives the required $\chi^{p,p'}_{a,b,c}(L)$. However, in the other cases, some paths are not accounted for. Theorems 8.10 and 8.12 show that these uncounted paths may be viewed as elements of $\mathcal{P}^{\hat{p},\hat{p}'}_{\hat{a},\hat{b},\hat{c}}(L)$ for certain values $\hat{p},\hat{p}',\hat{a},\hat{b},\hat{c}$. Thus $\chi^{\hat{p},\hat{p}'}_{\hat{a},\hat{b},\hat{c}}(L)$ sometimes appears in the fermionic expression for $\chi^{p,p'}_{a,b,c}(L)$, and in this way, the fermionic expressions that we give are recursive. Theorem 8.10 proves the expression (1.37), and Theorem 8.12 proves the expression (1.43).

In Section 9, we use the fermionic expressions for $\chi^{p,p'}_{a,b,c}(L)$ to obtain fermionic expressions for all characters $\chi^{p,p'}_{r,s}$. In the first instance, in Section 9.1, we set $a = s$, and choose b and c such that (1.10) is satisfied. Taking the $L \to \infty$ limit of the expressions for $\chi^{p,p'}_{a,b,c}(L)$ then yields fermionic expressions for $\chi^{p,p'}_{r,s}$. These resulting expressions, stated in Theorems 9.3 and 9.6, still make use of the Takahashi trees for a and b. However, in each of the two cases, certain subsets of $\mathcal{U}(b)$ must be omitted: they correspond to terms of $\chi^{p,p'}_{a,b,c}(L)$ that are zero in the $L \to \infty$ limit, and yet their inclusion would yield an incorrect expression for $\chi^{p,p'}_{r,s}$.

This undesirable feature of the expressions for $\chi^{p,p'}_{r,s}$, as well as the inconveniences of finding suitable b and c, and considering two separate cases, is ameliorated in Section 9.2. Here it is shown that the required subset of $\mathcal{U}(b)$ is (apart from inconsequential differences) equal to the set $\widetilde{\mathcal{U}}(r)$ that is obtained from the truncated Takahashi tree for r which was described in Section 1.8. The final result, Theorem 9.16, proves the fermionic expression (1.16) stated in Section 1.4 for an arbitrary character $\chi^{p,p'}_{r,s}$.

We discuss our results in Section 10.

2.2. The band structure

Consider again the path picture introduced in Section 1.1. The regions of the picture between adjacent heights will be known as *bands*. There are $p' - 2$ of these. The hth band is that between heights h and $h + 1$.

We now assign a parity to each band: the hth band is said to be an *even* band if $\lfloor hp/p' \rfloor = \lfloor (h+1)p/p' \rfloor$, and an *odd* band if $\lfloor hp/p' \rfloor \neq \lfloor (h+1)p/p' \rfloor$. The array of odd and even bands so obtained will be referred to as the (p,p')-model. It may immediately be deduced that the (p,p')-model has $p' - p - 1$ even bands and $p - 1$ odd bands. In addition, it is easily shown that for $1 \leq r < p$, the band lying between heights $\lfloor rp'/p \rfloor$ and $\lfloor rp'/p \rfloor + 1$ is odd: it will be referred to as the rth odd band.

When drawing the (p,p')-model, we distinguish the bands by shading the odd bands. For example, we obtain Fig. 2.1 when we impose the shading for the $(3,8)$-model on the path picture of Fig. 1.1.

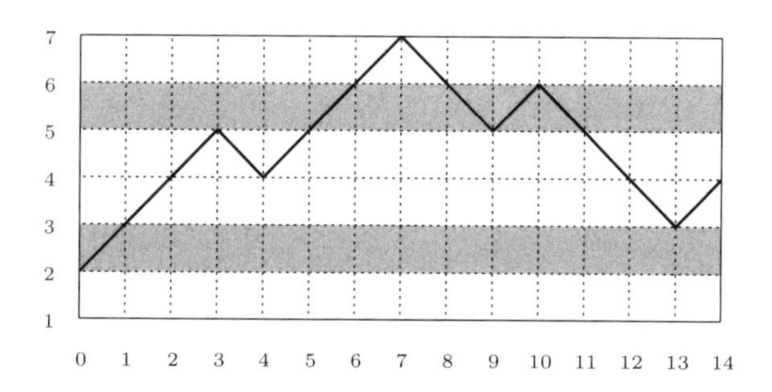

FIGURE 2.1. Typical path with shaded bands.

We note that the band structure is up-down symmetrical. Furthermore, if $p' > 2p$ then both the 1st and $(p'-2)$th bands are even, and there are no two adjacent odd bands.

Note 2.1. *It will be useful to consider there being a band (the* 0*th) below the bottom edge of the* (p,p')*-model, and a band (the* $(p'-1)$*th) above the top edge of the* (p,p')*-model. If* $p'>2p$ *these two bands are both designated as even, and if* $p'<2p$ *they are both designated as odd. Thus they have the same parity as both the* 1*st and* $(p'-2)$*th bands.*

For $2 \le a \le p'-2$, we say that a is *interfacial* if $\lfloor (a+1)p/p' \rfloor = \lfloor (a-1)p/p' \rfloor + 1$. In addition, 0 and p' are defined to be always interfacial, and 1 and $p'-1$ are defined to be always non-interfacial. Thus, for $1 \le a \le p'-1$, a is interfacial if and only if a lies between an odd and even band in the (p,p')-model. Thus for the case of the $(3,8)$-model depicted in Fig. 2.1, a is interfacial for $a = 0,2,3,5,6,8$. We define $\rho^{p,p'}(a) = \lfloor (a+1)p/p' \rfloor$ so that if a is interfacial with $1 \le a < p'$ then the odd band that it borders is the $\rho^{p,p'}(a)$th odd band. Note that $\rho^{p,p'}(0) = 0$ and $\rho^{p,p'}(p') = p$.

It is easily seen that the $(p'-p,p')$-model differs from the (p,p')-model in that each band has changed parity. It follows that if $0 < a < p'$ and a is interfacial in the (p,p')-model then a is also interfacial in the $(p'-p,p')$-model.

For $2 \le a \le p'-2$, we say that a is *multifacial* if $\lfloor (a+1)p/p' \rfloor = \lfloor (a-1)p/p' \rfloor + 2$. In addition, 1 and $p'-1$ are defined to be multifacial if and only if $p' < 2p$. Then a is multifacial if and only if a lies between two odd bands. Since there are no two adjacent odd bands when $p' > 2p$, multifacial a may only occur if $p' < 2p$.

Some basic results relating to the above notions are derived in Appendix C.

2.3. Weighting function

Given a path h of length L, for $1 \le i < L$, the values of h_{i-1}, h_i and h_{i+1} determine the shape of the ith vertex. The four possible shapes are given in Fig. 2.2.

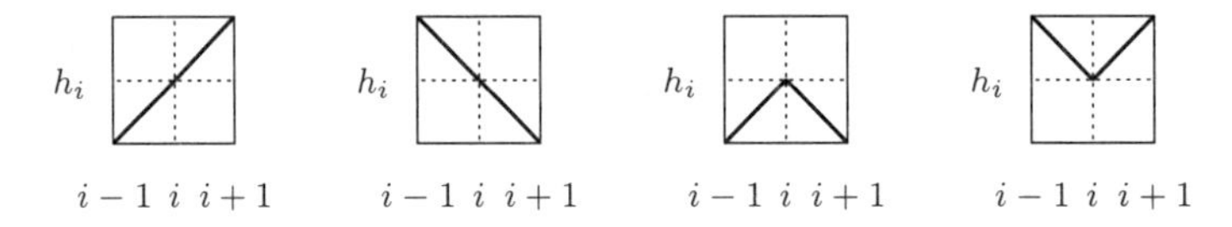

FIGURE 2.2. Vertex shapes.

The four types of vertices shown in Fig. 2.2 are referred to as a *straight-up* vertex, a *straight-down* vertex, a *peak-up* vertex and a *peak-down* vertex respectively. Each vertex is also assigned a parity: the parity of the ith vertex is defined to be the parity of the band that lies between heights h_i and h_{i+1}.

For paths $h \in \mathcal{P}^{p,p'}_{a,b,c}(L)$, we define $h_{L+1} = c$, whereupon the shape and parity of the vertex at $i = L$ is defined as above. This also applies if we extend the definition of $\mathcal{P}^{p,p'}_{a,b,c}(L)$ to encompass the cases $c = 0$ and $c = p'$.

The weight function for the paths is best specified in terms of a (x,y)-coordinate system which is inclined at 45° to the original (i,h)-coordinate system and whose origin is at the path's initial point at $(i=0, h=a)$. Specifically,

$$x = \frac{i-(h-a)}{2}, \qquad y = \frac{i+(h-a)}{2}.$$

Note that at each step in the path, either x or y is incremented and the other is constant. In this system, the path depicted in Fig. 2.1 has its first few vertices at $(0,1)$, $(0,2)$, $(0,3)$, $(1,3)$, $(1,4)$, $(1,5)$, $(1,6)$, $(2,6)$, $\ldots$

Now, for $1 \leq i \leq L$, we define the weight $c_i = c(h_{i-1}, h_i, h_{i+1})$ of the ith vertex according to its shape, its parity and its (x, y)-coordinate, as specified in Table 2.1.

Vertex	c_i	Vertex	c_i
	x	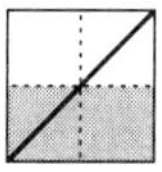	0
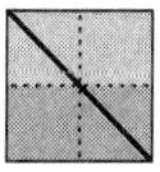	y	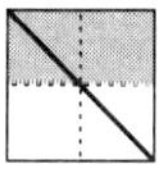	0
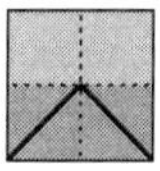	0	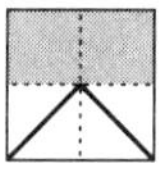	x
	0	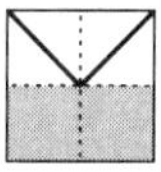	y

TABLE 2.1. Vertex weights.

In Table 2.1, the lightly shaded bands can be either even or odd bands (including the 0th or $(p'-1)$th bands as specified in Note 2.1). Note that for each vertex shape, only one parity case contributes non-zero weight in general. We shall refer to those four vertices, with assigned parity, for which in general, the weight is non-zero, as *scoring* vertices. The other four vertices will be termed *non-scoring*.

We now define:

$$wt(h) = \sum_{i=1}^{L} c_i. \tag{2.1}$$

To illustrate this procedure, consider again the path $h \in \mathcal{P}^{3,8}_{2,4,3}(14)$ depicted in Fig. 2.1. With $h_{15} = c = 3$, Table 2.1 indicates that there are scoring vertices at $i = 3, 4, 5, 7, 8, 13$ and 14. This leads to:

$$wt(h) = 0 + 3 + 1 + 1 + 6 + 7 + 6 = 24.$$

The generating function $\chi^{p,p'}_{a,b,c}(L)$ for the set of paths $\mathcal{P}^{p,p'}_{a,b,c}(L)$ is defined to be:

$$\chi^{p,p'}_{a,b,c}(L; q) = \sum_{h \in \mathcal{P}^{p,p'}_{a,b,c}(L)} q^{wt(h)}. \tag{2.2}$$

Often, we drop the base q from the notation so that $\chi^{p,p'}_{a,b,c}(L) = \chi^{p,p'}_{a,b,c}(L; q)$. The same will be done for other functions without comment.

The expression (1.47) for $\chi^{1,2}_{1,1,0}(L)$ and $\chi^{1,2}_{1,1,2}(L)$ is immediately obtained from (2.2).

Appendix B proves the bosonic expression for $\chi^{p,p'}_{a,b,c}(L)$ given in (1.9). With the 0th and $(p'-1)$th bands as specified in Note 2.1, this result extends to the cases $c=0$ and $c=p'$ (when $b=1$ and $b=p'-1$ respectively), provided that we extend the definition (1.10) as follows:

$$r = \begin{cases} \lfloor pc/p' \rfloor + (b-c+1)/2 & \text{if } 1 \le c < p'; \\ 1 & \text{if } c=0 \text{ and } p' > 2p; \\ 0 & \text{if } c=0 \text{ and } p' < 2p; \\ p-1 & \text{if } c=p' \text{ and } p' > 2p; \\ p & \text{if } c=p' \text{ and } p' < 2p. \end{cases} \tag{2.3}$$

2.4. Winged generating functions

For $h \in \mathcal{P}^{p,p'}_{a,b,c}(L)$, the description of the previous section, in effect, uses the values of b and c to specify a path *post-segment* that extends from (L,b) and $(L+1,c)$. In this section, we define another set of paths $\mathcal{P}^{p,p'}_{a,b,e,f}(L)$ for which the values of e and f serve to specify both a *pre-segment* and a post-segment.

Let p and p' be positive coprime integers for which $1 \le p < p'$. Then, given $a,b,L \in \mathbb{Z}_{\ge 0}$ such that $1 \le a,b \le p'-1$, $L+a-b \equiv 0 \pmod 2$, and $e,f \in \{0,1\}$, a path $h \in \mathcal{P}^{p,p'}_{a,b,e,f}(L)$ is a sequence $h_0,h_1,h_2,\ldots,h_L$, of integers such that:

(1) $1 \le h_i \le p'-1$ for $0 \le i \le L$,
(2) $h_{i+1} = h_i \pm 1$ for $0 \le i < L$,
(3) $h_0 = a$, $h_L = b$.

For $h \in \mathcal{P}^{p,p'}_{a,b,e,f}(L)$, we define $e(h)=e$ and $f(h)=f$.

As with the elements of $\mathcal{P}^{p,p'}_{a,b,c}(L)$, each $h \in \mathcal{P}^{p,p'}_{a,b,e,f}(L)$ can be depicted on the (p,p')-model of length L, by connecting the points (i,h_i) and $(i+1,h_{i+1})$ for $0 \le i < L$. For these values of i, the shape and parity of the ith vertex is determined as in Section 2.3. In addition, f specifies the direction of a path post-segment which starts at (L,b) and is in the NE (resp. SE) direction if $f=0$ (resp. $f=1$). Similarly, e specifies the direction of a path pre-segment which ends at $(0,a)$ and is in the SE (resp. NE) direction if $e=0$ (resp. $e=1$). The pre- and post-segments enable a shape and a parity to be assigned to both the zeroth and the Lth vertices of h. If a path h is displayed without its pre- and post-segments, their directions may be inferred from $e(h)$ and $f(h)$.

We now define a weight $\tilde{wt}(h)$, for $h \in \mathcal{P}^{p,p'}_{a,b,e,f}(L)$. For $1 \le i < L$, set $\tilde{c}_i = c(h_{i-1},h_i,h_{i+1})$ using Table 2.1 as in Section 2.3. Then set:

$$\tilde{c}_L = \begin{cases} x & \text{if } h_L - h_{L-1} = 1 \text{ and } f(h)=1; \\ y & \text{if } h_L - h_{L-1} = -1 \text{ and } f(h)=0; \\ 0 & \text{otherwise,} \end{cases} \tag{2.4}$$

where (x,y) is the coordinate of the Lth vertex of h. We then designate this vertex as scoring if it is a peak vertex ($h_L = h_{L-1} - (-1)^{f(h)}$), and as non-scoring otherwise. We define:

$$\tilde{wt}(h) = \sum_{i=1}^{L} \tilde{c}_i. \tag{2.5}$$

Consider the corresponding path $h' \in \mathcal{P}^{p,p'}_{a,b,c}(L)$ with $c = b + (-1)^f$, defined by $h'_i = h_i$ for $0 \le i \le L$. From Table 2.1, we see that $\tilde{wt}(h) = wt(h')$ if the post-segment of h lies in an even band.

Define the generating function:

$$\tilde{\chi}^{p,p'}_{a,b,e,f}(L;q) = \sum_{h \in \mathcal{P}^{p,p'}_{a,b,e,f}(L)} q^{\tilde{wt}(h)}, \tag{2.6}$$

where $\tilde{wt}(h)$ is given by (2.5). Of course, $\tilde{\chi}^{p,p'}_{a,b,0,f}(L) = \tilde{\chi}^{p,p'}_{a,b,1,f}(L)$.

2.5. Striking sequence of a path

In this section, for each $h \in \mathcal{P}^{p,p'}_{a,b,e,f}(L)$, we define $\pi(h)$, $d(h)$, $L(h)$, $m(h)$, $\alpha(h)$ and $\beta(h)$. Define $\pi(h) \in \{0,1\}$ to be the parity of the band between heights h_0 and h_1 (if $L(h) = 0$, we set $h_1 = h_0 + (-1)^{f(h)}$). Thus, for the path h shown in Fig. 2.1, we have $\pi(h) = 1$. Next, define $d(h) = 0$ when $h_1 - h_0 = 1$ and $d(h) = 1$ when $h_1 - h_0 = -1$. We then see that if $e(h) + d(h) + \pi(h) \equiv 0 \pmod 2$ then the 0th vertex is a scoring vertex, and if $e(h) + d(h) + \pi(h) \equiv 1 \pmod 2$ then it is a non-scoring vertex.

Now consider each path $h \in \mathcal{P}^{p,p'}_{a,b,e,f}(L)$ as a sequence of straight lines, alternating in direction between NE and SE. Then, reading from the left, let the lines be of lengths $w_1, w_2, w_3, \ldots, w_l$, for some l, with $w_i > 0$ for $1 \le i \le l$. Thence $w_1 + w_2 + \cdots + w_l = L(h)$, where $L(h) = L$ is the length of h.

For each of these lines, the last vertex will be considered to be part of the line but the first will not. Then, the ith of these lines contains w_i vertices, the first $w_i - 1$ of which are straight vertices. Then write $w_i = a_i + b_i$ so that b_i is the number of scoring vertices in the ith line. The striking sequence of h is then the array:

$$\begin{pmatrix} a_1 & a_2 & a_3 & \cdots & a_l \\ b_1 & b_2 & b_3 & \cdots & b_l \end{pmatrix}^{(e(h),f(h),d(h))}.$$

With $\pi = \pi(h)$, $e = e(h)$, $f = f(h)$ and $d = d(h)$, we define:

$$m(h) = \begin{cases} (e + d + \pi) \bmod 2 + \sum_{i=1}^{l} a_i & \text{if } L > 0; \\ (e + f) \bmod 2 & \text{if } L = 0, \end{cases}$$

whence $m(h)$ is the number of non-scoring vertices possessed by h (altogether, h has $L(h)+1$ vertices). We also define $\alpha(h) = (-1)^d((w_1 + w_3 + \cdots) - (w_2 + w_4 + \cdots))$ and for $L > 0$,

$$\beta(h) = \begin{cases} (-1)^d((b_1 + b_3 + \cdots) - (b_2 + b_4 + \cdots)) \\ \qquad\qquad\qquad \text{if } e + d + \pi \equiv 0 \pmod 2; \\ (-1)^d((b_1 + b_3 + \cdots) - (b_2 + b_4 + \cdots)) + (-1)^e \\ \qquad\qquad\qquad \text{otherwise.} \end{cases}$$

For $L = 0$, we set $\beta(h) = f - e$.

For example, for the path shown in Fig. 2.1 for which $d(h) = 0$ and $\pi(h) = 1$, the striking sequence is:

$$\begin{pmatrix} 2 & 0 & 1 & 1 & 1 & 2 & 0 \\ 1 & 1 & 2 & 1 & 0 & 1 & 1 \end{pmatrix}^{(e,1,0)}.$$

In this case, $m(h) = 8 - e$, $\alpha(h) = 2$, and $\beta(h) = 2 - e$.

We note that given the startpoint $h_0 = a$ of the path, the path can be reconstructed from its striking sequence[1]. In particular, $h_L = b = a + \alpha(h)$. In addition, the nature of the final vertex may be deduced from a_l and b_l.[2]

Lemma 2.2. *Let the path h have the striking sequence* $\begin{pmatrix} a_1 & a_2 & a_3 & \cdots & a_l \\ b_1 & b_2 & b_3 & \cdots & b_l \end{pmatrix}^{(e,f,d)}$, *with* $w_i = a_i + b_i$ *for* $1 \le i \le l$. *Then:*

$$\tilde{wt}(h) = \sum_{i=1}^{l} b_i(w_{i-1} + w_{i-3} + \cdots + w_{1+i \bmod 2}).$$

Proof: For $L = 0$, both sides are clearly 0. So assume $L > 0$. First consider $d = 0$. For i odd, the ith line is in the NE direction and its x-coordinate is $w_2 + w_4 + \cdots + w_{i-1}$. By the prescription of the previous section, and the definition of b_i, this line thus contributes $b_i(w_2 + w_4 + \cdots + w_{i-1})$ to the weight $\tilde{wt}(h)$ of h. Similarly, for i even, the ith line is in the SE direction and contributes $b_i(w_1 + w_3 + \cdots + w_{i-1})$ to $\tilde{wt}(h)$. The lemma then follows for $d = 0$. The case $d = 1$ is similar. □

2.6. Path parameters

We make the following definitions:

$$\begin{aligned} \alpha^{p,p'}_{a,b} &= b - a; \\ \beta^{p,p'}_{a,b,e,f} &= \left\lfloor \frac{bp}{p'} \right\rfloor - \left\lfloor \frac{ap}{p'} \right\rfloor + f - e; \\ \delta^{p,p'}_{a,e} &= \begin{cases} 0 & \text{if } \left\lfloor \frac{(a+(-1)^e)p}{p'} \right\rfloor = \left\lfloor \frac{ap}{p'} \right\rfloor; \\ 1 & \text{if } \left\lfloor \frac{(a+(-1)^e)p}{p'} \right\rfloor \ne \left\lfloor \frac{ap}{p'} \right\rfloor. \end{cases} \end{aligned}$$

(The superscripts of $\alpha^{p,p'}_{a,b}$ are superfluous, of course.) It may be seen that the value of $\delta^{p,p'}_{a,e}$ gives the parity of the band in which the path pre-segment resides.

Lemma 2.3. *Let $h \in \mathcal{P}^{p,p'}_{a,b,e,f}(L)$. Then $\alpha(h) = \alpha^{p,p'}_{a,b}$ and $\beta(h) = \beta^{p,p'}_{a,b,e,f}$.*

Proof: That $\alpha(h) = \alpha^{p,p'}_{a,b}$ follows immediately from the definitions.

The second result is proved by induction on L. If $h \in \mathcal{P}^{p,p'}_{a,b,e,f}(0)$ then $a = b$, whence $\beta^{p,p'}_{a,b,e,f} = f - e = \beta(h)$, immediately from the definitions.

For $L > 0$, let $h \in \mathcal{P}^{p,p'}_{a,b,e,f}(L)$ and assume that the result holds for all $h' \in \mathcal{P}^{p,p'}_{a,b',e,f'}(L-1)$. We consider a particular h' by setting $h'_i = h_i$ for $0 \le i < L$, $b' = h_{L-1}$ and choosing $f' \in \{0,1\}$ so that $f' = 0$ if either $b - b' = 1$ and the Lth segment of h lies in an even band, or $b - b' = -1$ and the Lth segment of h lies in an odd band; and $f' = 1$ otherwise. It may easily be checked that the $(L-1)$th

[1]We only need $w_1, w_2, \ldots, w_l$ together with d.

[2]Thus the value of f in the striking sequence is redundant — we retain it for convenience.

vertex of h' is scoring if and only if the $(L-1)$th vertex of h is scoring. Then, from the definition of $\beta(h)$, we find that:

$$\beta(h) = \begin{cases} \beta(h')+1 & \text{if } b-b' = 1 \text{ and } f=1; \\ \beta(h')-1 & \text{if } b-b' = -1 \text{ and } f=0; \\ \beta(h') & \text{otherwise.} \end{cases}$$

The induction hypothesis gives $\beta(h') = \lfloor b'p/p' \rfloor - \lfloor ap/p' \rfloor + f' - e$. Then when the Lth segment of h lies in an even band so that $\lfloor bp/p' \rfloor = \lfloor b'p/p' \rfloor$, consideration of the four cases of $b-b' = \pm 1$ and $f \in \{0,1\}$ shows that $\beta(h) = \lfloor bp/p' \rfloor - \lfloor ap/p' \rfloor + f - e$. When the Lth segment of h lies in an odd band so that $\lfloor bp/p' \rfloor = \lfloor b'p/p' \rfloor + b - b'$, consideration of the four cases of $b - b' = \pm 1$ and $f \in \{0,1\}$ again shows that $\beta(h) = \lfloor bp/p' \rfloor - \lfloor ap/p' \rfloor + f - e$. The result follows by induction. □

2.7. Scoring generating functions

We now define a generating function for paths that have a particular number of non-scoring vertices. First define the subset $\mathcal{P}^{p,p'}_{a,b,e,f}(L,m)$ of $\mathcal{P}^{p,p'}_{a,b,e,f}(L)$ to comprise those paths h for which $m(h) = m$. Then define:

$$\tilde{\chi}^{p,p'}_{a,b,e,f}(L,m;q) = \sum_{h \in \mathcal{P}^{p,p'}_{a,b,e,f}(L,m)} q^{\tilde{wt}(h)}. \tag{2.7}$$

Lemma 2.4. *Let $\beta = \beta^{p,p'}_{a,b,e,f}$. Then*

$$\tilde{\chi}^{p,p'}_{a,b,e,f}(L) = \sum_{\substack{m \equiv L+\beta \\ (\mathrm{mod}\, 2)}} \tilde{\chi}^{p,p'}_{a,b,e,f}(L,m).$$

Proof: Let $h \in \mathcal{P}^{p,p'}_{a,b,e,f}(L)$. We claim that $m(h) + L(h) + \beta(h) \equiv 0 \,(\mathrm{mod}\, 2)$. This will follow from showing that $L(h) - m(h) + (-1)^{d(h)}\beta(h)$ is even. If h has striking sequence $\begin{pmatrix} a_1 & a_2 & a_3 & \cdots & a_l \\ b_1 & b_2 & b_3 & \cdots & b_l \end{pmatrix}^{(e,f,d)}$, then $L(h) - m(h) = (b_1 + b_2 + \cdots + b_l) - (e + d + \pi) \bmod 2$, where $\pi = \pi(h)$. For $e + d + \pi \equiv 0 \,(\mathrm{mod}\, 2)$, we immediately obtain $L(h) - m(h) + (-1)^d \beta(h) = 2(b_1 + b_3 + \ldots)$. For $e + d + \pi \not\equiv 0 \,(\mathrm{mod}\, 2)$, we obtain $L(h) - m(h) + (-1)^d\beta(h) = 2(b_1 + b_3 + \ldots) - 1 + (-1)^{d+e}$, whence the claim is proved in all cases. The lemma then follows, once it is noted, via Lemma 2.3, that $\beta(h) = \beta^{p,p'}_{a,b,e,f}$. □

Note 2.5. *Since each element of $\mathcal{P}^{p,p'}_{a,b,e,f}(L,m)$ has $L+1$ vertices, it follows that $\chi^{p,p'}_{a,b,e,f}(L,m)$ is non-zero only if $0 \le m \le L+1$. Therefore the sum in Lemma 2.4 may be further restricted to $0 \le m \le L+1$.*

Later, we shall define generating functions for certain subsets of $\mathcal{P}^{p,p'}_{a,b,e,f}(L,m)$.

2.8. A seed

The following result provides a seed on which the results of later sections will act.

Lemma 2.6. *If $L \geq 0$ is even then:*

$$\tilde{\chi}^{1,3}_{1,1,0,0}(L,m) = \tilde{\chi}^{1,3}_{2,2,1,1}(L,m) = \delta_{m,0} q^{\frac{1}{4}L^2}.$$

If $L > 0$ is odd then:

$$\tilde{\chi}^{1,3}_{1,2,0,1}(L,m) = \tilde{\chi}^{1,3}_{2,1,1,0}(L,m) = \delta_{m,0} q^{\frac{1}{4}(L^2-1)}.$$

Proof: The $(1,3)$-model comprises one even band. Thus when L is even, there is precisely one $h \in \mathcal{P}^{1,3}_{1,1,0,0}(L)$. It has $h_i = 1$ for i even, and $h_i = 2$ for i odd. We see that h has striking sequence $\begin{pmatrix} 0 & 0 & 0 & \cdots & 0 \\ 1 & 1 & 1 & \cdots & 1 \end{pmatrix}^{(0,0,0)}$ and $m(h) = 0$. Lemma 2.2 then yields $\tilde{wt}(h) = 0+1+1+2+2+3+\cdots+(\frac{1}{2}L-1)+\frac{1}{2}L = (L/2)^2$, as required.

The other expressions follow in a similar way. □

2.9. Partitions

A partition $\lambda = (\lambda_1, \lambda_2, \ldots, \lambda_k)$ is a sequence of k integer parts $\lambda_1, \lambda_2, \ldots, \lambda_k$, satisfying $\lambda_1 \geq \lambda_2 \geq \cdots \geq \lambda_k > 0$. For $i > k$, define $\lambda_i = 0$. The weight $\mathrm{wt}(\lambda)$ of λ is given by $\mathrm{wt}(\lambda) = \sum_{i=1}^{k} \lambda_i$.

We define $\mathcal{Y}(k,m)$ to be the set of all partitions λ with at most k parts, and for which $\lambda_1 \leq m$. A proof of the following well known result may be found in [**4**].

Lemma 2.7. *The generating function,*

$$\sum_{\lambda \in \mathcal{Y}(k,m)} q^{\mathrm{wt}(\lambda)} = \begin{bmatrix} m+k \\ m \end{bmatrix}_q.$$

We also require the following easily proved result.

Lemma 2.8.

$$\begin{bmatrix} m+k \\ m \end{bmatrix}_{q^{-1}} = q^{-mk} \begin{bmatrix} m+k \\ m \end{bmatrix}_q.$$

3. The $\mathcal{B}$-transform

In this section, we introduce the $\mathcal{B}$-transform which maps the set of paths $\mathcal{P}^{p,p'}_{a,b,e,f}(L)$ into the sets $\mathcal{P}^{p,p'+p}_{a',b',e,f}(L')$ for certain a', b' and various L'.

The band structure of the $(p,p'+p)$-model is easily obtained from that of the (p,p')-model. Indeed, according to Section 2.2, for $1 \le r < p$, the rth odd band of the $(p,p'+p)$-model lies between heights $\lfloor r(p'+p)/p \rfloor = \lfloor rp'/p \rfloor + r$ and $\lfloor r(p'+p)/p \rfloor + 1 = \lfloor rp'/p \rfloor + r + 1$. Thus the height of the rth odd band in the $(p,p'+p)$-model is r greater than that in the (p,p')-model. Therefore, the $(p,p'+p)$-model may be obtained from the (p,p')-model by increasing the distance between neighbouring odd bands by one unit and appending an extra even band to both the top and the bottom of the grid. For example, compare the $(3,8)$-model of Fig. 2.1 with the $(3,11)$-model of Fig. 3.1.

The $\mathcal{B}$-transform has three components, which will be termed *path-dilation*, *particle-insertion*, and *particle-motion*. These three components will also be known as the $\mathcal{B}_1$-, $\mathcal{B}_2$- and $\mathcal{B}_3$-transforms respectively. In fact, particle-insertion is dependent on a parameter $k \in \mathbb{Z}_{\ge 0}$, and particle-motion is dependent on a partition λ that has certain restrictions. Consequently, we sometimes refer to particle-insertion and particle-motion as $\mathcal{B}_2(k)$- and $\mathcal{B}_3(\lambda)$-transforms respectively. Then, combining the $\mathcal{B}_1$-, $\mathcal{B}_2(k)$- and $\mathcal{B}_3(\lambda)$-transforms produces the $\mathcal{B}(k,\lambda)$-transform.

3.1. Path-dilation: the $\mathcal{B}_1$-transform

The $\mathcal{B}_1$-transform acts on each path $h \in \mathcal{P}^{p,p'}_{a,b,e,f}(L)$ to yield a new path $h^{(0)} \in \mathcal{P}^{p,p'+p}_{a',b',e,f}(L^{(0)})$, for certain a', b' and $L^{(0)}$. First, the starting point a' of the new path $h^{(0)}$ is specified to be:

$$a' = a + \left\lfloor \frac{ap}{p'} \right\rfloor + e. \tag{3.1}$$

If $r = \lfloor ap/p' \rfloor$ then r is the number of odd bands below $h = a$ in the (p,p')-model. Since the height of the rth odd band in the $(p,p'+p)$-model is r greater than that in the (p,p')-model, we thus see that under path-dilation, the height of the startpoint above the next lowermost odd band (or if there isn't one, the bottom of the grid) has either increased by one or remained constant.

We define $d(h^{(0)}) = d(h)$. The above definition specifies that $e(h^{(0)}) = e(h)$ and $f(h^{(0)}) = f(h)$.

In the case that $L = 0$ and $e = f$, we specify $h^{(0)}$ by setting $L^{(0)} = L(h^{(0)}) = 0$. When $L = 0$ and $e \ne f$, we leave the action of the $\mathcal{B}_1$-transform on h undefined (it will not be used in this case). Thus in Lemmas 3.2, 3.5, 3.6, 3.9, 4.3, 5.7, 5.9, and Corollary 3.3, we implicitly exclude consideration of the case $L = 0$ and $e \ne f$. However, it must be considered in the proofs of Corollaries 5.8 and 5.10.

In the case $L > 0$ consider, as in Section 2.5, h to comprise l straight lines that alternate in direction, the ith of which is of length w_i and possesses b_i scoring vertices. $h^{(0)}$ is then defined to comprise l straight lines that alternate in direction (since $d(h^{(0)}) = d(h)$, the direction of the first line in $h^{(0)}$ is the same as that in h), the ith of which has length

$$w_i' = \begin{cases} w_i + b_i & \text{if } i \geq 2 \text{ or } e(h) + d(h) + \pi(h) \equiv 0 \,(\mathrm{mod}\, 2); \\ w_1 + b_1 + 2\pi(h) - 1 & \text{if } i = 1 \text{ and } e(h) + d(h) + \pi(h) \not\equiv 0 \,(\mathrm{mod}\, 2). \end{cases}$$

In particular, this determines $L^{(0)} = L(h^{(0)})$ and $b' = h^{(0)}_{L^{(0)}}$.

As an example, consider the path h shown in Fig. 2.1, regarded as an element of $\mathcal{P}^{3,8}_{2,4,e,1}(14)$. Here $d(h) = 0$, $\pi(h) = 1$ and $\lfloor ap/p' \rfloor = 0$. Thus when $e = 0$, the action of path-dilation on h produces the path given in Fig. 3.1.

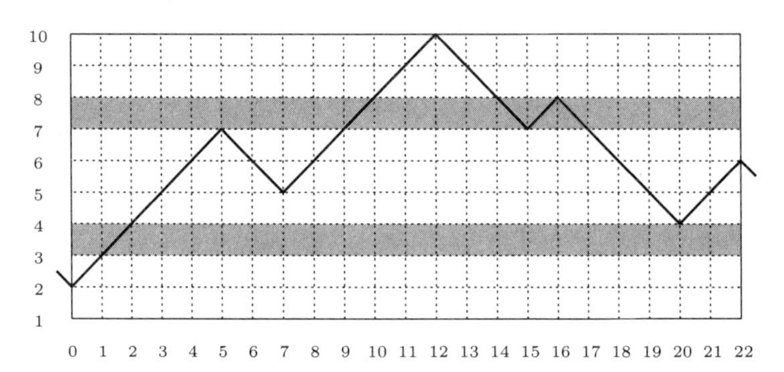

FIGURE 3.1. Result of $\mathcal{B}_1$-transform when $e = 0$ and $f = 1$.

This path is an element of $\mathcal{P}^{3,11}_{2,6,e,1}(22)$.

In contrast, when $e = 1$, the action of path-dilation on h produces the element of $\mathcal{P}^{3,11}_{3,6,e,1}(21)$ given in Fig. 3.2.

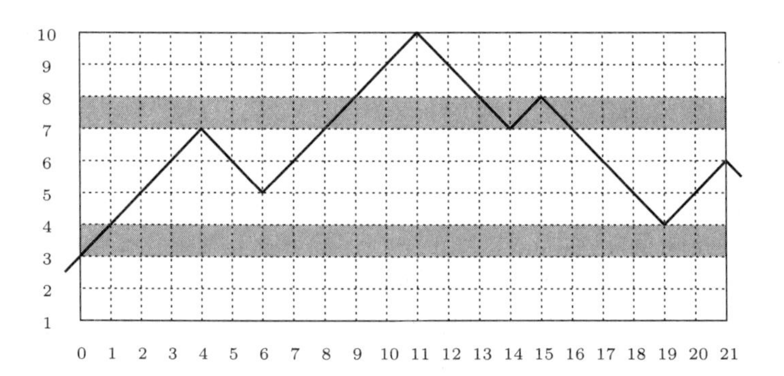

FIGURE 3.2. Result of $\mathcal{B}_1$-transform when $e = 1$ and $f = 1$.

The situation at the startpoint may be considered as falling into one of eight cases, corresponding to $e(h), d(h), \pi(h) \in \{0, 1\}$.[1] In Table 3.1, we illustrate the four cases that arise when $d(h) = 0$ (the four cases for $d(h) = 1$ may be obtained from

[1]These cases may be seen to correspond to the eight cases of vertex type as listed in Table 2.1.

these by an up-down reflection and changing the value of $e(h)$).[2] From Table 3.1 we see that the pre-segment of $h^{(0)}$ always lies in an even band. This also follows from Lemma C.1.

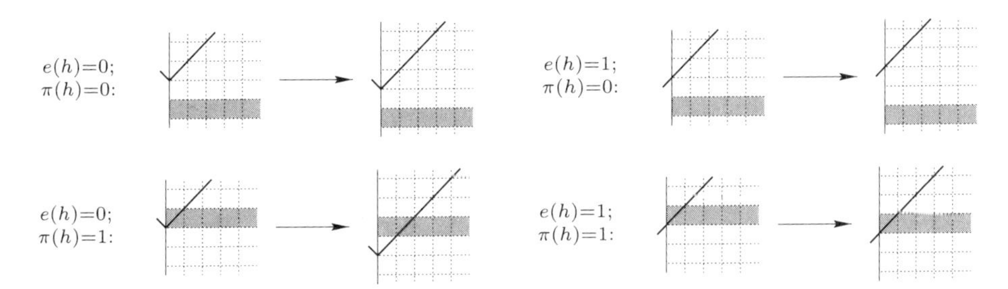

TABLE 3.1. $\mathcal{B}_1$-transforms at the startpoint.

Note 3.1. *The action of path-dilation on* $h \in \mathcal{P}^{p,p'}_{a,b,e,f}(L)$ *yields a path* $h^{(0)} \in \mathcal{P}^{p,p'+p}_{a',b',e,f}(L^{(0)})$ *that has, including the vertex at* $i = 0$, *no adjacent scoring vertices, except in the case where* $\pi(h) = 1$ and $e(h) = d(h)$, *when a single pair of scoring vertices occurs in* $h^{(0)}$ *at* $i = 0$ *and* $i = 1$.

Also note that $\pi(h^{(0)}) = \pi(h)$ *unless* $\pi(h) = 1$ and $e(h) = d(h)$, *in which case* $\pi(h^{(0)}) = 0$.

Now compare the ith line of $h^{(0)}$ (which has length w'_i) with the ith line of h (which has length w_i). If the lines in question are in the NE direction, we claim that the height of the final vertex of that in $h^{(0)}$ above the next lowermost odd band (or if there isn't one, the bottom of the model at $a = 1$) is one greater than that in h.

If the lines in question are in the SE direction, we claim that the height of the final vertex of that in $h^{(0)}$ below the next highermost odd band (or if there isn't one, the top of the model) is one greater than that in h. In particular, if either the first or last segment of the ith line is in an odd band, then the corresponding segment of $h^{(0)}$ lies in the same odd band.

We also claim that if that of h has a straight vertex that passes into the kth odd band in the (p,p')-model then that of $h^{(0)}$ has a straight vertex that passes into the kth odd band in the $(p,p'+p)$-model.

Comparing Figs. 3.1 and 3.2 with Fig. 2.1 demonstrates these claims.

These claims follow because in passing from the (p,p')-model to the $(p,p'+p)$-model, the distance between neighbouring odd bands has increased by one, and because the length of each line has increased by one for every scoring vertex with a small adjustment made to the length of the first line in certain cases. In effect, a new straight vertex has been inserted immediately prior to each scoring vertex and, if $e(h)+d(h)+\pi(h) \not\equiv 0 \pmod 2$, the length of the resulting first line has been adjusted by $2\pi(h) - 1$.

---

[2]The examples here are such that $w_1 \geq 3$ and $c_1 \geq 3$

Lemma 3.2. *Let $h \in \mathcal{P}^{p,p'}_{a,b,e,f}(L)$ have striking sequence $\begin{pmatrix} a_1 & a_2 & a_3 & \cdots & a_l \\ b_1 & b_2 & b_3 & \cdots & b_l \end{pmatrix}^{(e,f,d)}$, and let $h^{(0)} \in \mathcal{P}^{p,p'+p}_{a',b',e,f}(L^{(0)})$ be obtained from the action of the $\mathcal{B}_1$-transform on h. Let $\pi = \pi(h)$. If $e + d + \pi \equiv 0\,(mod 2)$ then $h^{(0)}$ has striking sequence:*

$$\begin{pmatrix} a_1+b_1 & a_2+b_2 & a_3+b_3 & \cdots & a_l+b_l \\ b_1 & b_2 & b_3 & \cdots & b_l \end{pmatrix}^{(e,f,d)},$$

and if $e + d + \pi \not\equiv 0\,(mod 2)$ then $h^{(0)}$ has striking sequence:

$$\begin{pmatrix} a_1+b_1+\pi-1 & a_2+b_2 & a_3+b_3 & \cdots & a_l+b_l \\ b_1+\pi & b_2 & b_3 & \cdots & b_l \end{pmatrix}^{(e,f,d)}.$$

Moreover, if $m = m(h)$:

- $m(h^{(0)}) = L$;
- $L^{(0)} = \begin{cases} 2L - m + 2 & \text{if } \pi = 1 \text{ and } e = d, \\ 2L - m & \text{otherwise;} \end{cases}$
- $\alpha(h^{(0)}) = \alpha(h) + \beta(h)$;
- $\beta(h^{(0)}) = \beta(h)$.

Proof: The form of the striking sequence for $h^{(0)}$ follows because, for $i > 1$, every scoring vertex in the ith line of h accounts for an extra non-scoring vertex in that line. The same is true when $i = 1$, except in the case $(e(h) + d(h) + \pi(h)) \equiv 1$ (throughout this proof, we take all equivalences, modulo 2) when the length of the new 1st line becomes $a_1 + 2b_1 + 2\pi - 1$. That there are $b_1 + \pi$ scoring vertices in this case, follows from examining Table 3.1.

Let $\pi' = \pi(h^{(0)})$. By definition, $e(h^{(0)}) = e$ and $d(h^{(0)}) = d$.

If $(e + d + \pi) \equiv 0$ then $(e + d + \pi') \equiv 0$ by Note 3.1. Thereupon $m^{(0)} = \sum_{i=1}^{l}(a_i + b_i) = L$. Additionally, $L^{(0)} = \sum_{i=1}^{l}(a_i + 2b_i) = 2L - \sum_{i=1}^{l} a_i = 2L - m$. That $\beta(h^{(0)}) = \beta(h)$ and $\alpha(h^{(0)}) = \alpha(h) + \beta(h)$ both follow immediately in this case.

On the other hand, if $(e + d + \pi) \not\equiv 0$ then $\pi = 0 \implies e \neq d$ and $\pi = 1 \implies e = d$. In each instance, Note 3.1 implies that $\pi' = 0$. Thereupon, $m^{(0)} = (e+d+\pi') \bmod 2 + \pi - 1 + \sum_{i=1}^{l}(a_i + b_i) = \sum_{i=1}^{l}(a_i + b_i) = L$. Additionally, $L^{(0)} = 2\pi - 1 + \sum_{i=1}^{l}(a_i + 2b_i) = 2L - (1 + \sum_{i=1}^{l} a_i) + 2\pi = 2L - m + 2\pi$. This is the required value. Now in this case, $\beta(h) = (-1)^d((b_1+b_3+\cdots)-(b_2+b_4+\cdots)) + (-1)^e$. When $\pi = 0$ so that $(e+d+\pi') \equiv 1$ then $\beta(h^{(0)}) = \beta(h)$ follows immediately. When $\pi = 1$, we have $\beta(h^{(0)}) = (-1)^d((b_1+1+b_3+\cdots) - (b_2+b_4+\cdots))$. $\beta(h^{(0)}) = \beta(h)$ now follows in this case because $(e + d + \pi) \not\equiv 0$ implies that $e = d$. Finally, $\alpha(h^{(0)}) = \alpha(h) + (-1)^d((b_1 + b_3 + \cdots) - (b_2 + b_4 + \cdots)) + (-1)^d(2\pi - 1)$. Since $(-1)^d(2\pi - 1) = -(-1)^d(-1)^\pi = (-1)^e$, the lemma then follows. □

Corollary 3.3. *Let $h \in \mathcal{P}^{p,p'}_{a,b,e,f}(L)$ and $h^{(0)} \in \mathcal{P}^{p,p'+p}_{a',b',e,f}(L^{(0)})$ be the path obtained by the action of the $\mathcal{B}_1$-transform on h. Then $a' = a + \lfloor ap/p' \rfloor + e$ and $b' = b + \lfloor bp/p' \rfloor + f$.*

Proof: $a' = a + \lfloor ap/p' \rfloor + e$ is by definition. Lemma 3.2 gives $\alpha(h^{(0)}) = \alpha(h) + \beta(h)$, whence Lemma 2.3 implies that $\alpha^{p,p'+p}_{a',b'} = \alpha^{p,p'}_{a,b} + \beta^{p,p'}_{a,b,e,f}$. Expanding this gives $b' - a' = b - a + \lfloor bp/p' \rfloor - \lfloor ap/p' \rfloor + f - e$, whence $b' = b + \lfloor bp/p' \rfloor + f$. □

The above result implies that the $\mathcal{B}_1$-transform maps $\mathcal{P}^{p,p'}_{a,b,e,f}(L)$ into a set of paths whose startpoints are equal and whose endpoints are equal. However, the lengths of these paths are not necessarily equal. We also see that the transformation of the endpoint is analogous to that which occurs at the startpoint. In particular, Lemma C.1 implies that $\delta^{p,p'+p}_{b',f} = 0$ so that the path post-segment of $h^{(0)}$ always resides in an even band. For the four cases where $h_L = h_{L-1} - 1$, the $\mathcal{B}_1$-transform affects the endpoint as in Table 3.2 (the value $\pi'(h)$ is the parity of the band in which the Lth segment of h lies).

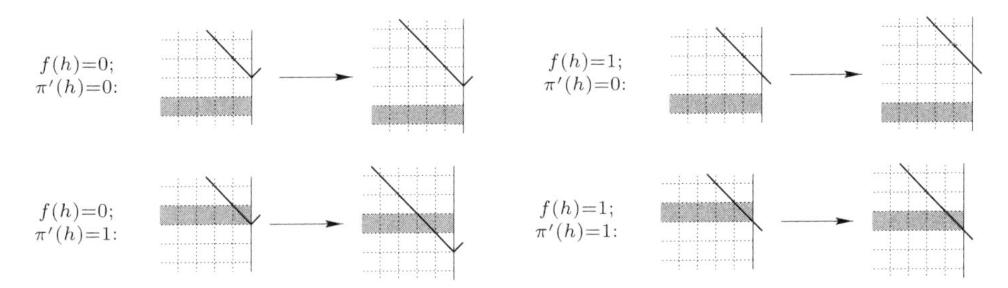

TABLE 3.2. $\mathcal{B}_1$-transforms at the endpoint.

Lemma 3.4. *Let $1 \le p < p'$, $1 \le a, b < p'$, $e, f \in \{0,1\}$, $a' = a + \lfloor ap/p' \rfloor + e$, and $b' = b + \lfloor bp/p' \rfloor + f$. Then $\alpha^{p,p'+p}_{a',b'} = \alpha^{p,p'}_{a,b} + \beta^{p,p'}_{a,b,e,f}$ and $\beta^{p,p'+p}_{a',b',e,f} = \beta^{p,p'}_{a,b,e,f}$.*

Proof: Lemma C.1 implies that $\lfloor a'p/(p'+p) \rfloor = \lfloor ap/p' \rfloor$, $\lfloor b'p/(p'+p) \rfloor = \lfloor bp/p' \rfloor$. The results then follow immediately from the definitions. □

Lemma 3.5. *Let $h \in \mathcal{P}^{p,p'}_{a,b,e,f}(L)$ and $h^{(0)} \in \mathcal{P}^{p,p'+p}_{a',b',e,f}(L^{(0)})$ be the path obtained by the action of the $\mathcal{B}_1$-transform on h. Then:*

$$\tilde{wt}(h^{(0)}) = \tilde{wt}(h) + \frac{1}{4}\left((L^{(0)} - m^{(0)})^2 - \beta^2\right),$$

where $m^{(0)} = m(h^{(0)})$ and $\beta = \beta^{p,p'}_{a,b,e,f}$.

Proof: Let h have striking sequence $\begin{pmatrix} a_1 & a_2 & a_3 & \cdots & a_l \\ b_1 & b_2 & b_3 & \cdots & b_l \end{pmatrix}^{(e,f,d)}$, and let $\pi = \pi(h)$. If $(e + d + \pi) \equiv 0 \pmod 2$, then Lemmas 3.2 and 2.2 show that:

$$\tilde{wt}(h^{(0)}) - \tilde{wt}(h) = (b_1 + b_3 + b_5 + \cdots)(b_2 + b_4 + b_6 + \cdots).$$

Via Lemma 3.2, we obtain $L^{(0)} - m^{(0)} = L - m(h) = b_1 + b_2 + \cdots + b_l$. Then since $\beta(h) = \pm((b_1 + b_3 + b_5 + \cdots) - (b_2 + b_4 + b_6 + \cdots))$, it follows that:

$$\tilde{wt}(h^{(0)}) - \tilde{wt}(h) = \frac{1}{4}((L^{(0)} - m^{(0)})^2 - \beta(h)^2).$$

If $(e + d + \pi) \not\equiv 0 \pmod 2$, then Lemmas 3.2 and 2.2 show that:

$$\begin{aligned} \tilde{wt}(h^{(0)}) - \tilde{wt}(h) &= (2\pi - 1 + b_1 + b_3 + b_5 + \cdots)(b_2 + b_4 + b_6 + \cdots) \\ &= \frac{1}{4}((L^{(0)} - m^{(0)})^2 - \beta(h)^2), \end{aligned}$$

the second equality resulting because $L^{(0)} - m^{(0)} = L - m(h) + 2\pi = b_1 + b_2 + \cdots + b_l + 2\pi - 1$ and

$$\begin{aligned}\beta(h) &= (-1)^d((b_1 + b_3 + b_5 + \cdots) - (b_2 + b_4 + b_6 + \cdots)) + (-1)^e \\ &= \pm((2\pi - 1 + b_1 + b_3 + b_5 + \cdots) - (b_2 + b_4 + b_6 + \cdots)),\end{aligned}$$

on using $(-1)^{e+d} = -(-1)^\pi = 2\pi - 1$.

Finally, Lemma 2.3 gives $\beta(h) = \beta^{p,p'}_{a,b,e,f} = \beta$. □

3.2. Particle insertion: the $\mathcal{B}_2$-transform

Let $p' > 2p$ so that the (p,p')-model has no two neighbouring odd bands, and let a', e be such that $\delta^{p,p'}_{a',e} = 0$. Then if $h^{(0)} \in \mathcal{P}^{p,p'}_{a',b',e,f}(L^{(0)})$, the pre-segment of $h^{(0)}$ lies in an even band. By *inserting a particle* into $h^{(0)}$, we mean displacing $h^{(0)}$ two positions to the right and inserting two segments: the leftmost of these is in the NE (resp. SE) direction if $e = 0$ (resp. $e = 1$), and the rightmost is in the opposite direction, which is thus the direction of the pre-segment of $h^{(0)}$. In this way, we obtain a path $h^{(1)}$ of length $L^{(0)} + 2$. We assign $e(h^{(1)}) = e$ and $f(h^{(1)}) = f$. Note also that $d(h^{(1)}) = e$ and $\pi(h^{(1)}) = 0$.

Thereupon, we may repeat this process of particle insertion. After inserting k particles into $h^{(0)}$, we obtain a path $h^{(k)} \in \mathcal{P}^{p,p'}_{a',b',e,f}(L^{(0)} + 2k)$. We say that $h^{(k)}$ has been obtained by the action of a $\mathcal{B}_2(k)$-transform on $h^{(0)}$.

In the case of the element of $\mathcal{P}^{3,11}_{3,6,1,1}(21)$ shown in Fig. 3.2, the insertion of two particles produces the element of $\mathcal{P}^{3,11}_{3,6,1,1}(25)$ shown in Fig. 3.3.

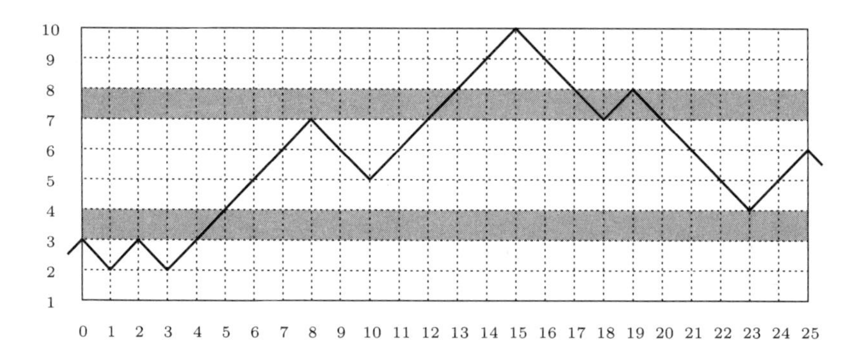

FIGURE 3.3. Result of $\mathcal{B}_2(k)$-transform.

Lemma 3.6. *Let $h \in \mathcal{P}^{p,p'}_{a,b,e,f}(L)$. Apply a $\mathcal{B}_1$-transform to h to obtain the path $h^{(0)} \in \mathcal{P}^{p,p'+p}_{a',b',e,f}(L^{(0)})$. Then obtain $h^{(k)} \in \mathcal{P}^{p,p'+p}_{a',b',e,f}(L^{(k)})$ by applying a $\mathcal{B}_2(k)$-transform to $h^{(0)}$. If $m^{(k)} = m(h^{(k)})$, then $L^{(k)} = L^{(0)} + 2k$, $m^{(k)} = m^{(0)}$ and*

$$\tilde{wt}(h^{(k)}) = \tilde{wt}(h) + \frac{1}{4}((L^{(k)} - m^{(k)})^2 - \beta^2),$$

where $\beta = \beta^{p,p'}_{a,b,e,f}$.

Proof: That $L^{(k)} = L^{(0)} + 2k$ follows immediately from the definition of a $\mathcal{B}_2(k)$-transform. Lemma 3.5 yields:

$$\tilde{wt}(h^{(0)}) = \tilde{wt}(h) + \frac{1}{4}\left((L^{(0)} - m(h^{(0)}))^2 - \beta^2\right). \tag{3.2}$$

Let the striking sequence of $h^{(0)}$ be $\left(\begin{smallmatrix} a_1 & a_2 & \cdots & a_l \\ b_1 & b_2 & \cdots & b_l \end{smallmatrix}\right)^{(e,f,d)}$, and let $\pi = \pi(h^{(0)})$.

If $e = d$, we are restricted to the case $\pi = 0$, since $\delta^{p,p'+p}_{a',e} = 0$ by Lemma C.1. The striking sequence of $h^{(1)}$ is then $\left(\begin{smallmatrix} 0 & 0 & a_1 & a_2 & \cdots & a_l \\ 1 & 1 & b_1 & b_2 & \cdots & b_l \end{smallmatrix}\right)^{(e,f,e)}$. Thereupon $m(h^{(1)}) = \sum_{i=1}^{l} a_i = m(h^{(0)})$. In this case, Lemma 2.2 shows that $\tilde{wt}(h^{(1)}) - \tilde{wt}(h^{(0)}) = 1 + b_1 + b_2 + \cdots + b_l = L^{(0)} - m^{(0)} + 1$.

If $e \neq d$, the striking sequence of $h^{(1)}$ is $\left(\begin{smallmatrix} 0 & a_1+1-\pi & a_2 & \cdots & a_l \\ 1 & b_1+\pi & b_2 & \cdots & b_l \end{smallmatrix}\right)^{(e,f,e)}$. Then $m(h^{(1)}) = 1 - \pi + \sum_{i=1}^{l} a_i$ which equals $m(h^{(0)}) = (e + d + \pi) \bmod 2 + \sum_{i=1}^{l} a_i$ for both $\pi = 0$ and $\pi = 1$. Here, Lemma 2.2 shows that $\tilde{wt}(h^{(1)}) - \tilde{wt}(h^{(0)}) = \pi + b_1 + b_2 + \cdots + b_l$. Since $L^{(0)} - m^{(0)} = -(e + d + \pi) \bmod 2 + b_1 + b_2 + \cdots + b_l$, we once more have $\tilde{wt}(h^{(1)}) - \tilde{wt}(h^{(0)}) = L^{(0)} - m^{(0)} + 1$.

Repeated application of these results, yields $m(h^{(k)}) = m(h^{(0)})$ and

$$\tilde{wt}(h^{(k)}) = \tilde{wt}(h^{(0)}) + k\left(L^{(0)} - m(h^{(0)})\right) + k^2.$$

Then, on using (3.2) and $L^{(k)} = L^{(0)} + 2k$, the lemma follows. □

3.3. Particle moves

In this section, we once more restrict to the case $p' > 2p$ so that the (p, p')-model has no two neighbouring odd bands, and consider only paths $h \in \mathcal{P}^{p,p'}_{a',b',e,f}(L')$ for which either $d(h) = e$ or $\pi(h) = 0$.

We specify four types of local deformations of a path. These deformations will be known as *particle moves*. In each of the four cases, a particular sequence of contiguous segments of a path is changed to a different sequence, the remainder of the path being unchanged. The moves are as follows — the path portion to the left of the arrow is changed to that on the right:

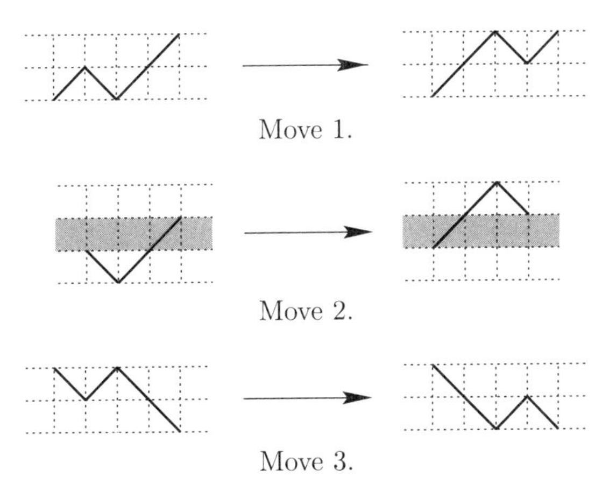

Move 1.

Move 2.

Move 3.

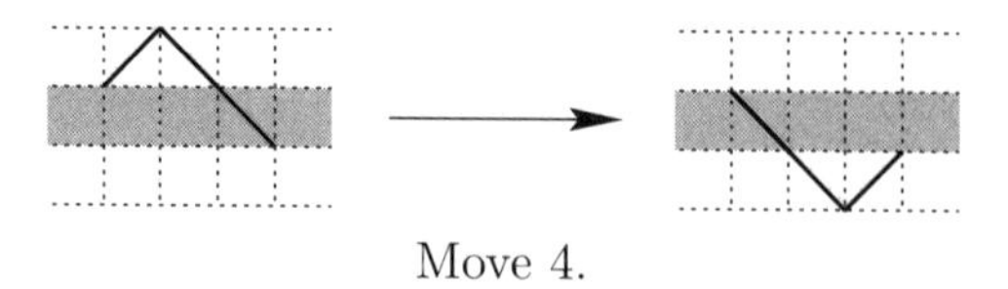

Move 4.

Since $p' > 2p$, each odd band is straddled by a pair of even bands. Thus, there is no impediment to enacting moves 2 and 4 for paths in $\mathcal{P}^{p,p'}_{a',b',e,f}(L)$. Note that moves 1 and 2 are reflections of moves 3 and 4.

In addition to the four moves described above, we permit certain additional deformations of a path close to its left and right extremities in certain circumstances. Each of these moves will be referred to as an *edge-move*. They, together with their validities, are as follows:

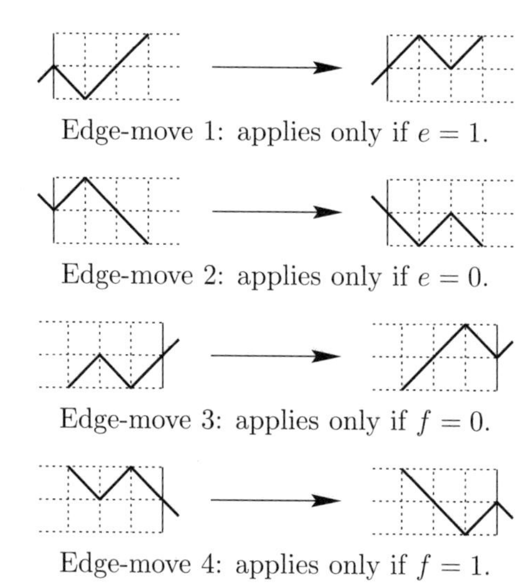

Edge-move 1: applies only if $e = 1$.

Edge-move 2: applies only if $e = 0$.

Edge-move 3: applies only if $f = 0$.

Edge-move 4: applies only if $f = 1$.

With the pre- and post-segments drawn, we see that these edge-moves may be considered as instances of moves 1 and 3 described beforehand.

Lemma 3.7. *Let the path $\hat{h}$ differ from the path h in that one change has been made according to one of the four moves described above, or to one of the four edge-moves described above (subject to their restrictions). Then*

$$\tilde{wt}(\hat{h}) = \tilde{wt}(h) + 1.$$

Additionally, $L(\hat{h}) = L(h)$ and $m(\hat{h}) = m(h)$.

Proof: For each of the four moves and four edge-moves, take the (x, y)-coordinate of the leftmost point of the depicted portion of h to be (x_0, y_0). Now consider the contribution to the weight of the three vertices in question before and after the move (although the vertex immediately before those considered may change, its contribution doesn't). In each of the eight cases, the contribution is $x_0 + y_0 + 1$ before the move and $x_0 + y_0 + 2$ afterwards. Thus $\tilde{wt}(\hat{h}) = \tilde{wt}(h) + 1$. The other statements are immediate on inspecting all eight moves. □

Now observe that for each of the eight moves specified above, the sequence of path segments before the move consists of an adjacent pair of scoring vertices followed by a non-scoring vertex. The specified move replaces this combination with a non-scoring vertex followed by two scoring vertices. As anticipated above, the pair of adjacent scoring vertices is viewed as a particle. Thus each of the above eight moves describes a particle moving to the right by one step.

When $p' > 2p$, so that there are no two adjacent odd bands in the (p,p')-model, and noting that $\delta^{p,p'}_{b',f} = 0$, we see that each sequence comprising two scoring vertices followed by a non-scoring vertex is present amongst the eight configurations prior to a move, except for the case depicted in Fig. 3.4 and its up-down reflection.

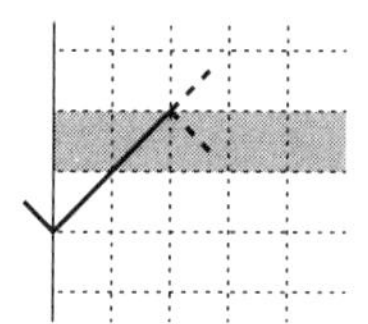

FIGURE 3.4. Not a particle

Only in these cases, where the 0th and 1st segments are scoring and the first two segments are in the same direction, do we *not* refer to the adjacent pair of scoring vertices as a particle.

Also note that when $p' > 2p$ and $\delta^{p,p'}_{a',e} = \delta^{p,p'}_{b',f} = 0$, each sequence of a non-scoring vertex followed by two scoring vertices appears amongst the eight configurations that result from a move. In such cases, the move may thus be reversed.

3.4. The $\mathcal{B}_3$-transform

Since in each of the moves described in Section 3.3, a pair of scoring vertices shifts to the right by one step, we see that a succession of such moves is possible until the pair is followed by another scoring vertex. If this itself is followed by yet another scoring vertex, we forbid further movement. However, if it is followed by a non-scoring vertex, further movement is allowed after considering the latter two of the three consecutive scoring vertices to be the particle (instead of the first two).

As above, let $h^{(k)}$ be a path resulting from a $\mathcal{B}_2(k)$-transform acting on a path that itself is the image of a $\mathcal{B}_1$-transform. We now consider moving the k particles that have been inserted.

Lemma 3.8. *Let $\delta^{p,p'}_{b',f} = 0$. There is a bijection between the set of paths obtained by moving the particles in $h^{(k)}$ and $\mathcal{Y}(k,m)$, where $m = m(h^{(k)})$. This bijection is such that if $\lambda \in \mathcal{Y}(k,m)$ is the bijective image of a particular h then*

$$\tilde{wt}(h) = \tilde{wt}(h^{(k)}) + \mathrm{wt}\,(\lambda).$$

Additionally, $L(h) = L(h^{(k)})$ and $m(h) = m(h^{(k)})$.

Proof: Since each particle moves by traversing a non-scoring vertex, and there are m of these to the right of the rightmost particle in $h^{(k)}$, and there are no consecutive scoring vertices to its right, this particle can make λ_1 moves to the right, with $0 \le \lambda_1 \le m$. Similarly, the next rightmost particle can make λ_2 moves to the right with $0 \le \lambda_2 \le \lambda_1$. Here, the upper restriction arises because the two scoring vertices would then be adjacent to those of the first particle. Continuing in this way, we obtain that all possible final positions of the particles are indexed by $\lambda = (\lambda_1, \lambda_2, \dots, \lambda_k)$ with $m \geq \lambda_1 \geq \lambda_2 \geq \cdots \geq \lambda_k \geq 0$, that is, by partitions of at most k parts with no part exceeding m. Moreover, since by Lemma 3.7 the weight increases by one for each move, the weight increase after the sequence of moves specified by a particular λ is equal to $\mathrm{wt}\,(\lambda)$. The final statement also follows from Lemma 3.7. □

We say that a path obtained by moving the particles in $h^{(k)}$ according to the partition λ has been obtained by the action of a $\mathcal{B}_3(\lambda)$-transform.

Having defined $\mathcal{B}_1$, $\mathcal{B}_2(k)$ for $k \geq 0$ and $\mathcal{B}_3(\lambda)$ for λ a partition with at most k parts, we now define a $\mathcal{B}(k,\lambda)$-transform as the composition $\mathcal{B}(k,\lambda) = \mathcal{B}_3(\lambda) \circ \mathcal{B}_2(k) \circ \mathcal{B}_1$.

Lemma 3.9. *Let $h' \in \mathcal{P}^{p,p'+p}_{a',b',e,f}(L')$ be obtained from $h \in \mathcal{P}^{p,p'}_{a,b,e,f}(L)$ by the action of the $\mathcal{B}(k,\lambda)$-transform. If $\pi = \pi(h)$ and $m = m(h)$ then:*

- $L' = \begin{cases} 2L - m + 2k + 2 & \text{if } \pi = 1 \text{ and } e = d, \\ 2L - m + 2k & \text{otherwise}; \end{cases}$
- $m(h') = L$;
- $\tilde{wt}(h') = \tilde{wt}(h) + \frac{1}{4}\left((L'-L)^2 - \beta^2\right) + \mathrm{wt}\,(\lambda)$,

where $\beta = \beta^{p,p'}_{a,b,e,f}$.

Proof: These results follow immediately from Lemmas 3.2, 3.6 and 3.8. □

Note 3.10. *Since particle insertion and the particle moves don't change the startpoint, endpoint or value $e(h)$ or $f(h)$ of a path h, then in view of Lemma C.1 and Corollary 3.3, we see that the action of a $\mathcal{B}$-transform on $h \in \mathcal{P}^{p,p'}_{a,b,e,f}(L)$ yields a path $h' \in \mathcal{P}^{p,p'+p}_{a',b',e,f}(L')$, where $a' = a + \lfloor ap/p' \rfloor + e$, $b' = b + \lfloor bp/p' \rfloor + f$, and $\delta^{p,p'+p}_{a',e} = \delta^{p,p'+p}_{b',f} = 0$.*

3.5. Particle content of a path

We again restrict to the case $p' > 2p$ so that the (p,p')-model has no two neighbouring odd bands, and let $h' \in \mathcal{P}^{p,p'}_{a',b',e,f}(L')$. In the following lemma, we once more restrict to the cases for which $\delta^{p,p'}_{a',e} = \delta^{p,p'}_{b',f} = 0$, and thus only consider the cases for which the pre-segment and the post-segment of h' lie in even bands.

Lemma 3.11. *For $1 \leq p < p'$ with $p' > 2p$, let $1 \leq a', b' < p'$ and $e, f \in \{0,1\}$, with $\delta^{p,p'}_{a',e} = \delta^{p,p'}_{b',f} = 0$. Set $a = a' - \lfloor a'p/p' \rfloor - e$ and $b = b' - \lfloor b'p/p' \rfloor - f$. If $h' \in \mathcal{P}^{p,p'}_{a',b',e,f}(L')$, then there is a unique triple (h,k,λ) where $h \in \mathcal{P}^{p,p'-p}_{a,b,e,f}(L)$ for some L, such that the action of a $\mathcal{B}(k,\lambda)$-transform on h results in h'.*

Proof: This is proved by reversing the constructions described in the previous sections. Locate the leftmost pair of consecutive scoring vertices in h', and move them leftward by reversing the particle moves, until they occupy the 0th and 1st positions. This is possible in all cases when $\delta^{p,p'}_{a',e} = \delta^{p,p'}_{b',f} = 0$. Now ignoring these two vertices, do the same with the next leftmost pair of consecutive scoring vertices, moving them leftward until they occupy the second and third positions. Continue in this way until all consecutive scoring vertices occupy the leftmost positions of the path. Denote this path by $h^{(\cdot)}$. At the leftmost end of $h^{(\cdot)}$, there will be a number of even segments (possibly zero) alternating in direction. Let this number be $2k$ or $2k+1$ according to whether is it even or odd. Clearly h' results from $h^{(\cdot)}$ by a $\mathcal{B}_3(\lambda)$-transform for a particular λ with at most k parts.

Removing the first $2k$ segments of $h^{(\cdot)}$ yields a path $h^{(0)} \in \mathcal{P}^{p,p'}_{a',b',e,f}$. This path thus has no two consecutive scoring vertices, except possibly at the 0th and 1st positions, and then only if the first vertex is a straight vertex (as in Fig. 3.4). Moreover, $h^{(k)}$ arises by the action of a $\mathcal{B}_2(k)$-transform on $h^{(0)}$.

Ignoring for the moment the case where there are scoring vertices at the 0th and 1st positions, $h^{(0)}$ has by construction no pair of consecutive scoring vertices. Therefore, beyond the 0th vertex, we may remove a non-scoring vertex before every scoring vertex (and increment the length of the first line if $\pi(h') = 1$ and $e(h') \neq f(h')$) to obtain a path $h \in \mathcal{P}^{p,p'-p}_{a,b,e,f}(L)$ for some L. Since $\lfloor a'p/p' \rfloor = \lfloor a'p/(p'-p) \rfloor$, and $\lfloor b'p/p' \rfloor = \lfloor b'p/(p'-p) \rfloor$ by Lemma C.1, and thus $a' = a + \lfloor ap/(p'-p) \rfloor + e$ and $b' = b + \lfloor bp/(p'-p) \rfloor + f$, it follows that $h^{(0)}$ arises by the action of a $\mathcal{B}_1$-transform on h.

On examining the third case depicted in Table 3.1, we see that the case where $h^{(0)}$ has a pair of scoring vertices at the 0th and 1st positions, arises similarly from a particular $h \in \mathcal{P}^{p,p'-p}_{a,b,e,f}(L)$ for some L. The lemma is then proved. □

The value of k obtained above will be referred to as the particle content of h'.

We could now proceed as in Lemma 3.13 and Corollary 3.14 of [**25**] to, for appropriate parameters, give a weight-respecting bijection between $\bigcup_k \mathcal{P}^{p,p'}_{a,b,e,f}(m_1, 2k + 2m_1 - m_0) \times \mathcal{Y}(k, m_1)$ and $\mathcal{P}^{p,p'+p}_{a',b',e,f}(m_0, m_1)$, and state the corresponding relationship between the generating functions. However, something more general is required in this paper. This will be derived in Lemma 5.7 and Corollary 5.8 for certain restricted paths once the notation is developed in Section 5.

4. The $\mathcal{D}$-transform

4.1. The pure $\mathcal{D}$-transform

The $\mathcal{D}$*-transform* is defined to act on each $h \in \mathcal{P}^{p,p'}_{a,b,e,f}(L)$ to yield a path $\hat{h} \in \mathcal{P}^{p'-p,p'}_{a,b,1-e,1-f}(L)$ with exactly the same sequence of integer heights, i.e., $\hat{h}_i = h_i$ for $0 \le i \le L$. Note that, by definition, $e(\hat{h}) = 1 - e(h)$ and $f(\hat{h}) = 1 - f(h)$.

Since the band structure of the $(p'-p, p')$-model is obtained from that of the (p, p')-model simply by replacing odd bands by even bands and vice-versa, then, apart from the vertex at $i = 0$, each scoring vertex maps to a non-scoring vertex and vice-versa. That $e(h)$ and $e(\hat{h})$ differ implies that the vertex at $i = 0$ is both scoring or both non-scoring in h and $\hat{h}$.

Lemma 4.1. *Let* $\hat{h} \in \mathcal{P}^{p'-p,p'}_{a,b,1-e,1-f}(L)$ *be obtained from* $h \in \mathcal{P}^{p,p'}_{a,b,e,f}(L)$ *by the action of the* $\mathcal{D}$*-transform. Then* $\pi(\hat{h}) = 1 - \pi(h)$*. Moreover, if* $m = m(h)$ *then:*

- $L(\hat{h}) = L$;
- $m(\hat{h}) = \begin{cases} L - m & \text{if } e + d + \pi(h) \equiv 0 \,(mod 2), \\ L - m + 2 & \text{if } e + d + \pi(h) \not\equiv 0 \,(mod 2); \end{cases}$
- $\tilde{wt}(\hat{h}) = \frac{1}{4}\left(L^2 - \alpha(h)^2\right) - \tilde{wt}(h)$.

Proof: Let h have striking sequence $\begin{pmatrix} a_1 & a_2 & a_3 & \cdots & a_l \\ b_1 & b_2 & b_3 & \cdots & b_l \end{pmatrix}^{(e,f,d)}$. Since, beyond the zeroth vertex, the $\mathcal{D}$-transform exchanges scoring vertices for non-scoring vertices and vice-versa, it follows that the striking sequence for $\hat{h}$ is $\begin{pmatrix} b_1 & b_2 & b_3 & \cdots & b_l \\ a_1 & a_2 & a_3 & \cdots & a_l \end{pmatrix}^{(1-e,1-f,d)}$. It is immediate that $L(\hat{h}) = L$, $\pi(\hat{h}) = 1 - \pi(h)$, $e(\hat{h}) = 1 - e(h)$ and $d(\hat{h}) = d(h)$. Then $m(\hat{h}) = (e(\hat{h}) + d(\hat{h}) + \pi(\hat{h})) \bmod 2 + \sum_{i=1}^{l} b_i = (e + d + \pi(h)) \bmod 2 + L - \sum_{i=1}^{l} a_i = 2((e + d + \pi(h)) \bmod 2) + L - m(h)$.

Now let $w_i = a_i + b_i$ for $1 \le i \le l$. Then, using Lemma 2.2, we obtain:

$$\begin{aligned} \tilde{wt}(h) + \tilde{wt}(\hat{h}) &= \sum_{i=1}^{l} b_i(w_{i-1} + w_{i-3} + \cdots + w_{1+i \bmod 2}) \\ &\quad + \sum_{i=1}^{l} a_i(w_{i-1} + w_{i-3} + \cdots + w_{1+i \bmod 2}) \\ &= \sum_{i=1}^{l} w_i(w_{i-1} + w_{i-3} + \cdots + w_{1+i \bmod 2}) \\ &= (w_1 + w_3 + w_5 + \cdots)(w_2 + w_4 + w_6 + \cdots). \end{aligned}$$

The lemma then follows because $(w_1 + w_3 + w_5 + \cdots) + (w_2 + w_4 + w_6 + \cdots) = L$ and $(w_1 + w_3 + w_5 + \cdots) - (w_2 + w_4 + w_6 + \cdots) = \pm\alpha(h)$. $\square$

Lemma 4.2. *If $1 \le p < p'$ with p coprime to p', $1 \le a, b < p'$ and $e, f \in \{0,1\}$ then $\alpha^{p'-p,p'}_{a,b} = \alpha^{p,p'}_{a,b}$ and $\beta^{p'-p,p'}_{a,b,1-e,1-f} + \beta^{p,p'}_{a,b,e,f} = \alpha^{p,p'}_{a,b}$.*

Proof: Lemma C.2 gives $\lfloor ap/p' \rfloor + \lfloor a(p'-p)/p' \rfloor = a - 1$ and likewise, $\lfloor bp/p' \rfloor + \lfloor b(p'-p)/p' \rfloor = b - 1$. The required results then follow immediately. $\square$

4.2. The $\mathcal{BD}$-pair

It will often be convenient to consider the combined action of a $\mathcal{D}$-transform followed immediately by a $\mathcal{B}$-transform. Such a pair will naturally be referred to as a $\mathcal{BD}$-transform and maps a path $h \in \mathcal{P}^{p'-p,p'}_{a,b,e,f}(L)$ to a path $h' \in \mathcal{P}^{p,p'+p}_{a',b',1-e,1-f}(L')$, where a', b', L' are determined by our previous results.

In what follows, the $\mathcal{BD}$-transform will always follow a $\mathcal{B}$-transform. Thus we restrict consideration to where $2(p'-p) < p'$.

Lemma 4.3. *With $p' < 2p$, let $h \in \mathcal{P}^{p'-p,p'}_{a,b,e,f}(L)$. Let $h' \in \mathcal{P}^{p,p'+p}_{a',b',1-e,1-f}(L')$ result from the action of a $\mathcal{D}$-transform on h, followed by a $\mathcal{B}(k,\lambda)$-transform. Then:*

- $L' = \begin{cases} L + m(h) + 2k - 2 & \text{if } \pi(h) = 1 \text{ and } e = d(h), \\ L + m(h) + 2k & \text{otherwise;} \end{cases}$
- $m(h') = L$;
- $\tilde{wt}(h') = \frac{1}{4}\left(L^2 + (L'-L)^2 - \alpha^2 - \beta^2\right) + \mathrm{wt}(\lambda) - \tilde{wt}(h)$,

where $\alpha = \alpha^{p,p'}_{a,b}$ and $\beta = \beta^{p,p'}_{a,b,1-e,1-f}$.

Proof: Let $\hat{h}$ result from the action of the $\mathcal{D}$-transform on h, and let $d = d(h)$, $\pi = \pi(h)$, $\hat{e} = e(\hat{h})$ $\hat{d} = d(\hat{h})$, $\hat{\pi} = \pi(\hat{h})$. Then we immediately have $\hat{d} = d$, $\hat{e} = 1 - e$, and $\hat{\pi} = 1 - \pi$.

In the case where $\pi = 0$ and $e \ne d$, we then have, using Lemmas 3.9 and 4.1, $L' = 2L(\hat{h}) - m(\hat{h}) + 2k + 2 = 2L - (L - m(h) + 2) + 2k + 2 = L + m(h) + 2k$.

In the case where $\pi = 1$ and $e = d$, we then have, using Lemmas 3.9 and 4.1, $L' = 2L(\hat{h}) - m(\hat{h}) + 2k + 2 = 2L - (L - m(h) + 2) + 2k = L + m(h) + 2k - 2$.

In the other cases, $e + d + \pi \equiv 0 \pmod 2$ and so $\hat{e} + \hat{d} + \hat{\pi} \equiv 0 \pmod 2$. Lemmas 3.9 and 4.1 yield $L' = 2L(\hat{h}) - m(\hat{h}) + 2k = 2L - (L - m(h)) + 2k = L + m(h) + 2k$.

The expressions for $m(h')$ and $\tilde{wt}(h')$ follow immediately from Lemmas 3.9, 4.1 and 2.3. $\square$

We could now proceed as in Lemma 4.5 and Corollary 4.6 of [**25**] to show that the $\mathcal{BD}$-transform implies, for certain parameters, a weight respecting bijection between $\bigcup_k \mathcal{P}^{p'-p,p'}_{a,b,e,f}(m_1, m_0 - m_1 - 2k) \times \mathcal{Y}(k, m_1)$ and $\mathcal{P}^{p,p'+p}_{a',b',1-e,1-f}(m_0, m_1)$, and consequently a relationship between the corresponding generating functions. However again, something more general is required in the current paper. This is developed in Section 5, resulting in Lemma 5.9 and Corollary 5.10.

We finish this section by examining how the $\mathcal{BD}$-transform affects some of the parameters.

Lemma 4.4. *Let* $1 \leq p < p' < 2p$ *with* p *coprime to* p', $1 \leq a, b < p'$ *and* $e, f \in \{0,1\}$, *and set* $a' = a + 1 - e + \lfloor ap/p' \rfloor$ *and* $b' = b + 1 - f + \lfloor bp/p' \rfloor$. *Then* $\alpha^{p,p'+p}_{a',b'} = 2\alpha^{p'-p,p'}_{a,b} - \beta^{p'-p,p'}_{a,b,e,f}$ *and* $\beta^{p,p'+p}_{a',b',1-e,1-f} = \alpha^{p'-p,p'}_{a,b} - \beta^{p'-p,p'}_{a,b,e,f}$.

Proof: By using first Lemma 3.4 and then Lemma 4.2, we obtain: $\alpha^{p,p'+p}_{a',b'} = \alpha^{p,p'}_{a,b} + \beta^{p,p'}_{a,b,1-e,1-f} = 2\alpha^{p'-p,p'}_{a,b} - \beta^{p'-p,p'}_{a,b,e,f}$. By a similar route, we obtain: $\beta^{p,p'+p}_{a',b',1-e,1-f} = \beta^{p,p'}_{a,b,1-e,1-f} = \alpha^{p'-p,p'}_{a,b} - \beta^{p'-p,p'}_{a,b,e,f}$. □

5. Mazy runs

In this section, we examine paths that are constrained to attain certain heights in a certain order. We will find that the subset of $\mathcal{P}^{p,p'}_{a,b,e,f}(L)$ that corresponds to an individual term $F_{a,b}(\boldsymbol{u}^L, \boldsymbol{u}^R, L)$ as defined in Section 1.14, may be characterised in such a way.

For $1 \le a < p'$, let a be interfacial in the (p,p')-model, and let $r = \rho^{p,p'}(a)$ so that a borders the rth odd band. Then $a = \lfloor p'r/p \rfloor$ if and only if $r = \lfloor pa/p' \rfloor + 1$, and $a = \lfloor p'r/p \rfloor + 1$ if and only if $r = \lfloor pa/p' \rfloor$. In the former case, we write $\omega^{p,p'}(a) = r^-$, and in the latter case we write $\omega^{p,p'}(a) = r^+$. It will also be convenient to define $\omega^{p,p'}(a) = \infty$ when a is not interfacial (simply so that certain statements make sense). When p and p' are implicit, we write $\omega(a) = \omega^{p,p'}(a)$.

5.1. The passing sequence

For each path $h \in \mathcal{P}^{p,p'}_{a,b,e,f}(L)$, the passing sequence $\omega(h)$ of h is a word that indicates how the path h meanders between the interfacial values. It is a word in the elements of $\mathcal{R}^{p,p'} = \{1^-, 1^+, 2^-, 2^+, \ldots, (p-1)^-, (p-1)^+\}$ and is obtained as follows. First start with an empty word. For $i = 0, 1, \ldots, L$, if h_i is interfacial, append $\omega(h_i)$ to the word. Then replace each subsequence of consecutive identical symbols within this word by a single instance of that symbol. We define $\omega(h)$ to be this resulting word.

For example, in the case of the path $h \in \mathcal{P}^{3,8}_{2,4,e,1}(14)$ depicted in Fig. 2.1, we obtain the passing sequence $\omega(h) = 1^-1^+2^-2^+2^-2^+2^-1^+$.

In the case $p' < 2p$, it will be useful to note that not all the elements of $\mathcal{R}^{p,p'}$ can occur in a passing sequence, because some of them correspond to multifacial heights. The set of elements that cannot occur in this $p' < 2p$ case is thus given by:

$$\overline{\mathcal{R}}^{p,p'} = \{(r-1)^+, r^- : 1 < r < p', \lfloor (r-1)p'/p \rfloor = \lfloor rp'/p \rfloor - 1\} \cup \{1^-, (p-1)^+\}.$$

The utility of the passing function lies in its near invariance under the $\mathcal{B}$-transform, and its simple transformation under the $\mathcal{BD}$-transform.

Lemma 5.1. *Let $p' > 2p$ and let $h' \in \mathcal{P}^{p,p'+p}_{a',b',e,f}(L')$ be obtained from $h \in \mathcal{P}^{p,p'}_{a,b,e,f}(L)$ through the action of a $\mathcal{B}(k,\lambda)$-transform. Then either $\omega(h') = \omega(h)$, $\omega(a)\omega(h') = \omega(h)$, $\omega(h')\omega(b) = \omega(h)$, or $\omega(a)\omega(h')\omega(b) = \omega(h)$.*

Moreover, if either $\omega(a)\omega(h') = \omega(h)$ or $\omega(a)\omega(h')\omega(b) = \omega(h)$ then

$$\delta^{p,p'}_{a,e} = 1 \textit{ and } h_1 = a - (-1)^e,$$

and if either $\omega(h') = \omega(h)\omega(b)$ or $\omega(h') = \omega(a)\omega(h)\omega(b)$ then

$$\delta^{p,p'}_{b,f} = 1 \textit{ and } h_{L-1} = b - (-1)^f.$$

Proof: Let $h^{(0)}$ result from the action of the $\mathcal{B}_1$-transform on h. The discussion of the $\mathcal{B}_1$-transform immediately following Note 3.1 together with Tables 3.1 and 3.2 show that either $\omega(h^{(0)}) = \omega(h)$, $\omega(a)\omega(h^{(0)}) = \omega(h)$, $\omega(h^{(0)})\omega(b) = \omega(h)$, or $\omega(a)\omega(h^{(0)})\omega(b) = \omega(h)$. In addition (see Table 3.1) $\omega(a)\omega(h^{(0)}) = \omega(h)$ or $\omega(a)\omega(h^{(0)})\omega(b) = \omega(h)$, only if $\delta^{p,p'}_{a,e} = 1$ and $h_1 = a - (-1)^e$. Similarly (see Table 3.2) $\omega(h^{(0)})\omega(b) = \omega(h)$ or $\omega(a)\omega(h^{(0)})\omega(b) = \omega(h)$, only if $\delta^{p,p'}_{b,f} = 1$ and $h_1 = b - (-1)^f$.

Now let $h^{(k)}$ be the result of applying the $\mathcal{B}_2(k)$-transform of Section 3.2 to $h^{(0)}$. Noting that because, when $p' > 2p$, neighbouring odd bands in the $(p, p'+p)$-model are separated by at least two even bands, we see that inserting particles into $h^{(0)}$ can change the passing sequence only if $a' + (-1)^e$ is interfacial and $h^{(0)}_1 = a' - (-1)^e$. This occurs only in that case where $\delta^{p,p'}_{a,e} = 1$ and $h_1 = a - (-1)^e$. In this case, $\omega(h^{(k)}) = \omega(a)\omega(h^{(0)}) = \omega(h)$. Otherwise $\omega(h^{(k)}) = \omega(h^{(0)})$.

Now let h' be the result of applying the $\mathcal{B}_3(\lambda)$-transform to $h^{(k)}$, By inspection, again noting that neighbouring odd bands are separated by at least two even bands, each of the moves from Section 3.3 does not change the passing sequence. This is not necessarily true of the edge-moves. Edge-moves 1 and 2 can change the passing sequence only if $a' + (-1)^e$ is interfacial, $\omega(h^{(k)})$ begins with $\omega(a)$, and $h_1 = a - (-1)^e$. In this case, the effect of the edge-move is to remove the $\omega(a)$. Edge-moves 3 and 4 can change the passing sequence only if $b' + (-1)^f$ is interfacial, $\omega(h^{(k)})$ does not end with $\omega(b)$, and $h_1 = b - (-1)^f$. In this case, the effect of the edge-move is to append $\omega(b)$. Because $a' + (-1)^e$ is interfacial implies that $\delta^{p,p'}_{a,e} = 1$, and $b' + (-1)^f$ is interfacial implies that $\delta^{p,p'}_{b,f} = 1$, the lemma then follows. □

For the case $p' < 2p$, the above result doesn't hold and $\omega(h')$ can be very different from $\omega(h)$. However, in this case, we can make do with the following result. For the sake of clarification, by a subword ω' of $\omega = \omega_1\omega_2\cdots\omega_k$, we mean a sequence $\omega' = \omega_{i_1}\omega_{i_2}\cdots\omega_{i_{k'}}$ where $1 \le i_1 < i_2 < \cdots < i_{k'} \le k$.

Lemma 5.2. *Let $p' < 2p$ and let $h' \in \mathcal{P}^{p,p'+p}_{a',b',e,f}(L')$ be obtained from the path $h \in \mathcal{P}^{p,p'}_{a,b,e,f}(L)$ through the action of a $\mathcal{B}(k,\lambda)$-transform. Let $\omega' = \omega'_1\omega'_2\cdots\omega'_{k'}$ be the longest subword of $\omega(h')$ for which $\omega'_i \ne \omega'_{i+1}$ for $1 \le i < k'$, that comprises the symbols in $\mathcal{R}^{p,p'}\backslash\overline{\mathcal{R}}^{p,p'}$. Then either $\omega(h') = \omega(h)$, $\omega(a)\omega(h') = \omega(h)$, $\omega(h')\omega(b) = \omega(h)$, or $\omega(a)\omega(h')\omega(b) = \omega(h)$.*

Moreover, if either $\omega(a)\omega(h') = \omega(h)$ or $\omega(a)\omega(h')\omega(b) = \omega(h)$ then $\delta^{p,p'}_{a,e} = 1$ and $h_1 = a - (-1)^e$, and if either $\omega(h')\omega(b) = \omega(h)$ or $\omega(a)\omega(h')\omega(b) = \omega(h)$ then $\delta^{p,p'}_{b,f} = 1$ and $h_{L-1} = b - (-1)^f$.

Proof: Note first that the heights in the $(p, p'+p)$-model that correspond to the symbols from $\mathcal{R}^{p,p'}\backslash\overline{\mathcal{R}}^{p,p'}$ are either separated by a single odd band or at least two even bands. We then compare the actions of the $\mathcal{B}_1$-, $\mathcal{B}_2(k)$-, and $\mathcal{B}_3(k,\lambda)$-transforms to those described in the proof of Lemma 5.1. As there, $\omega(a)$ might get removed from the start of the passing sequence if a is interfacial in the (p,p')-model, $\delta^{p,p'}_{a,e} = 1$ and $h_1 = a - (-1)^e$. Similarly $\omega(b)$ might get removed from the end of of the passing sequence if b is interfacial in the (p,p')-model, $\delta^{p,p'}_{b,f} = 1$ and $h_{L-1} = b - (-1)^f$. Apart from these changes, the action of the $\mathcal{B}_1$-transform can only insert elements from $\overline{\mathcal{R}}^{p,p'}$ into the passing sequence. The same is true of the $\mathcal{B}_2(k)$-transform. In the case of the $\mathcal{B}_3(k)$-transform, we see that only those symbols that correspond to heights separated by a single even band might get inserted or removed. The lemma then follows. $\square$

Since heights that are interfacial in the $(p'-p,p')$-model are also interfacial in the (p,p')-model and vice-versa, we see that the effect of a $\mathcal{D}$-transform on a passing sequence is just to change the name of each symbol. So as to be able to describe this change, for $1 \le r < (p'-p)$ set $R = \lfloor rp/(p'-p) \rfloor$ and define:

$$\sigma(r^-) = R^+;$$
$$\sigma(r^+) = (R+1)^-,$$

Then, if $\omega(h) = \omega_1\omega_2\cdots\omega_k$, define $\sigma(\omega(h)) = \sigma(\omega_1)\sigma(\omega_2)\cdots\sigma(\omega_k)$.

Lemma 5.3. *Let $p' < 2p$ and let $h' \in \mathcal{P}^{p,p'+p}_{a',b',1-e,1-f}(L')$ be obtained from the path $h \in \mathcal{P}^{p'-p,p'}_{a,b,e,f}(L)$ through the action of a $\mathcal{BD}$-transform. If $\omega' = \omega'_1\omega'_2\cdots\omega'_{k'}$ is the longest subword of $\omega(h')$ for which $\omega'_i \ne \omega'_{i+1}$ for $1 \le i < k'$, and which comprises the symbols in $\{\sigma(x) : x \in \mathcal{R}^{p'-p,p'}\}$, then either $\omega' = \sigma(\omega(h))$, $\omega(a)\omega' = \sigma(\omega(h))$, $\omega'\omega(b) = \sigma(\omega(h))$, or $\omega(a)\omega'\omega(b) = \sigma(\omega(h))$.*

Moreover, if either $\omega(a)\omega' = \sigma(\omega(h))$ or $\omega(a)\omega'\omega(b) = \sigma(\omega(h))$ then a is interfacial in the (p,p')-model, $\delta^{p,p'}_{a,e} = 1$ and $h_1 = a + (-1)^e$, and if either $\omega'\omega(b) = \sigma(\omega(h))$ or $\omega(a)\omega'\omega(b) = \sigma(\omega(h))$ then b is interfacial in the (p,p')-model, $\delta^{p,p'}_{b,f} = 1$ and $h_{L-1} = b + (-1)^f$.

Proof: For $1 \le r < (p'-p)$, the lowermost edge of the rth odd band in the $(p'-p,p')$-model is at height $a = \lfloor rp'/(p'-p) \rfloor$. Since $p' > 2(p'-p)$, no two odd bands in the $(p'-p,p')$-model are adjacent and so a is the height of the uppermost edge of the Rth even band since $R = \lfloor rp'/(p'-p) \rfloor - r$. Thus, in the (p,p')-model, the uppermost edge of the Rth odd band is at height a. We also see that the uppermost edge of the rth odd band in the $(p'-p,p')$-model is at height $a+1$ which is also the height of the lowermost edge of the $(R+1)$th odd band of the (p,p')-model.

Thus if $\hat{h} \in \mathcal{P}^{p,p'}_{a,b,1-e,1-f}(L)$ results from a $\mathcal{D}$-transform acting on h, then we immediately obtain $\omega(\hat{h}) = \sigma(\omega(h))$. The lemma then follows on applying Lemma 5.2 to $\hat{h}$. $\square$

5.2. The passing function

Let $p' > 2p$, $1 \le a, b < p'$, $e, f \in \{0,1\}$ and $L \ge 0$. We now define certain subsets of $\mathcal{P}^{p,p'}_{a,b,e,f}(L)$ that depend on d^L-dimensional vectors $\boldsymbol{\mu} = (\mu_1, \mu_2, \ldots, \mu_{d^L})$ and $\boldsymbol{\mu}^* = (\mu^*_1, \mu^*_2, \ldots, \mu^*_{d^L})$, and d^R-dimensional vectors $\boldsymbol{\nu} = (\nu_1, \nu_2, \ldots, \nu_{d^R})$ and $\boldsymbol{\nu}^* = (\nu^*_1, \nu^*_2, \ldots, \nu^*_{d^R})$, where $d^L, d^R \ge 0$. We say that a path $h \in \mathcal{P}^{p,p'}_{a,b,e,f}(L)$ is *mazy-compliant* (with $\boldsymbol{\mu}, \boldsymbol{\mu}^*, \boldsymbol{\nu}, \boldsymbol{\nu}^*$) if the following five conditions are satisfied:

(1) for $1 \le j \le d^L$, there exists i with $0 \le i \le L$ such that $h_i = \mu^*_j$;
(2) for $1 \le j \le d^L$, if there exists i' with $0 \le i' \le L$ such that $h_{i'} = \mu_j$, then there exists i with $0 \le i < i'$ such that $h_i = \mu^*_j$;
(3) for $1 \le j \le d^R$, there exists i with $0 \le i \le L$ such that $h_i = \nu^*_j$;

(4) for $1 \le j \le d^R$, if there exists i' with $0 \le i' \le L$ such that $h_{i'} = \nu_j$, then there exists i with $i' < i \le L$ such that $h_i = \nu_j^*$;
(5) if $d^L, d^R > 0$ and i is the smallest value such that $0 \le i \le L$ and $h_i = \mu_1^*$ and i' is the largest value such that $0 \le i' \le L$ and $h_{i'} = \nu_1^*$, then $i \le i'$.

Loosely speaking, these are the paths which, for $1 \le j \le d^L$, attain μ_j^*, and do so before they attain any μ_j; which, for $1 \le k \le d^R$, attain ν_k^*, and do so after they attain any ν_k; and which, if $d^L, d^R > 0$, attain ν_1^* after they attain the first μ_1^*.

The set $\mathcal{P}^{p,p'}_{a,b,e,f}(L)\left\{\begin{smallmatrix}\boldsymbol{\mu}\ ;\ \boldsymbol{\nu}\\ \boldsymbol{\mu}^*;\ \boldsymbol{\nu}^*\end{smallmatrix}\right\}$ is defined to be the subset of $\mathcal{P}^{p,p'}_{a,b,e,f}(L)$ comprising those paths h which are mazy-compliant with $\boldsymbol{\mu}, \boldsymbol{\mu}^*, \boldsymbol{\nu}, \boldsymbol{\nu}^*$. The generating function for these paths is defined to be:

$$\tilde{\chi}^{p,p'}_{a,b,e,f}(L;q)\left\{\begin{matrix}\boldsymbol{\mu}\ ;\boldsymbol{\nu}\\ \boldsymbol{\mu}^*;\boldsymbol{\nu}^*\end{matrix}\right\} = \sum_{h\in\mathcal{P}^{p,p'}_{a,b,e,f}(L)\left\{\begin{smallmatrix}\boldsymbol{\mu}\ ;\ \boldsymbol{\nu}\\ \boldsymbol{\mu}^*;\ \boldsymbol{\nu}^*\end{smallmatrix}\right\}} q^{\tilde{wt}(h)}. \tag{5.1}$$

We also define $\mathcal{P}^{p,p'}_{a,b,e,f}(L,m)\left\{\begin{smallmatrix}\boldsymbol{\mu}\ ;\ \boldsymbol{\nu}\\ \boldsymbol{\mu}^*;\ \boldsymbol{\nu}^*\end{smallmatrix}\right\}$ to be the subset of $\mathcal{P}^{p,p'}_{a,b,e,f}(L,m)$ comprising those paths h which are mazy-compliant with $\boldsymbol{\mu}, \boldsymbol{\mu}^*, \boldsymbol{\nu}, \boldsymbol{\nu}^*$. We then define the generating function:

$$\tilde{\chi}^{p,p'}_{a,b,e,f}(L,m;q)\left\{\begin{matrix}\boldsymbol{\mu}\ ;\boldsymbol{\nu}\\ \boldsymbol{\mu}^*;\boldsymbol{\nu}^*\end{matrix}\right\} = \sum_{h\in\mathcal{P}^{p,p'}_{a,b,e,f}(L,m)\left\{\begin{smallmatrix}\boldsymbol{\mu}\ ;\ \boldsymbol{\nu}\\ \boldsymbol{\mu}^*;\ \boldsymbol{\nu}^*\end{smallmatrix}\right\}} q^{\tilde{wt}(h)}.$$

The following result is proved in exactly the same way as Lemma 2.4:

Lemma 5.4. *Let* $\beta = \beta^{p,p'}_{a,b,e,f}$. *Then:*

$$\tilde{\chi}^{p,p'}_{a,b,e,f}(L;q)\left\{\begin{matrix}\boldsymbol{\mu}\ ;\boldsymbol{\nu}\\ \boldsymbol{\mu}^*;\boldsymbol{\nu}^*\end{matrix}\right\} = \sum_{\substack{m\equiv L+\beta\\ (\mathrm{mod}\,2)}} \tilde{\chi}^{p,p'}_{a,b,e,f}(L,m;q)\left\{\begin{matrix}\boldsymbol{\mu}\ ;\boldsymbol{\nu}\\ \boldsymbol{\mu}^*;\boldsymbol{\nu}^*\end{matrix}\right\}.$$

Our vectors $\boldsymbol{\mu}$, $\boldsymbol{\mu}^*$, $\boldsymbol{\nu}$, $\boldsymbol{\nu}^*$, will satisfy certain constraints. We say that $\boldsymbol{\mu}$ and $\boldsymbol{\mu}^*$ are a *mazy-pair* in the (p,p')-model sandwiching a if they satisfy the following three conditions:

(1) $0 \le \mu_j, \mu_j^* \le p'$ for $1 \le j \le d^L$ (we permit $\mu_j^* = 0$, $\mu_j^* = p'$ although the set of paths constrained by such parameters will be empty);
(2) $\mu_j \ne \mu_j^*$ for $1 \le j \le d^L$;
(3) a is strictly between[1] μ_{d^L} and $\mu^*_{d^L}$, and for $1 \le j < d^L$, both μ_{j+1} and μ^*_{j+1} are between μ_j and μ_j^*.

If in addition,

4. μ_j and μ_j^* are interfacial in the (p,p')-model for $1 \le j \le d^L$,

we say that $\boldsymbol{\mu}$ and $\boldsymbol{\mu}^*$ are an *interfacial mazy-pair* in the (p,p')-model sandwiching a. If $\boldsymbol{\mu}$ and $\boldsymbol{\mu}^*$ are a mazy-pair (resp. interfacial mazy-pair) in the (p,p')-model sandwiching a, and $\boldsymbol{\nu}$ and $\boldsymbol{\nu}^*$ are a mazy-pair (resp. interfacial mazy-pair) in the (p,p')-model sandwiching b, then we say that $\boldsymbol{\mu}$, $\boldsymbol{\mu}^*$, $\boldsymbol{\nu}$, $\boldsymbol{\nu}^*$, are a *mazy-four* (resp. *interfacial mazy-four*) in the (p,p')-model sandwiching (a,b).

---

[1]By x between y and z, we mean that either $y \le x \le z$ or $z \le x \le y$. By x strictly between y and z, we mean that either $y < x < z$ or $z < x < y$.

Lemma 5.5. *Let $\boldsymbol{\mu}, \boldsymbol{\mu}^*, \boldsymbol{\nu}, \boldsymbol{\nu}^*$ be a mazy-four in the (p,p')-model sandwiching (a,b). If $d^L \ge 2$ and $\mu_{d^L} = \mu_{d^L-1}$, then:*

$$\tilde{\chi}^{p,p'}_{a,b,e,f}(L,m)\left\{\begin{matrix}\mu_1,\ldots,\mu_{d^L-1},\mu_{d^L};\boldsymbol{\nu}\\ \mu^*_1,\ldots,\mu^*_{d^L-1},\mu^*_{d^L};\boldsymbol{\nu}^*\end{matrix}\right\} = \tilde{\chi}^{p,p'}_{a,b,e,f}(L,m)\left\{\begin{matrix}\mu_1,\ldots,\mu_{d^L-1};\boldsymbol{\nu}\\ \mu^*_1,\ldots,\mu^*_{d^L-1};\boldsymbol{\nu}^*\end{matrix}\right\}.$$

If $d^R \ge 2$ and $\nu_{d^R} = \nu_{d^R-1}$, then

$$\tilde{\chi}^{p,p'}_{a,b,e,f}(L,m)\left\{\begin{matrix}\boldsymbol{\mu}\;;\nu_1,\ldots,\nu_{d^R-1},\nu_{d^R}\\ \boldsymbol{\mu}^*;\nu^*_1,\ldots,\nu^*_{d^R-1},\nu^*_{d^R}\end{matrix}\right\} = \tilde{\chi}^{p,p'}_{a,b,e,f}(L,m)\left\{\begin{matrix}\boldsymbol{\mu}\;;\nu_1,\ldots,\nu_{d^R-1}\\ \boldsymbol{\mu}^*;\nu^*_1,\ldots,\nu^*_{d^R-1}\end{matrix}\right\}.$$

Proof: Immediately from the definition, $\mathcal{P}^{p,p'}_{a,b,e,f}(L,m)\left\{\begin{smallmatrix}\mu_1,\ldots,\mu_{d^L-1},\mu_{d^L};\boldsymbol{\nu}\\ \mu^*_1,\ldots,\mu^*_{d^L-1},\mu^*_{d^L};\boldsymbol{\nu}^*\end{smallmatrix}\right\} \subseteq \mathcal{P}^{p,p'}_{a,b,e,f}(L,m)\left\{\begin{smallmatrix}\mu_1,\ldots,\mu_{d^L-1};\boldsymbol{\nu}\\ \mu^*_1,\ldots,\mu^*_{d^L-1};\boldsymbol{\nu}^*\end{smallmatrix}\right\}$. Now consider a path h belonging to the second of these sets. Since a is between $\mu_{d^L} = \mu_{d^L-1}$ and $\mu^*_{d^L}$, and $\mu^*_{d^L}$ is between $\mu_{d^L} = \mu_{d^L-1}$ and $\mu^*_{d^L-1}$, then $\mu^*_{d^L}$ is between a and $\mu^*_{d^L-1}$. Thus if h attains $\mu^*_{d^L-1}$ then it necessarily attains $\mu^*_{d^L}$, and if it attains $\mu^*_{d^L-1}$ before it attains $\mu_{d^L} = \mu_{d^L-1}$, then it necessarily attains $\mu^*_{d^L}$ before it attains $\mu_{d^L} = \mu_{d^L-1}$. Thus h belongs to the first of the above sets, whence the first part of the lemma follows. The second expression is proved in an analogous way. □

Lemma 5.6. *Let $\boldsymbol{\mu}, \boldsymbol{\mu}^*, \boldsymbol{\nu}, \boldsymbol{\nu}^*$ be a mazy-four in the (p,p')-model sandwiching (a,b). If $1 \le j < d^L$ and $\mu^*_j = \mu^*_{j+1}$, then:*

$$\begin{aligned}\tilde{\chi}^{p,p'}_{a,b,e,f}(L)&\left\{\begin{matrix}\mu_1,\ldots,\mu_{j-1},\mu_j,\mu_{j+1},\ldots,\mu_{d^L};\boldsymbol{\nu}\\ \mu^*_1,\ldots,\mu^*_{j-1},\mu^*_j,\mu^*_{j+1},\ldots,\mu^*_{d^L};\boldsymbol{\nu}^*\end{matrix}\right\}\\ &= \tilde{\chi}^{p,p'}_{a,b,e,f}(L)\left\{\begin{matrix}\mu_1,\ldots,\mu_{j-1},\mu_{j+1},\ldots,\mu_{d^L};\boldsymbol{\nu}\\ \mu^*_1,\ldots,\mu^*_{j-1},\mu^*_{j+1},\ldots,\mu^*_{d^L};\boldsymbol{\nu}^*\end{matrix}\right\}.\end{aligned}$$

If $1 \le j < d^R$ and $\nu^*_j = \nu^*_{j+1}$, then

$$\begin{aligned}\tilde{\chi}^{p,p'}_{a,b,e,f}(L)&\left\{\begin{matrix}\boldsymbol{\mu}\;;\nu_1,\ldots,\nu_{j-1},\nu_j,\nu_{j+1},\ldots,\nu_{d^R}\\ \boldsymbol{\mu}^*;\nu^*_1,\ldots,\nu^*_{j-1},\nu^*_j,\nu^*_{j+1},\ldots,\nu^*_{d^R}\end{matrix}\right\}\\ &= \tilde{\chi}^{p,p'}_{a,b,e,f}(L)\left\{\begin{matrix}\boldsymbol{\mu}\;;\nu_1,\ldots,\nu_{j-1},\nu_{j+1},\ldots,\nu_{d^R}\\ \boldsymbol{\mu}^*;\nu^*_1,\ldots,\nu^*_{j-1},\nu^*_{j+1},\ldots,\nu^*_{d^R}\end{matrix}\right\}.\end{aligned}$$

Proof: Immediately from the definition, $\mathcal{P}^{p,p'}_{a,b,e,f}(L)\left\{\begin{smallmatrix}\mu_1,\ldots,\mu_{j-1},\mu_j,\mu_{j+1},\ldots,\mu_{d^R};\boldsymbol{\nu}\\ \mu^*_1,\ldots,\mu^*_{j-1},\mu^*_j,\mu^*_{j+1},\ldots,\mu^*_{d^R};\boldsymbol{\nu}^*\end{smallmatrix}\right\} \subseteq \mathcal{P}^{p,p'}_{a,b,e,f}(L)\left\{\begin{smallmatrix}\mu_1,\ldots,\mu_{j-1},\mu_{j+1},\ldots,\mu_{d^R};\boldsymbol{\nu}\\ \mu^*_1,\ldots,\mu^*_{j-1},\mu^*_{j+1},\ldots,\mu^*_{d^R};\boldsymbol{\nu}^*\end{smallmatrix}\right\}$. Now consider a path h belonging to the second of these sets. Since μ_{j+1} is between $\mu^*_j = \mu^*_{j+1}$ and μ_j, then if h attains μ^*_{j+1} and does so before it attains any μ_{j+1}, then it necessarily attains μ^*_j and does so before it attains any μ_j. Thus h belongs to the first of the above sets, whence the first part of the lemma follows. The second expression is proved in an analogous way. □

5.3. Transforming the passing function

In this section, we determine how the generating functions defined in the previous section behave under the $\mathcal{B}$-transform and the $\mathcal{BD}$-transform.

Lemma 5.7. *For $1 \le p < p'$ with $p' > 2p$, let $1 \le a, b < p'$, $e, f \in \{0, 1\}$, and $m_0, m_1 \ge 0$. Let $\boldsymbol{\mu}, \boldsymbol{\mu}^*, \boldsymbol{\nu}, \boldsymbol{\nu}^*$ be an interfacial mazy-four in the (p, p')-model sandwiching (a, b). If $\delta^{p,p'}_{a,e} = 1$ we restrict to $d^L > 0$, and likewise if $\delta^{p,p'}_{b,f} = 1$ we restrict to $d^R > 0$. Set $a' = a + e + \lfloor ap/p' \rfloor$ and $b' = b + f + \lfloor bp/p' \rfloor$. Define the vectors $\boldsymbol{\mu}', \boldsymbol{\mu}^{*\prime}, \boldsymbol{\nu}', \boldsymbol{\nu}^{*\prime}$, by setting $\mu'_j = \mu_j + \rho^{p,p'}(\mu_j)$ and $\mu^{*\prime}_j = \mu^*_j + \rho^{p,p'}(\mu^*_j)$ for $1 \le j \le d^L$; and setting $\nu'_j = \nu_j + \rho^{p,p'}(\nu_j)$ and $\nu^{*\prime}_j = \nu^*_j + \rho^{p,p'}(\nu^*_j)$ for $1 \le j \le d^R$.*

If $\delta^{p,p'}_{a,e} = 0$, then the map $(h, k, \lambda) \mapsto h'$ effected by the action of a $\mathcal{B}(k, \lambda)$-transform on h, is a bijection between

$$\bigcup_k \mathcal{P}^{p,p'}_{a,b,e,f}(m_1, 2k + 2m_1 - m_0) \left\{ \begin{matrix} \boldsymbol{\mu} \ ; \boldsymbol{\nu} \\ \boldsymbol{\mu}^*; \boldsymbol{\nu}^* \end{matrix} \right\} \times \mathcal{Y}(k, m_1)$$

and $\mathcal{P}^{p,p'+p}_{a',b',e,f}(m_0, m_1) \left\{ \begin{matrix} \boldsymbol{\mu}' \ ; \boldsymbol{\nu}' \\ \boldsymbol{\mu}^{*\prime}; \boldsymbol{\nu}^{*\prime} \end{matrix} \right\}$.

If $\delta^{p,p'}_{a,e} = 1$ and $\mu_{d^L} = a - (-1)^e$, then the map $(h, k, \lambda) \mapsto h'$ effected by the action of a $\mathcal{B}(k, \lambda)$-transform on h, is a bijection between

$$\bigcup_k \mathcal{P}^{p,p'}_{a,b,e,f}(m_1, 2k + 2m_1 - m_0 + 2) \left\{ \begin{matrix} \boldsymbol{\mu} \ ; \boldsymbol{\nu} \\ \boldsymbol{\mu}^*; \boldsymbol{\nu}^* \end{matrix} \right\} \times \mathcal{Y}(k, m_1)$$

and $\mathcal{P}^{p,p'+p}_{a',b',e,f}(m_0, m_1) \left\{ \begin{matrix} \boldsymbol{\mu}' \ ; \boldsymbol{\nu}' \\ \boldsymbol{\mu}^{*\prime}; \boldsymbol{\nu}^{*\prime} \end{matrix} \right\}$.

In both cases,

$$\tilde{wt}(h') = \tilde{wt}(h) + \frac{1}{4}\left((m_0 - m_1)^2 - \beta^2\right) + \mathrm{wt}\,(\lambda),$$

where $\beta = \beta^{p,p'}_{a,b,e,f}$.

Additionally, $\boldsymbol{\mu}', \boldsymbol{\mu}^{*\prime}, \boldsymbol{\nu}', \boldsymbol{\nu}^{*\prime}$ are an interfacial mazy-four in the $(p, p'+p)$-model sandwiching (a', b').

Proof: It is immediate that $\boldsymbol{\mu}', \boldsymbol{\mu}^{*\prime}, \boldsymbol{\nu}', \boldsymbol{\nu}^{*\prime}$ are a mazy-four in the $(p, p' + p)$-model sandwiching (a', b'). Let $1 \le j \le d^L$. Since μ^*_j is interfacial in the (p, p')-model, Lemma C.4(1) implies that $\mu^{*\prime}_j = \mu^*_j + \rho^{p,p'}(\mu^*_j)$ is interfacial in the $(p, p'+p)$-model, The analogous results hold for μ_j for $1 \le j \le d^L$ and for ν^*_j and ν_j for $1 \le j \le d^R$. It follows that $\boldsymbol{\mu}', \boldsymbol{\mu}^{*\prime}, \boldsymbol{\nu}', \boldsymbol{\nu}^{*\prime}$ are an interfacial mazy-four in the $(p, p' + p)$-model sandwiching (a', b').

Lemma C.4(1) also shows that $\omega^{p,p'+p}(\mu^{*\prime}_j) = \omega^{p,p'+p}(\mu^*_j + \rho^{p,p'}(\mu^*_j)) = \omega^{p,p'}(\mu^*_j)$, with again, the analogous results holding for μ_j for $1 \le j \le d^L$ and for ν^*_j and ν_j for $1 \le j \le d^R$.

Let $h \in \mathcal{P}^{p,p'}_{a,b,e,f}(m_1, m) \left\{ \begin{matrix} \boldsymbol{\mu} \ ; \boldsymbol{\nu} \\ \boldsymbol{\mu}^*; \boldsymbol{\nu}^* \end{matrix} \right\}$ and let h' result from the action of the $\mathcal{B}(k, \lambda)$-transform on h. Lemma 5.1 implies that either $\omega(h') = \omega(h)$, $\omega(a)\omega(h') = \omega(h)$, $\omega(h')\omega(b) = \omega(h)$, or $\omega(a)\omega(h')\omega(b) = \omega(h)$. In the case $\omega(h') = \omega(h)$, it follows immediately that h' is mazy-compliant with $\boldsymbol{\mu}', \boldsymbol{\mu}^{*\prime}, \boldsymbol{\nu}', \boldsymbol{\nu}^{*\prime}$.

In the case $\omega(a)\omega(h') = \omega(h)$, then necessarily a is interfacial in the (p, p')-model. Since $\boldsymbol{\mu}, \boldsymbol{\mu}^*$ are a mazy-pair sandwiching a, $a \ne \mu^*_j$ for $1 \le j \le d^L$. It follows that h' satisfies conditions 1 and 2 for being mazy-compliant with $\boldsymbol{\mu}', \boldsymbol{\mu}^{*\prime}, \boldsymbol{\nu}', \boldsymbol{\nu}^{*\prime}$. For the other conditions, note first that Lemma 5.1 implies that $\delta^{p,p'}_{a,e} = 1$ so that $d^L > 0$, whereupon if $d^R > 0$, there exists $i' \ge i > 0$ such that $h_i = \mu^*_1$ and $h_{i'} = \nu^*_1$. Since $\boldsymbol{\nu}, \boldsymbol{\nu}^*$ are a mazy-pair sandwiching b, it follows that if $1 \le j \le d^R$, then there exists $i'' \ge i'$ such that $h_{i''} = \nu^*_j$. Then from $\omega(a)\omega(h') = \omega(h)$, it follows that h' satisfies the remaining criteria for being mazy-compliant with $\boldsymbol{\mu}', \boldsymbol{\mu}^{*\prime}, \boldsymbol{\nu}', \boldsymbol{\nu}^{*\prime}$.

The cases $\omega(h')\omega(b) = \omega(h)$ and $\omega(a)\omega(h')\omega(b) = \omega(h)$ are dealt with in a similar way. Thereupon, making use of Lemma 3.9, $h' \in \mathcal{P}^{p,p'+p}_{a',b',e,f}(L', m_1)\left\{\begin{smallmatrix}\boldsymbol{\mu}' & ;\boldsymbol{\nu}' \\ \boldsymbol{\mu}^{*\prime} & ;\boldsymbol{\nu}^{*\prime}\end{smallmatrix}\right\}$, for some L'.

In the case $\delta^{p,p'}_{a,e} = 0$, if $\pi(h) = 1$ then $d(h) \ne e$. Lemma 3.9 thus gives $L' = 2m_1 - m + 2k$.

In the case $\delta^{p,p'}_{a,e} = 1$, we have $\mu_{d^L} = a-(-1)^e$ which implies that $h_1 = a+(-1)^e$. Thence, $\pi(h) = 1$ and $d(h) = e$. Lemma 3.9 thus gives $L' = 2m_1 - m + 2k + 2$.

Now consider $h' \in \mathcal{P}^{p,p'+p}_{a',b',e,f}(L', m_1)\left\{\begin{smallmatrix}\boldsymbol{\mu}' & ;\boldsymbol{\nu}' \\ \boldsymbol{\mu}^{*\prime} & ;\boldsymbol{\nu}^{*\prime}\end{smallmatrix}\right\}$. Lemma C.1 shows that $\delta^{p,p'+p}_{a',e} = \delta^{p,p'+p}_{b',f} = 0$, whereupon Lemma 3.11 shows that there is a unique triple (h, k, λ) with $h \in \mathcal{P}^{p,p'}_{a,b,e,f}(m_1, m)$ and λ is a partition having at most k parts, the greatest of which does not exceed m_1, such that h' arises from the action of a $\mathcal{B}(k, \lambda)$-transform on h. Now employ Lemma 5.1. Whether $\omega(h') = \omega(h)$, $\omega(a)\omega(h') = \omega(h)$, $\omega(h')\omega(b) = \omega(h)$, or $\omega(a)\omega(h')\omega(b) = \omega(h)$, it is readily seen that h is mazy-compliant with $\boldsymbol{\mu}, \boldsymbol{\mu}^*, \boldsymbol{\nu}, \boldsymbol{\nu}^*$. The bijection follows on taking $m_0 = L'$.

The expression for $\tilde{wt}(h)$ also results from Lemma 3.9. □

Corollary 5.8. *Let all parameters be as in the first paragraph of Lemma 5.7. If either $\delta^{p,p'}_{a,e} = 0$, or $\delta^{p,p'}_{a,e} = 1$ and $\mu_{d^L} = a - (-1)^e$, then:*

$$\tilde{\chi}^{p,p'+p}_{a',b',e,f}(m_0, m_1)\left\{\begin{matrix}\boldsymbol{\mu}' & ;\boldsymbol{\nu}' \\ \boldsymbol{\mu}^{*\prime} & ;\boldsymbol{\nu}^{*\prime}\end{matrix}\right\}$$
$$= q^{\frac{1}{4}((m_0-m_1)^2-\beta^2)} \sum_{\substack{m \equiv m_0 \\ (\mathrm{mod}\, 2)}} \begin{bmatrix}\frac{1}{2}(m_0 + m) \\ m_1\end{bmatrix}_q \tilde{\chi}^{p,p'}_{a,b,e,f}(m_1, m + 2\delta^{p,p'}_{a,e})\left\{\begin{matrix}\boldsymbol{\mu} & ;\boldsymbol{\nu} \\ \boldsymbol{\mu}^* & ;\boldsymbol{\nu}^*\end{matrix}\right\},$$

where $\beta = \beta^{p,p'}_{a,b,e,f}$.

In addition, $\alpha^{p,p'+p}_{a',b'} = \alpha^{p,p'}_{a,b} + \beta^{p,p'}_{a,b,e,f}$ and $\beta^{p,p'+p}_{a',b',e,f} = \beta^{p,p'}_{a,b,e,f}$.

Proof: Apart from the case for which $m_1 = 0$ and $e \ne f$, the first statement follows immediately from Lemma 5.7 on setting $m = 2k + 2m_1 - m_0$, once it is noted, via Lemma 2.7, that $\begin{bmatrix}k+m_1 \\ m_1\end{bmatrix}_q$ is the generating function for $\mathcal{Y}(k, m_1)$.

In the case $m_1 = 0$ and $e \ne f$, after noting that $\delta^{p,p'}_{b',f} = 0$, it is readily seen that if a path h' is to contribute to the left side, then m_0 is odd, $|a' - b'| = 1$, the band between a' and b' is even and h' alternates between heights a' and b'. Also note that $a' - e = b' - f$ from which we obtain $a = b$. There is only one path h' satisfying the above. Via the same calculation as in the proof of 2.6, $\tilde{wt}(h') = \frac{1}{4}(m_0^2 - 1)$. Then, since $\tilde{\chi}^{p,p'}_{a,b,e,f}(0, m)\{\} = \delta_{m,1}$, the expression certainly holds if $d^L = d^R = 0$. If either $d^L > 0$ or $d^R > 0$ then the right side is zero since each contributing path must attain $\mu^*_{d^L} \ne a$ or $\nu^*_{d^R} \ne b$ respectively. Correspondingly, the left side is zero in the case $d^L > 0$ since because $\mu^*_{d^L}$ is interfacial and $\mu^{*\prime}_{d^L} = \mu^*_{d^L} + \rho^{p,p'}(\mu^*_{d^L})$, it follows that $\mu^{*\prime}_{d^L} \neq a'$ and $\mu^{*\prime}_{d^L} \neq b'$. That the left side is zero in the $d^R > 0$ case follows similarly.

By the above reasoning, the left side is zero if either m_0 is even or $a \neq b$. It is easily checked that the right side is also zero in these cases.

The second statement is Lemma 3.4. □

We also require $\mathcal{BD}$-transform analogues of Lemma 5.7 and Corollary 5.8. However, the restrictions that applied in the cases where $\delta^{p,p'}_{a,e} = 1$ or $\delta^{p,p'}_{b,f} = 1$ are not sufficient for what is required later.

Lemma 5.9. *For $1 \le p < p'$ with $p' < 2p$, let $1 \le a, b < p'$, $e, f \in \{0,1\}$, and $m_0, m_1 \ge 0$. Let $\boldsymbol{\mu}, \boldsymbol{\mu}^*, \boldsymbol{\nu}, \boldsymbol{\nu}^*$ be an interfacial mazy-four in the (p,p')-model sandwiching (a,b). If $\delta^{p'-p,p'}_{a,e} = 1$ we restrict to $d^L > 0$, and likewise if $\delta^{p'-p,p'}_{b,f} = 1$ we restrict to $d^R > 0$. Set $a' = a+1-e+\lfloor ap/p' \rfloor$ and $b' = b+1-f+\lfloor bp/p' \rfloor$. Define the vectors $\boldsymbol{\mu}', \boldsymbol{\mu}^{*\prime}, \boldsymbol{\nu}', \boldsymbol{\nu}^{*\prime}$, by setting $\mu'_j = \mu_j + \rho^{p,p'}(\mu_j)$ and $\mu^{*\prime}_j = \mu^*_j + \rho^{p,p'}(\mu^*_j)$ for $1 \le j \le d^L$; and setting $\nu'_j = \nu_j + \rho^{p,p'}(\nu_j)$ and $\nu^{*\prime}_j = \nu^*_j + \rho^{p,p'}(\nu^*_j)$ for $1 \le j \le d^R$.*

If $\delta^{p'-p,p'}_{a,e} = 0$, then the map $(h,k,\lambda) \mapsto h'$ effected by the action of a $\mathcal{D}$-transform on h followed by a $\mathcal{B}(k,\lambda)$-transform, is a bijection between

$$\bigcup_k \mathcal{P}^{p'-p,p'}_{a,b,e,f}(m_1, m_0 - m_1 - 2k) \left\{ \begin{matrix} \boldsymbol{\mu}\ ;\boldsymbol{\nu} \\ \boldsymbol{\mu}^*;\boldsymbol{\nu}^* \end{matrix} \right\} \times \mathcal{Y}(k, m_1)$$

and $\mathcal{P}^{p,p'+p}_{a',b',1-e,1-f}(m_0, m_1) \left\{ \begin{matrix} \boldsymbol{\mu}'\ ;\boldsymbol{\nu}' \\ \boldsymbol{\mu}^{*\prime};\boldsymbol{\nu}^{*\prime} \end{matrix} \right\}$.

If $\delta^{p'-p,p'}_{a,e} = 1$ and $\mu_{d^L} = a - (-1)^e$, then the map $(h,k,\lambda) \mapsto h'$ effected by the action of a $\mathcal{B}(k,\lambda)$-transform on h, is a bijection between

$$\bigcup_k \mathcal{P}^{p'-p,p'}_{a,b,e,f}(m_1, m_0 - m_1 - 2k + 2) \left\{ \begin{matrix} \boldsymbol{\mu}\ ;\boldsymbol{\nu} \\ \boldsymbol{\mu}^*;\boldsymbol{\nu}^* \end{matrix} \right\} \times \mathcal{Y}(k, m_1)$$

and $\mathcal{P}^{p,p'+p}_{a',b',1-e,1-f}(m_0, m_1) \left\{ \begin{matrix} \boldsymbol{\mu}'\ ;\boldsymbol{\nu}' \\ \boldsymbol{\mu}^{*\prime};\boldsymbol{\nu}^{*\prime} \end{matrix} \right\}$.

In both cases,

$$\tilde{wt}(h') = \frac{1}{4}\left(m_1^2 + (m_0 - m_1)^2 - \alpha^2 - \beta^2\right) + \mathrm{wt}\,(\lambda) - \tilde{wt}(h),$$

where $\alpha = \alpha^{p,p'}_{a,b}$ and $\beta = \beta^{p,p'}_{a,b,1-e,1-f}$.

Additionally, $\boldsymbol{\mu}', \boldsymbol{\mu}^{*\prime}, \boldsymbol{\nu}', \boldsymbol{\nu}^{*\prime}$ are an interfacial mazy-four in the $(p, p'+p)$-model sandwiching (a', b').

Proof: It is immediate that $\boldsymbol{\mu}', \boldsymbol{\mu}^{*\prime}, \boldsymbol{\nu}', \boldsymbol{\nu}^{*\prime}$ are a mazy-four in the $(p, p'+p)$-model sandwiching (a', b'). Let $1 \le j \le d^L$. Since μ^*_j is interfacial in the $(p'-p, p')$-model, Lemma C.4 implies that μ^*_j is interfacial in the (p,p')-model and then also that $\mu^{*\prime}_j = \mu^*_j + \rho^{p,p'}(\mu^*_j)$ is interfacial in the $(p, p'+p)$-model. The analogous results hold for μ_j for $1 \le j \le d^L$ and for ν^*_j and ν_j for $1 \le j \le d^R$. It follows that $\boldsymbol{\mu}', \boldsymbol{\mu}^{*\prime}, \boldsymbol{\nu}', \boldsymbol{\nu}^{*\prime}$ are an interfacial mazy-four in the $(p, p'+p)$-model sandwiching (a', b').

Lemma C.4 also shows that if $R = \lfloor rp/(p'-p) \rfloor$ then $\omega^{p'-p,p'}(\mu^*_j) = r^- \implies \omega^{p,p'}(\mu^*_j) = R^+ \implies \omega^{p,p'+p}(\mu^{*\prime}_j) = R^+$, and $\omega^{p'-p,p'}(\mu^*_j) = r^+ \implies \omega^{p,p'}(\mu^*_j) =$

$(R+1)^- \implies \omega^{p,p'+p}(\mu_j^{*\prime}) = (R+1)^-$. The analogous results hold for μ_j for $1 \le j \le d^L$, and for ν_j^* and ν_j for $1 \le j \le d^R$.

Let $h \in \mathcal{P}^{p'-p,p'}_{a,b,e,f}(m_1,m)\left\{\begin{smallmatrix}\boldsymbol{\mu}\;;\boldsymbol{\nu}\\ \boldsymbol{\mu}^*;\boldsymbol{\nu}^*\end{smallmatrix}\right\}$, and let $\hat{h}$ result from the action of a $\mathcal{D}$-transform on h, and let h' result from the action of a $\mathcal{B}(k,\lambda)$-transform on $\hat{h}$. Lemma 5.3 then implies that the longest subword $\omega' = \omega'_1\omega'_2\cdots\omega'_{k'}$ of $\omega(h')$ that comprises the symbols in $\{\sigma(x) : x \in \mathcal{R}^{p'-p,p'}\}$, and for which $\omega'_i \ne \omega'_{i+1}$ for $1 \le i < k'$, is such that either $\omega' = \sigma(\omega(h))$, $\omega(a)\omega' = \sigma(\omega(h))$, $\omega'\omega(b) = \sigma(\omega(h))$, or $\omega(a)\omega'\omega(b) = \sigma(\omega(h))$. In the case $\omega' = \omega(h)$, it follows immediately that h' is mazy-compliant with $\boldsymbol{\mu}', \boldsymbol{\mu}^{*\prime}, \boldsymbol{\nu}', \boldsymbol{\nu}^{*\prime}$. The other cases are dealt with as in the proof of Lemma 5.7. Thereupon, making use of Lemma 4.3, $h' \in \mathcal{P}^{p,p'+p}_{a',b',1-e,1-f}(L',m_1)\left\{\begin{smallmatrix}\boldsymbol{\mu}'\;;\boldsymbol{\nu}'\\ \boldsymbol{\mu}^{*\prime};\boldsymbol{\nu}^{*\prime}\end{smallmatrix}\right\}$, for some L'.

To determine L', first consider $\delta^{p'-p,p'}_{a,e} = 0$. Here, if $\pi(h) = 1$ then $d(h) \ne e$. Lemma 4.3 thus gives $L' = m + m_1 + 2k$.

In the case $\delta^{p'-p,p'}_{a,e} = 1$, we have $\mu_{d^L} = a - (-1)^e$ which implies that $h_1 = a + (-1)^e$. Thence, $\pi(h) = 1$ and $d(h) = e$. Lemma 4.3 thus gives $L' = m+m_1+2k+2$.

Now consider $h' \in \mathcal{P}^{p,p'+p}_{a',b',e,f}(L',m_1)\left\{\begin{smallmatrix}\boldsymbol{\mu}'\;;\boldsymbol{\nu}'\\ \boldsymbol{\mu}^{*\prime};\boldsymbol{\nu}^{*\prime}\end{smallmatrix}\right\}$ and let ω' be the longest subword $\omega' = \omega'_1\omega'_2\cdots\omega'_{k'}$ of $\omega(h')$ that comprises the symbols in $\{\sigma(x) : x \in \mathcal{R}^{p'-p,p'}\}$, and for which $\omega'_i \ne \omega'_{i+1}$ for $1 \le i < k'$. Lemma C.1 shows that $\delta^{p,p'+p}_{a',e} = \delta^{p,p'+p}_{b',f} = 0$, whereupon Lemma 3.11 shows that there is a unique triple $(\hat{h},k,\lambda)$ with $\hat{h} \in \mathcal{P}^{p,p'}_{a,b,e,f}(m_1,m)$ and λ is a partition having at most k parts, the greatest of which does not exceed m_1, such that h' arises from the action of a $\mathcal{B}(k,\lambda)$-transform on $\hat{h}$. The $\mathcal{D}$-transform maps $\hat{h}$ to a unique $h \in \mathcal{P}^{p'-p,p'}_{a,b,e,f}(m_1,m)$, so that h' arises from a unique h by a $\mathcal{BD}$-transform. Lemma 5.3 implies that $\omega' = \sigma(\omega(h))$, $\omega(a)\omega' = \sigma(\omega(h))$, $\omega'\omega(b) = \sigma(\omega(h))$, or $\omega(a)\omega'\omega(b) = \sigma(\omega(h))$. In either case, it follows that h is mazy-compliant with $\boldsymbol{\mu}, \boldsymbol{\mu}^*, \boldsymbol{\nu}, \boldsymbol{\nu}^*$. Therefore, $h \in \mathcal{P}^{p'-p,p'}_{a,b,e,f}(m_1,m)\left\{\begin{smallmatrix}\boldsymbol{\mu}\;;\boldsymbol{\nu}\\ \boldsymbol{\mu}^*;\boldsymbol{\nu}^*\end{smallmatrix}\right\}$, and the bijection follows on taking $m_0 = L'$.

Lemma 4.3 also yields the expression for $\tilde{wt}(h)$. □

Corollary 5.10. *Let all parameters be as in the premise of Lemma 5.9. If either* $\delta^{p'-p,p'}_{a,e} = 0$, *or* $\delta^{p'-p,p'}_{a,e} = 1$ *and* $\mu_{d^L} = a - (-1)^e$, *then:*

$$\tilde{\chi}^{p,p'+p}_{a',b',1-e,1-f}(m_0,m_1;q)\left\{\begin{matrix}\boldsymbol{\mu}'\;;\boldsymbol{\nu}'\\ \boldsymbol{\mu}^{*\prime};\boldsymbol{\nu}^{*\prime}\end{matrix}\right\}$$

$$= q^{\frac14(m_1^2+(m_0-m_1)^2-\alpha^2-\beta^2)} \sum_{\substack{m\equiv m_0-m_1\\ (\text{mod } 2)}} \left(\begin{bmatrix} \frac12(m_0+m_1-m)\\ m_1 \end{bmatrix}_q \right.$$

$$\left. \times\ \tilde{\chi}^{p'-p,p'}_{a,b,e,f}(m_1, m+2\delta^{p'-p,p'}_{a,e}; q^{-1})\left\{\begin{matrix}\boldsymbol{\mu}\;;\boldsymbol{\nu}\\ \boldsymbol{\mu}^*;\boldsymbol{\nu}^*\end{matrix}\right\} \right),$$

where $\alpha = \alpha^{p,p'}_{a,b}$ *and* $\beta = \beta^{p,p'}_{a,b,1-e,1-f}$.

In addition, $\alpha^{p,p'+p}_{a',b'} = 2\alpha^{p'-p,p'}_{a,b} - \beta^{p'-p,p'}_{a,b,e,f}$ *and* $\beta^{p,p'+p}_{a',b',1-e,1-f} = \alpha^{p'-p,p'}_{a,b} - \beta^{p'-p,p'}_{a,b,e,f}$.

Proof: Apart from the case for which $m_1 = 0$ and $e \neq f$, the first statement follows immediately from Lemma 5.9 on setting $m = m_0 - m_1 - 2k$, once it is noted, via Lemma 2.7, that $\begin{bmatrix} k+m_1 \\ m_1 \end{bmatrix}_q$ is the generating function for $\mathcal{Y}(k, m_1)$. The case $m_1 = 0$ and $e \neq f$ is dealt with exactly as in the proof of Corollary 5.8.

The second statement is Lemma 4.4. □

5.4. Mazy runs in the original weighting

Although not required until Section 7.4, here we define generating functions for paths restricted in the same way as in Section 5.2, but which have the original weight function (2.1) applied to them.

With $\boldsymbol{\mu} = (\mu_1, \mu_2, \ldots, \mu_{d^L})$, $\boldsymbol{\mu}^* = (\mu_1^*, \mu_2^*, \ldots, \mu_{d^L}^*)$, $\boldsymbol{\nu} = (\nu_1, \nu_2, \ldots, \nu_{d^R})$, and $\boldsymbol{\nu}^* = (\nu_1^*, \nu_2^*, \ldots, \nu_{d^R}^*)$ for $d^L, d^R \geq 0$, we say that a path $h \in \mathcal{P}^{p,p'}_{a,b,c}(L)$ is *mazy-compliant* (with $\boldsymbol{\mu}, \boldsymbol{\mu}^*, \boldsymbol{\nu}, \boldsymbol{\nu}^*$) if the five conditions of Section 5.2 hold. The set $\mathcal{P}^{p,p'}_{a,b,c}(L) \left\{ \begin{smallmatrix} \boldsymbol{\mu} & ; & \boldsymbol{\nu} \\ \boldsymbol{\mu}^* & ; & \boldsymbol{\nu}^* \end{smallmatrix} \right\}$ is defined to be the subset of $\mathcal{P}^{p,p'}_{a,b,c}(L)$ comprising those paths h which are mazy-compliant with $\boldsymbol{\mu}, \boldsymbol{\mu}^*, \boldsymbol{\nu}, \boldsymbol{\nu}^*$. The generating function for these paths is defined to be:

$$\chi^{p,p'}_{a,b,c}(L; q) \left\{ \begin{matrix} \boldsymbol{\mu} \ ; \boldsymbol{\nu} \\ \boldsymbol{\mu}^* ; \boldsymbol{\nu}^* \end{matrix} \right\} = \sum_{h \in \mathcal{P}^{p,p'}_{a,b,c}(L) \left\{ \begin{smallmatrix} \boldsymbol{\mu} & ; & \boldsymbol{\nu} \\ \boldsymbol{\mu}^* & ; & \boldsymbol{\nu}^* \end{smallmatrix} \right\}} q^{wt(h)}. \tag{5.2}$$

Lemma 5.11. *Let $1 \leq a, b < p'$ with b non-interfacial in the (p, p')-model, and let $\Delta \in \{\pm 1\}$. Let $\boldsymbol{\mu}, \boldsymbol{\mu}^*, \boldsymbol{\nu}, \boldsymbol{\nu}^*$ be a mazy-four in the (p, p')-model sandwiching (a, b).*

If $d^R = 0$ or $b - \Delta \neq \nu_{d^R}^*$ set $\boldsymbol{\nu}^+ = \boldsymbol{\nu}$ and $\boldsymbol{\nu}^{+*} = \boldsymbol{\nu}^*$. Otherwise, if $d^R > 0$ and $b - \Delta = \nu_{d^R}^*$, set $d^+ \geq 0$ to be the smallest integer such that $b - \Delta = \nu_{d^+ + 1}^*$, and set $\boldsymbol{\nu}^+ = (\nu_1, \nu_2, \ldots, \nu_{d^+})$ and $\boldsymbol{\nu}^{+*} = (\nu_1^*, \nu_2^*, \ldots, \nu_{d^+}^*)$.

1) If $p' > 2p$ and either $d^R = 0$ or $b + \Delta \neq \nu_{d^R}$ then:

$$\begin{aligned}
\chi^{p,p'}_{a,b,b-\Delta}(L) \left\{ \begin{matrix} \boldsymbol{\mu} \ ; \boldsymbol{\nu} \\ \boldsymbol{\mu}^* ; \boldsymbol{\nu}^* \end{matrix} \right\} &= q^{\frac{1}{2}(L + \Delta(a-b))} \chi^{p,p'}_{a,b-\Delta,b}(L-1) \left\{ \begin{matrix} \boldsymbol{\mu} \ ; \boldsymbol{\nu}^+ \\ \boldsymbol{\mu}^* ; \boldsymbol{\nu}^{+*} \end{matrix} \right\} \\
&\quad + \chi^{p,p'}_{a,b+\Delta,b}(L-1) \left\{ \begin{matrix} \boldsymbol{\mu} \ ; \boldsymbol{\nu} \\ \boldsymbol{\mu}^* ; \boldsymbol{\nu}^* \end{matrix} \right\}, \\
\chi^{p,p'}_{a,b,b+\Delta}(L) \left\{ \begin{matrix} \boldsymbol{\mu} \ ; \boldsymbol{\nu} \\ \boldsymbol{\mu}^* ; \boldsymbol{\nu}^* \end{matrix} \right\} &= \chi^{p,p'}_{a,b-\Delta,b}(L-1) \left\{ \begin{matrix} \boldsymbol{\mu} \ ; \boldsymbol{\nu}^+ \\ \boldsymbol{\mu}^* ; \boldsymbol{\nu}^{+*} \end{matrix} \right\} \\
&\quad + q^{\frac{1}{2}(L - \Delta(a-b))} \chi^{p,p'}_{a,b+\Delta,b}(L-1) \left\{ \begin{matrix} \boldsymbol{\mu} \ ; \boldsymbol{\nu} \\ \boldsymbol{\mu}^* ; \boldsymbol{\nu}^* \end{matrix} \right\}.
\end{aligned}$$

2) If $p' > 2p$ and either both $d^R > 0$ and $b + \Delta = \nu_{d^R}$, or $b + \Delta \in \{0, p'\}$ then:

$$\begin{aligned}
\chi^{p,p'}_{a,b,b-\Delta}(L) \left\{ \begin{matrix} \boldsymbol{\mu} \ ; \boldsymbol{\nu} \\ \boldsymbol{\mu}^* ; \boldsymbol{\nu}^* \end{matrix} \right\} &= q^{\frac{1}{2}(L + \Delta(a-b))} \chi^{p,p'}_{a,b,b+\Delta}(L) \left\{ \begin{matrix} \boldsymbol{\mu} \ ; \boldsymbol{\nu} \\ \boldsymbol{\mu}^* ; \boldsymbol{\nu}^* \end{matrix} \right\} \\
&= q^{\frac{1}{2}(L + \Delta(a-b))} \chi^{p,p'}_{a,b-\Delta,b}(L-1) \left\{ \begin{matrix} \boldsymbol{\mu} \ ; \boldsymbol{\nu}^+ \\ \boldsymbol{\mu}^* ; \boldsymbol{\nu}^{+*} \end{matrix} \right\}.
\end{aligned}$$

3) If $p' < 2p$ and either $d^R = 0$ or $b + \Delta \neq \nu_{d^R}$ then:

$$\chi^{p,p'}_{a,b,b-\Delta}(L)\left\{\begin{matrix}\boldsymbol{\mu}\ ;\boldsymbol{\nu}\\ \boldsymbol{\mu}^*;\boldsymbol{\nu}^*\end{matrix}\right\} = \chi^{p,p'}_{a,b-\Delta,b}(L-1)\left\{\begin{matrix}\boldsymbol{\mu}\ ;\boldsymbol{\nu}^+\\ \boldsymbol{\mu}^*;\boldsymbol{\nu}^{+*}\end{matrix}\right\}$$
$$+ q^{\frac{1}{2}(L-\Delta(a-b))}\chi^{p,p'}_{a,b+\Delta,b}(L-1)\left\{\begin{matrix}\boldsymbol{\mu}\ ;\boldsymbol{\nu}\\ \boldsymbol{\mu}^*;\boldsymbol{\nu}^*\end{matrix}\right\},$$

$$\chi^{p,p'}_{a,b,b+\Delta}(L)\left\{\begin{matrix}\boldsymbol{\mu}\ ;\boldsymbol{\nu}\\ \boldsymbol{\mu}^*;\boldsymbol{\nu}^*\end{matrix}\right\} = q^{\frac{1}{2}(L+\Delta(a-b))}\chi^{p,p'}_{a,b-\Delta,b}(L-1)\left\{\begin{matrix}\boldsymbol{\mu}\ ;\boldsymbol{\nu}^+\\ \boldsymbol{\mu}^*;\boldsymbol{\nu}^{+*}\end{matrix}\right\}$$
$$+ \chi^{p,p'}_{a,b+\Delta,b}(L-1)\left\{\begin{matrix}\boldsymbol{\mu}\ ;\boldsymbol{\nu}\\ \boldsymbol{\mu}^*;\boldsymbol{\nu}^*\end{matrix}\right\}.$$

4) If $p' < 2p$ and either both $d^R > 0$ and $b + \Delta = \nu_{d^R}$, or $b + \Delta \in \{0, p'\}$ then:

$$\chi^{p,p'}_{a,b,b-\Delta}(L)\left\{\begin{matrix}\boldsymbol{\mu}\ ;\boldsymbol{\nu}\\ \boldsymbol{\mu}^*;\boldsymbol{\nu}^*\end{matrix}\right\} = q^{-\frac{1}{2}(L+\Delta(a-b))}\chi^{p,p'}_{a,b,b+\Delta}(L)\left\{\begin{matrix}\boldsymbol{\mu}\ ;\boldsymbol{\nu}\\ \boldsymbol{\mu}^*;\boldsymbol{\nu}^*\end{matrix}\right\}$$
$$= \chi^{p,p'}_{a,b-\Delta,b}(L-1)\left\{\begin{matrix}\boldsymbol{\mu}\ ;\boldsymbol{\nu}^+\\ \boldsymbol{\mu}^*;\boldsymbol{\nu}^{+*}\end{matrix}\right\}.$$

Proof: Consider $h \in \mathcal{P}^{p,p'}_{a,b,b-\Delta}(L)\left\{\begin{smallmatrix}\boldsymbol{\mu}\ ;\ \boldsymbol{\nu}\\ \boldsymbol{\mu}^*;\ \boldsymbol{\nu}^*\end{smallmatrix}\right\}$. If either $d^R = 0$ or $b + \Delta \neq \nu_{d^R}$, the first $(L-1)$ segments of h constitute a path which is either a member of $\mathcal{P}^{p,p'}_{a,b-\Delta,b}(L-1)\left\{\begin{smallmatrix}\boldsymbol{\mu}\ ;\ \boldsymbol{\nu}^+\\ \boldsymbol{\mu}^*;\ \boldsymbol{\nu}^{+*}\end{smallmatrix}\right\}$ or a member of $\mathcal{P}^{p,p'}_{a,b+\Delta,b}(L-1)\left\{\begin{smallmatrix}\boldsymbol{\mu}\ ;\ \boldsymbol{\nu}\\ \boldsymbol{\mu}^*;\ \boldsymbol{\nu}^*\end{smallmatrix}\right\}$. (In the first of these, we use $\boldsymbol{\nu}^+$ and $\boldsymbol{\nu}^{+*}$ instead of $\boldsymbol{\nu}$ and $\boldsymbol{\nu}^*$ when $b - \Delta = \nu^*_{d^R}$ because then $\boldsymbol{\nu}$ and $\boldsymbol{\nu}^*$ are not a mazy-pair sandwiching $b - \Delta$.) In either case $p' > 2p$ or $p' < 2p$, consideration of the weight of the Lth vertex of h, then yields the first identity between generating functions in 1) and 3). The second identity in these cases follows in a similar way.

Now consider h as above when either both $d^R > 0$ and $b + \Delta = \nu_{d^R}$, or $b + \Delta \in \{0, p'\}$. Necessarily $h_{L-1} = b - \Delta$. Thus when $p' > 2p$ the Lth vertex is scoring, and when $p' < 2p$ the Lth vertex is non-scoring. Changing its direction yields the first identity in both 2) and 4). The first $(L-1)$ segments of h yields a path which is a member of $\mathcal{P}^{p,p'}_{a,b-\Delta,b}(L-1)\left\{\begin{smallmatrix}\boldsymbol{\mu}\ ;\ \boldsymbol{\nu}^+\\ \boldsymbol{\mu}^*;\ \boldsymbol{\nu}^{+*}\end{smallmatrix}\right\}$. The second identity in 2) and 4) follows as in the first paragraph above. □

6. Extending and truncating paths

6.1. Extending paths

In this section, we specify a process by which a path $h \in \mathcal{P}^{p,p'}_{a,b,e,f}(L)$ may be extended by a single unit to its left, or by a single unit to its right. Consequently, the new path h' is of length $L' = L+1$. An extension on the right may follow one on the left to yield a path of length $L+2$.

Throughout this section, $\boldsymbol{\mu}, \boldsymbol{\mu}^*, \boldsymbol{\nu}, \boldsymbol{\nu}^*$ are a mazy-four in the (p,p')-model sandwiching (a,b), with $\boldsymbol{\mu}$, $\boldsymbol{\mu}^*$ each of dimension $d^L \geq 0$, and $\boldsymbol{\nu}$, $\boldsymbol{\nu}^*$ each of dimension $d^R \geq 0$.

We restrict path extension on the right to the cases where $\delta^{p,p'}_{b,f} = 0$ so that the post-segment of h lies in the even band. The extended path h' has endpoint $b' = b+\Delta$, where if b is interfacial we permit both $\Delta = \pm 1$, and if b is not interfacial we permit only $\Delta = (-1)^f$ (in this latter case, the $(L+1)$th segment of h' lies in the same direction as the post-segment of h). We specify that the $(L+1)$th vertex of h' is a scoring (peak) vertex. Thus, on setting $f' = 0$ if $b' = b-1$, and $f' = 1$ if $b' = b+1$, we have $h' \in \mathcal{P}^{p,p'}_{a,b',e,f'}(L+1)$ given by $h'_i = h_i$ for $0 \leq i \leq L$ and $h'_{L+1} = b'$.

The three diagrams in Fig. 6.1 depict the extending process when $f = 1$.

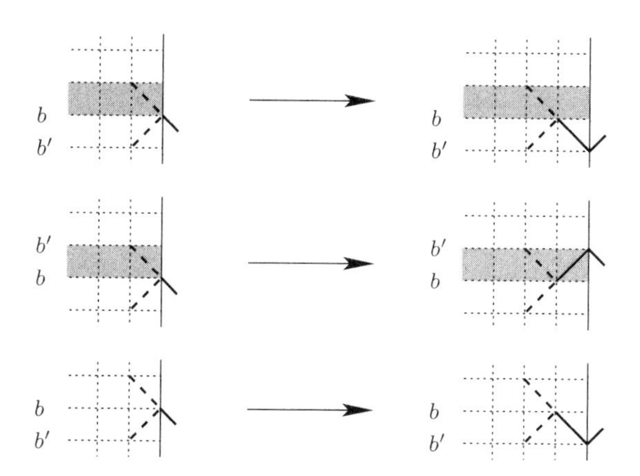

FIGURE 6.1. Path extension on the right.

The $f = 0$ cases may be obtained by reflecting these diagrams in a horizontal axis.

Lemma 6.1. *Let $1 \leq a, b < p'$ and $e, f \in \{0,1\}$ with $\delta^{p,p'}_{b,f} = 0$, and let $L \geq 0$. If b is interfacial let $\Delta = \pm 1$, and otherwise set $\Delta = (-1)^f$. Then set $b' = b+\Delta$*

and let $f' \in \{0,1\}$ be such that $\Delta = -(-1)^{f'}$. If $h' \in \mathcal{P}^{p,p'}_{a,b',e,f'}(L')$ is obtained from $h \in \mathcal{P}^{p,p'}_{a,b,e,f}(L)$ by the above process of path extension, then:

- $L' = L + 1$;
- $m(h') = m(h)$;
- $\tilde{wt}(h') = \tilde{wt}(h) + \frac{1}{2}(L - \Delta\alpha(h))$.

In addition, if $\lfloor bp/p' \rfloor = \lfloor b'p/p' \rfloor$ then $f' = 1 - f$, and if $\lfloor bp/p' \rfloor \neq \lfloor b'p/p' \rfloor$ then $f' = f$. Furthermore, $\alpha^{p,p'}_{a,b'} = \alpha^{p,p'}_{a,b} + \Delta$ and $\beta^{p,p'}_{a,b',e,f'} = \beta^{p,p'}_{a,b,e,f} + \Delta$.

Proof: Let h have striking sequence $\begin{pmatrix} a_1 & a_2 & a_3 & \cdots & a_l \\ b_1 & b_2 & b_3 & \cdots & b_l \end{pmatrix}^{(e,f,d)}$ and assume first that $L > 0$. With $\delta^{p,p'}_{b,f} = 0$, it is readily checked that the Lth vertex of h' is scoring if and only if the Lth vertex of h is scoring. This holds even if $L = 0$. Since the $(L+1)$th vertex of h' is scoring, it immediately follows that $m(h') = m(h)$.

If $L = 0$, then necessarily $\alpha(h) = 0$ whereupon the expression for $\tilde{wt}(h')$ holds. Now consider $L > 0$. If the extending segment is in the same direction as the Lth segment, h' has striking sequence $\begin{pmatrix} a_1 & a_2 & a_3 & \cdots & a_l \\ b_1 & b_2 & b_3 & \cdots & b_l+1 \end{pmatrix}^{(e,f',d)}$ and $\Delta = -(-1)^{d+l}$. If the extending segment is in the direction opposite to that of the Lth segment, h' has striking sequence $\begin{pmatrix} a_1 & a_2 & \cdots & a_l & 0 \\ b_1 & b_2 & \cdots & b_l & 1 \end{pmatrix}^{(e,f',d)}$ and $\Delta = (-1)^{d+l}$.

For $1 \leq i \leq l$, let $w_i = a_i + b_i$. We find $\alpha(h) = -(-1)^{d+l}((w_l + w_{l-2}\cdots) - (w_{l-1} + w_{l-3} + \cdots))$. In the first case above, Lemma 2.2 gives $\tilde{wt}(h') = \tilde{wt}(h) + (w_{l-1} + w_{l-3} + w_{l-5} + \cdots)$, whereupon we obtain $\tilde{wt}(h') = \tilde{wt}(h) + \frac{1}{2}(L(h) - \Delta\alpha(h))$. In the second case above, Lemma 2.2 gives $\tilde{wt}(h') = \tilde{wt}(h) + (w_l + w_{l-2} + w_{l-4} + \cdots)$, and we again obtain $\tilde{wt}(h') = \tilde{wt}(h) + \frac{1}{2}(L(h) - \Delta\alpha(h))$.

In the $L = 0$ case, let $d' = d(h')$. It is easily seen that $d' = f$ if and only if $\pi(h') = 0$. It then follows that h' has striking sequence $\binom{0}{1}^{(e,f',d')}$ and that $m(h') = m(h)$, $\alpha(h) = 0$ and $\tilde{wt}(h') = \tilde{wt}(h) = 0$ thus verifying the first statement in the $L = 0$ case.

The penultimate statement follows from the definitions, noting that $\lfloor b'p/p' \rfloor \neq \lfloor bp/p' \rfloor$ only if b is interfacial in the (p,p')-model and $\Delta = -(-1)^f$.

That $\alpha^{p,p'}_{a,b'} = \alpha^{p,p'}_{a,b} + \Delta$ is immediate. $\beta^{p,p'}_{a,b',e,f'} = \beta^{p,p'}_{a,b,e,f} + \Delta$ follows in the $\lfloor bp/p' \rfloor \neq \lfloor b'p/p' \rfloor$ case because then $\lfloor b'p/p' \rfloor = \lfloor bp/p' \rfloor + \Delta$ and $f' = f$, and in the $\lfloor bp/p' \rfloor = \lfloor b'p/p' \rfloor$ case because $f = 1 - f'$ and $\Delta = -(-1)^{f'} = 2f' - 1$. □

Lemma 6.2. *Let $1 \leq a, b < p'$ and $e, f \in \{0,1\}$ with $\delta^{p,p'}_{b,f} = 0$. If b is interfacial let $\Delta = \pm 1$, and otherwise set $\Delta = (-1)^f$. Then set $b' = b + \Delta$ and let $f' \in \{0,1\}$ be such that $\Delta = -(-1)^{f'}$. Define $\boldsymbol{\nu}' = (\nu_1, \ldots, \nu_{d^R}, b' + \Delta)$ and $\boldsymbol{\nu}^{*\prime} = (\nu^*_1, \ldots, \nu^*_{d^R}, b' - \Delta)$. If either $d^R = 0$ or both $b' \neq \nu_{d^R}$ and $b' \neq \nu^*_{d^R}$, then $\boldsymbol{\mu}, \boldsymbol{\mu}^*, \boldsymbol{\nu}', \boldsymbol{\nu}^{*\prime}$ are a mazy-four sandwiching (a, b') and*

$$\tilde{\chi}^{p,p'}_{a,b',e,f'}(L,m)\begin{Bmatrix} \boldsymbol{\mu} \,; \boldsymbol{\nu}' \\ \boldsymbol{\mu}^*; \boldsymbol{\nu}^{*\prime} \end{Bmatrix} = q^{\frac{1}{2}(L-1-\Delta\alpha)}\tilde{\chi}^{p,p'}_{a,b,e,f}(L-1,m)\begin{Bmatrix} \boldsymbol{\mu} \,; \boldsymbol{\nu} \\ \boldsymbol{\mu}^*; \boldsymbol{\nu}^* \end{Bmatrix},$$

where $\alpha = \alpha^{p,p'}_{a,b}$.

In addition, $\alpha^{p,p'}_{a,b'} = \alpha^{p,p'}_{a,b} + \Delta$, $\beta^{p,p'}_{a,b',e,f'} = \beta^{p,p'}_{a,b,e,f} + \Delta$, *and if* $f' = 1 - f$ *then* $\lfloor b'p/p' \rfloor = \lfloor bp/p' \rfloor$, *and if* $f' = f$ *then* $\lfloor b'p/p' \rfloor = \lfloor bp/p' \rfloor + \Delta$.

Proof: That $\boldsymbol{\mu}, \boldsymbol{\mu}^*, \boldsymbol{\nu}', \boldsymbol{\nu}^{*\prime}$ are a mazy-four sandwiching (a, b') follows immediately if $d^R = 0$. If $d^R > 0$, it follows after noting that since b is strictly between ν_{d^R} and $\nu^*_{d^R}$, and $b' \neq \nu_{d^R}$ and $b' \neq \nu^*_{d^R}$, then $b' + \Delta$ and $b' - \Delta$ are both between ν_{d^R} and $\nu^*_{d^R}$.

Let $L > 0$ and $h \in \mathcal{P}^{p,p'}_{a,b,e,f}(L-1, m)\left\{\begin{smallmatrix}\boldsymbol{\mu} & ;\boldsymbol{\nu} \\ \boldsymbol{\mu}^* & ;\boldsymbol{\nu}^*\end{smallmatrix}\right\}$. Extend this path on the right to obtain h' with $h'_L = b' = b + \Delta$. Clearly, h' attains $b = b' - \Delta$ and does so after it attains any $b' + \Delta$. It follows that h' is mazy-compliant with $\boldsymbol{\mu}, \boldsymbol{\mu}^*, \boldsymbol{\nu}', \boldsymbol{\nu}^{*\prime}$. Then, via Lemma 6.1, $h' \in \mathcal{P}^{p,p'}_{a,b',e,f'}(L, m)\left\{\begin{smallmatrix}\boldsymbol{\mu} & ;\boldsymbol{\nu}' \\ \boldsymbol{\mu}^* & ;\boldsymbol{\nu}^{*\prime}\end{smallmatrix}\right\}$. Conversely, any such h' arises from some $h \in \mathcal{P}^{p,p'}_{a,b,e,f}(L-1, m)\left\{\begin{smallmatrix}\boldsymbol{\mu} & ;\boldsymbol{\nu} \\ \boldsymbol{\mu}^* & ;\boldsymbol{\nu}^*\end{smallmatrix}\right\}$, in this way. For $L > 0$, the required result then follows from the expression for $\tilde{wt}(h')$ given in Lemma 6.1, and $\alpha(h) = \alpha^{p,p'}_{a,b}$ from Lemma 2.3. For $L \leq 0$, both sides are clearly equal to 0.

The final statement follows from Lemma 6.1. □

For $h \in \mathcal{P}^{p,p'}_{a,b,e,f}(L)$, we now define path extension to the left in a similar way. Here, we restrict path extension to the cases where $\delta^{p,p'}_{a,e} = 0$ so that the pre-segment of h lies in the even band. If $L = 0$, we also restrict to the case $\delta^{p,p'}_{b,f} = 0$. The extended path h' has startpoint $a' = a + \Delta$, where if a is interfacial we permit both $\Delta = \pm 1$, and if a is not interfacial we permit only $\Delta = (-1)^e$ (in this latter case, the 0th segment of h' lies in the same direction as the pre-segment of h). We specify that the 0th vertex of h' is a scoring (peak) vertex. Thus, on setting $e' = 0$ if $a' = a - 1$, and $e' = 1$ if $a' = a + 1$, we have $h' \in \mathcal{P}^{p,p'}_{a',b,e',f}(L+1)$ given by $h'_i = h_{i-1}$ for $1 \leq i \leq L+1$ and $h'_0 = b'$.

The three diagrams in Fig. 6.2 depict the extending process when $e = 1$.

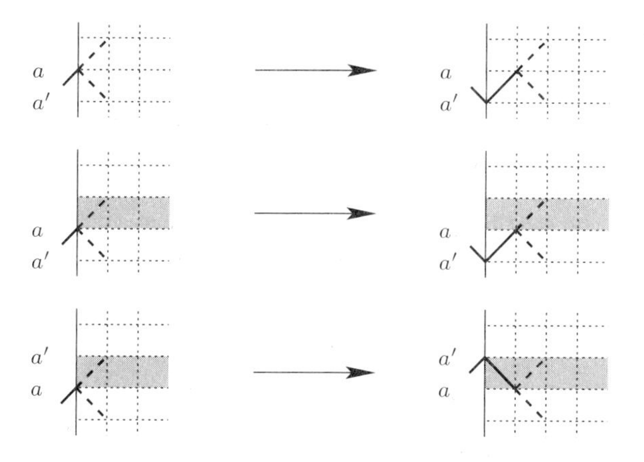

FIGURE 6.2. Path extension on the left.

The cases where $e = 0$ may be obtained by reflecting these diagrams in a horizontal axis.

Lemma 6.3. *Let $1 \le a, b < p'$ and $e, f \in \{0,1\}$ with $\delta^{p,p'}_{a,e} = 0$. Let $L \ge 0$, but if $\delta^{p,p'}_{b,f} = 1$ restrict to $L > 0$. If a is interfacial let $\Delta = \pm 1$, and otherwise set $\Delta = (-1)^e$. Then set $a' = a + \Delta$ and let $e' \in \{0,1\}$ be such that $\Delta = -(-1)^{e'}$. If $h' \in \mathcal{P}^{p,p'}_{a',b,e',f}(L')$ is obtained from $h \in \mathcal{P}^{p,p'}_{a,b,e,f}(L)$ by the above process of path extension, then:*

- $L' = L + 1$;
- $m(h') = \begin{cases} m(h) & \text{if } \lfloor ap/p' \rfloor = \lfloor a'p/p' \rfloor, \\ m(h) + 2 & \text{if } \lfloor ap/p' \rfloor \ne \lfloor a'p/p' \rfloor; \end{cases}$
- $\tilde{wt}(h') = \tilde{wt}(h) + \frac{1}{2}(L - m(h) + \Delta\beta(h))$.

In addition, if $\lfloor ap/p' \rfloor = \lfloor a'p/p' \rfloor$ then $e' = 1 - e$, and if $\lfloor ap/p' \rfloor \ne \lfloor a'p/p' \rfloor$ then $e' = e$. Furthermore, $\alpha^{p,p'}_{a',b} = \alpha^{p,p'}_{a,b} - \Delta$ and $\beta^{p,p'}_{a',b,e',f} = \beta^{p,p'}_{a,b,e,f} - \Delta$.

Proof: Let h have striking sequence $\begin{pmatrix} a_1 & a_2 & a_3 & \cdots & a_l \\ b_1 & b_2 & b_3 & \cdots & b_l \end{pmatrix}^{(e,f,d)}$, let $\pi = \pi(h)$, and assume first that $L > 0$. Consider the case $\lfloor ap/p' \rfloor = \lfloor a'p/p' \rfloor$, where necessarily $\Delta = (-1)^e$ (in the case of interfacial a, this follows from $\delta^{p,p'}_{a,e} = 0$). It follows immediately from the definition that $d(h') = e' = 1 - e$ and $\pi(h') = 0$.

In the subcase where $e = d$ then h' has striking sequence $\begin{pmatrix} 0 & a_1 & a_2 & \cdots & a_l \\ 1 & b_1 & b_2 & \cdots & b_l \end{pmatrix}^{(e',f,e')}$ and $\pi = 0$. We see that $m(h') = m(h)$ and $\tilde{wt}(h') = \tilde{wt}(h) + (b_1 + b_3 + \cdots)$. Since $\beta(h) = \Delta((b_1 + b_3 + \cdots) - (b_2 + b_4 + \cdots))$ and $m(h) = (a_1 + a_2 + a_3 + \cdots)$, we obtain $\tilde{wt}(h') = \tilde{wt}(h) + (L - m(h) + \Delta\beta(h))/2$.

In the subcase where $e \ne d$ and $\pi = 1$ then $\begin{pmatrix} a_1 & a_2 & \cdots & a_l \\ b_1{+}1 & b_2 & \cdots & b_l \end{pmatrix}^{(e',f,e')}$ is the striking sequence of h'. We see that $m(h') = m(h)$ and $\tilde{wt}(h') = \tilde{wt}(h) + (b_2 + b_4 + \cdots)$. Since $\beta(h) = -\Delta((b_1 + b_3 + \cdots) - (b_2 + b_4 + \cdots))$ and $m(h) = (a_1 + a_2 + a_3 + \cdots)$, we obtain $\tilde{wt}(h') = \tilde{wt}(h) + (L - m(h) + \Delta\beta(h))/2$.

In the subcase where $e \ne d$ and $\pi = 0$ then $\begin{pmatrix} a_1{+}1 & a_2 & \cdots & a_l \\ b_1 & b_2 & \cdots & b_l \end{pmatrix}^{(e',f,e')}$ is the striking sequence of h'. We see that $m(h') = m(h)$ and $\tilde{wt}(h') = \tilde{wt}(h) + (b_2 + b_4 + \cdots)$. Since $\beta(h) = -\Delta((b_1 + b_3 + \cdots) - (1 + b_2 + b_4 + \cdots))$ and $m(h) = (1 + a_1 + a_2 + a_3 + \cdots)$, we obtain $\tilde{wt}(h') = \tilde{wt}(h) + (L - m(h) + \Delta\beta(h))/2$.

Now consider the case $\lfloor ap/p' \rfloor \ne \lfloor a'p/p' \rfloor$, where necessarily a is interfacial and $\Delta = -(-1)^e$. It follows immediately from the definition that $d(h') = e' = e$ and $\pi(h') = 1$.

In the subcase where $e = d$ then h' has striking sequence $\begin{pmatrix} a_1{+}1 & a_2 & \cdots & a_l \\ b_1 & b_2 & \cdots & b_l \end{pmatrix}^{(e',f,e')}$ and $\pi = 0$. We see that $m(h') = m(h) + 2$ and $\tilde{wt}(h') = \tilde{wt}(h) + (b_2 + b_4 + \cdots)$. Since $\beta(h) = -\Delta((b_1 + b_3 + \cdots) - (b_2 + b_4 + \cdots))$ and $m(h) = (a_1 + a_2 + a_3 + \cdots)$, we obtain $\tilde{wt}(h') = \tilde{wt}(h) + (L - m(h) + \Delta\beta(h))/2$.

In the subcase where $e \ne d$ then h' has striking sequence $\begin{pmatrix} 1 & a_1 & a_2 & \cdots & a_l \\ 0 & b_1 & b_2 & \cdots & b_l \end{pmatrix}^{(e',f,e')}$ and $\pi = 1$. We see that $m(h') = m(h) + 2$ and $\tilde{wt}(h') = \tilde{wt}(h) + (b_1 + b_3 + \cdots)$. Since $\beta(h) = \Delta((b_1 + b_3 + \cdots) - (b_2 + b_4 + \cdots))$ and $m(h) = (a_1 + a_2 + a_3 + \cdots)$, we obtain $\tilde{wt}(h') = \tilde{wt}(h) + (L - m(h) + \Delta\beta(h))/2$.

Now consider $L=0$ when $\tilde{wt}(h)=0$. If $e=f$ then $m(h)=\beta(h)=0$. If in addition $\pi(h')=0$, then h' has striking sequence $\binom{0}{1}^{(1-e,f,1-e)}$ so that $m(h')=\tilde{wt}(h')=0$, as required. Otherwise, if in addition $\pi(h')=1$, then h' has striking sequence $\binom{1}{0}^{(e,f,e)}$ so that $m(h')=2$ and $\tilde{wt}(h')=0$, as required. For $e\neq f$, we have $m(h)=1$ and $\beta(h)=f-e$. The restriction that $\delta^{p,p'}_{b,f}=0$ forces $\pi(h')=0$, whereupon h' has striking sequence $\binom{1}{0}^{(1-e,f,1-e)}$, so that $\Delta=f-e=\beta(h)$, $m(h')=1$ and $\tilde{wt}(h')=0$, as required.

That $\alpha^{p,p'}_{a',b}=\alpha^{p,p'}_{a,b}-\Delta$ is immediate. $\beta^{p,p'}_{a',b,e',f}=\beta^{p,p'}_{a,b,e,f}-\Delta$ follows in the $\lfloor ap/p'\rfloor\neq\lfloor a'p/p'\rfloor$ case because then $\lfloor a'p/p'\rfloor=\lfloor ap/p'\rfloor+\Delta$ and $e'=e$, and in the $\lfloor ap/p'\rfloor=\lfloor a'p/p'\rfloor$ case because $e=1-e'$ and $\Delta=-(-1)^{e'}=2e'-1$. □

Lemma 6.4. *Let $1\leq a,b<p'$ and $e,f\in\{0,1\}$ with $\delta^{p,p'}_{a,e}=0$. In addition, if $\delta^{p,p'}_{b,f}=1$, restrict to $d^R>0$. If a is interfacial let $\Delta=\pm1$, and otherwise set $\Delta=(-1)^e$. Then set $a'=a+\Delta$ and let $e'\in\{0,1\}$ be such that $\Delta=-(-1)^{e'}$. Define $\boldsymbol{\mu}'=(\mu_1,\ldots,\mu_{d^L},a'+\Delta)$ and $\boldsymbol{\mu}^{*\prime}=(\mu^*_1,\ldots,\mu^*_{d^L},a'-\Delta)$. If either $d^L=0$ or both $a'\neq\nu_{d^L}$ and $a'\neq\nu^*_{d^L}$, then $\boldsymbol{\mu}',\boldsymbol{\mu}^{*\prime},\boldsymbol{\nu},\boldsymbol{\nu}^*$ are a mazy-four sandwiching (a',b) and*

$$\tilde{\chi}^{p,p'}_{a',b,e',f}(L,m+2\delta^{p,p'}_{a',e'})\left\{\begin{matrix}\boldsymbol{\mu}'\,;\boldsymbol{\nu}\\ \boldsymbol{\mu}^{*\prime};\boldsymbol{\nu}^*\end{matrix}\right\}=q^{\frac{1}{2}(L-1-m+\Delta\beta)}\tilde{\chi}^{p,p'}_{a,b,e,f}(L-1,m)\left\{\begin{matrix}\boldsymbol{\mu}\,;\boldsymbol{\nu}\\ \boldsymbol{\mu}^*;\boldsymbol{\nu}^*\end{matrix}\right\},$$

where $\beta=\beta^{p,p'}_{a,b,e,f}$.

In addition, $\alpha^{p,p'}_{a',b}=\alpha^{p,p'}_{a,b}-\Delta$, $\beta^{p,p'}_{a',b,e',f}=\beta^{p,p'}_{a,b,e,f}-\Delta$, and if $e'=1-e$ then $\lfloor a'p/p'\rfloor=\lfloor ap/p'\rfloor$, and if $e'=e$ then $\lfloor a'p/p'\rfloor=\lfloor ap/p'\rfloor+\Delta$.

Proof: That $\boldsymbol{\mu}',\boldsymbol{\mu}^{*\prime},\boldsymbol{\nu},\boldsymbol{\nu}^*$ are a mazy-four sandwiching (a',b) follows immediately if $d^L=0$. If $d^L>0$, it follows after noting that since a is strictly between μ_{d^L} and $\mu^*_{d^L}$, and $a'\neq\mu_{d^L}$ and $a'\neq\mu^*_{d^L}$, then $a'+\Delta$ and $a'-\Delta$ are both between μ_{d^L} and $\mu^*_{d^L}$.

Let $L>0$ and $h\in\mathcal{P}^{p,p'}_{a,b,e,f}(L-1,m)\left\{\begin{smallmatrix}\boldsymbol{\mu}\,;\boldsymbol{\nu}\\ \boldsymbol{\mu}^*;\boldsymbol{\nu}^*\end{smallmatrix}\right\}$. Extend h on the left to obtain h' with $h'_0=a'=a+\Delta$. Clearly, h' attains $a=a'-\Delta$ and does so before it attains any $a'+\Delta$. It follows that h' is mazy-compliant with $\boldsymbol{\mu}',\boldsymbol{\mu}^{*\prime},\boldsymbol{\nu},\boldsymbol{\nu}^*$. Then, via Lemma 6.3, $h'\in\mathcal{P}^{p,p'}_{a',b,e',f}(L,m)\left\{\begin{smallmatrix}\boldsymbol{\mu}'\,;\boldsymbol{\nu}\\ \boldsymbol{\mu}^{*\prime};\boldsymbol{\nu}^*\end{smallmatrix}\right\}$. Conversely, any such h' arises from some $h\in\mathcal{P}^{p,p'}_{a,b,e,f}(L-1,m)\left\{\begin{smallmatrix}\boldsymbol{\mu}\,;\boldsymbol{\nu}\\ \boldsymbol{\mu}^*;\boldsymbol{\nu}^*\end{smallmatrix}\right\}$ in this way. For $L>0$, the required result then follows from the expression for $\tilde{wt}(h')$ given in Lemma 6.3, and $\beta(h)=\beta^{p,p'}_{a,b,e,f}$ from Lemma 2.3. For $L\leq0$, both sides are clearly equal to 0.

The final statement also follows from Lemma 6.3. □

6.2. Truncating paths

In this section, we specify a process by which a path $h\in\mathcal{P}^{p,p'}_{a,b,e,f}(L)$, where $L>0$, may be shortened by removing just the first segment, or by removing just the Lth segment. Consequently, the new path h' is of length $L'=L-1$. A shortening on the right may follow one on the left to yield a path of length $L-2$.

Throughout this section, $\boldsymbol{\mu}, \boldsymbol{\mu}^*, \boldsymbol{\nu}, \boldsymbol{\nu}^*$ are a mazy-four in the (p,p')-model sandwiching (a,b), with $\boldsymbol{\mu}$, $\boldsymbol{\mu}^*$ each of dimension d^L, and $\boldsymbol{\nu}$, $\boldsymbol{\nu}^*$ each of dimension d^R.

Removing the Lth segment of $h \in \mathcal{P}^{p,p'}_{a,b,e,f}(L)$ is permitted only if $h_{L-1} = b+(-1)^f$ and $\delta^{p,p'}_{b,f} = 0$ so that the Lth segment of h lies in an even band and the Lth vertex is scoring. We specify that $f(h') = 1 - f(h)$ so that the post-segment of h' is in the same direction as the Lth segment of h. If $b' = h_{L-1}$ and $f' = 1-f$ then $h' \in \mathcal{P}^{p,p'}_{a,b',e,f'}(L-1)$ with $h'_i = h_i$ for $0 \le i \le L-1$.

Lemma 6.5. *Let $1 \le a,b < p'$ and $e,f \in \{0,1\}$ with $\delta^{p,p'}_{b,f} = 0$. For $L > 0$, let $h \in \mathcal{P}^{p,p'}_{a,b,e,f}(L)$ be such that $h_{L-1} = b+(-1)^f$ and let $h' \in \mathcal{P}^{p,p'}_{a,b',e,f'}(L')$ be obtained from h by the above process of path truncation. If $\Delta = b - b'$ then $\Delta = (-1)^{f'}$, $f = 1 - f'$, and*

- $L' = L - 1$;
- $m(h') = m(h)$;
- $\tilde{wt}(h') = \tilde{wt}(h) - \frac{1}{2}(L - \Delta\alpha(h))$.

In addition, $\alpha^{p,p'}_{a,b'} = \alpha^{p,p'}_{a,b} - \Delta$ and $\beta^{p,p'}_{a,b',e,f'} = \beta^{p,p'}_{a,b,e,f} - \Delta$.

Proof: That $\Delta = (-1)^{f'}$ and $f = 1 - f'$ and $L' = L - 1$ are immediate from the definition. Let h have striking sequence $\begin{pmatrix} a_1 & a_2 & a_3 & \cdots & a_l \\ b_1 & b_2 & b_3 & \cdots & b_l \end{pmatrix}^{(e,f,d)}$, whence $\Delta = -(-1)^{d+l}$. Since $h_{L-1} = b + (-1)^f$, the Lth vertex of h is scoring and thus $b_l \ge 1$. Since $\delta^{p,p'}_{b,f} = 0$ and $f' = 1 - f$, it follows that the $(L-1)$th vertex of h' is scoring if and only if the $(L-1)$th vertex of h is scoring. This holds even if $L = 1$. Thus $m(h') = m(h)$.

When $a_l + b_l > 1$, h' has striking sequence $\begin{pmatrix} a_1 & a_2 & a_3 & \cdots & a_l \\ b_1 & b_2 & b_3 & \cdots & b_l - 1 \end{pmatrix}^{(e,f',d)}$ and when $a_l + b_l > 1$, h' has striking sequence $\begin{pmatrix} a_1 & a_2 & a_3 & \cdots & a_{l-1} \\ b_1 & b_2 & b_3 & \cdots & b_{l-1} \end{pmatrix}^{(e,f',d)}$. It follows that $L' = L - 1$ and $m(h') = m(h)$.

For $1 \le i \le l$, let $w_i = a_i + b_i$. Lemma 2.2 gives $\tilde{wt}(h') = \tilde{wt}(h) - (w_{l-1} + w_{l-3} + w_{l-5} + \cdots)$, whereupon, since $\alpha(h) = \Delta((w_l + w_{l-2} \cdots) - (w_{l-1} + w_{l-3} + \cdots))$, we obtain $\tilde{wt}(h') = \tilde{wt}(h) - \frac{1}{2}(L(h) - \Delta\alpha(h))$.

That $\alpha^{p,p'}_{a,b'} = \alpha^{p,p'}_{a,b} - \Delta$ is immediate. Since $\delta^{p,p'}_{b,f} = 0$, we have $\lfloor b'p/p' \rfloor = \lfloor (b+(-1)^f)p/p' \rfloor = \lfloor bp/p' \rfloor$. Then $\beta^{p,p'}_{a,b',e,f'} = \lfloor b'p/p' \rfloor - \lfloor ap/p' \rfloor + f' - e = \lfloor bp/p' \rfloor - \lfloor ap/p' \rfloor + f - (1-2f') - e = \beta^{p,p'}_{a,b,e,f} - \Delta$, as required. $\square$

Lemma 6.6. *Let $1 \le a, b, b' < p'$ and $e, f \in \{0,1\}$ with $b - b' = -(-1)^f$ and $\delta^{p,p'}_{b,f} = 0$. Set $\Delta = b - b'$ and $f' = 1 - f$. If $d^R > 0$, let $\nu^*_{d^R} \ne b'$ and $\nu_{d^R} = b + \Delta$. If $d^R = 0$, restrict to $b \in \{1, p'-1\}$. Then $\boldsymbol{\mu}, \boldsymbol{\mu}^*, \boldsymbol{\nu}, \boldsymbol{\nu}^*$ are a mazy-four sandwiching (a,b'), and if $L \ge 0$ then:*

$$\tilde{\chi}^{p,p'}_{a,b',e,f'}(L,m)\begin{Bmatrix} \boldsymbol{\mu} \,;\boldsymbol{\nu} \\ \boldsymbol{\mu}^*;\boldsymbol{\nu}^* \end{Bmatrix} = q^{-\frac{1}{2}(L+1-\Delta\alpha)}\tilde{\chi}^{p,p'}_{a,b,e,f}(L+1,m)\begin{Bmatrix} \boldsymbol{\mu} \,;\boldsymbol{\nu} \\ \boldsymbol{\mu}^*;\boldsymbol{\nu}^* \end{Bmatrix},$$

where $\alpha = \alpha^{p,p'}_{a,b}$.

In addition, $\alpha^{p,p'}_{a,b'} = \alpha^{p,p'}_{a,b} - \Delta$, $\beta^{p,p'}_{a,b',e,f'} = \beta^{p,p'}_{a,b,e,f} - \Delta$ and $\lfloor b'p/p' \rfloor = \lfloor bp/p' \rfloor$.

Proof: If $d^R > 0$ then since b is strictly between ν_{d^R} and $\nu^*_{d^R}$, and $b' \neq \nu_{d^R}$ and $b' \neq \nu^*_{d^R}$, it follows that b' is strictly between ν_{d^R} and $\nu^*_{d^R}$. Thus $\boldsymbol{\mu}, \boldsymbol{\mu}^*, \boldsymbol{\nu}, \boldsymbol{\nu}^*$ are a mazy-four sandwiching (a, b').

Let $h \in \mathcal{P}^{p,p'}_{a,b,e,f}(L+1, m)\left\{ {\boldsymbol{\mu}\ ;\boldsymbol{\nu} \atop \boldsymbol{\mu}^*;\boldsymbol{\nu}^*} \right\}$. First consider $d^R > 0$. Since $\nu_{d^R} = b + \Delta$ and $\Delta = \pm 1$, and since h attains $\nu^*_{d^R}$ after any ν_{d^R}, then necessarily $h_{L-1} = b - \Delta = b'$. Then, on removing the final segment of h, we obtain a path h' for which, via Lemma 6.5, $h' \in \mathcal{P}^{p,p'}_{a,b',e,f'}(L, m)\left\{ {\boldsymbol{\mu}\ ;\boldsymbol{\nu} \atop \boldsymbol{\mu}^*;\boldsymbol{\nu}^*} \right\}$. Clearly, each h' in this latter set arises this way. If $d^R = 0$, the same argument may be used after replacing ν_{d^R} by 0 (resp. p') when $b = 1$ (resp. $p' - 1$) and necessarily $\Delta = -1$ (resp. $+1$). The lemma then follows on using the expression for $\tilde{wt}(h')$ given in Lemma 6.5, and $\alpha(h) = \alpha^{p,p'}_{a,b}$ from Lemma 2.3.

The final statement follows from Lemma 6.5 and noting that $\delta^{p,p'}_{b,f} = 0$ and $b' = b + (-1)^f$ implies $\lfloor b'p/p' \rfloor = \lfloor bp/p' \rfloor$. □

Removing the first segment of h is permitted only when $\pi(h) = 0$ and when $d(h) = e$ so that the 0th vertex of h is scoring. We specify that $e(h') = 1 - e$ so that the pre-segment of h' is in the same direction as the first segment of h. Let $a' = h_1$. Then, on setting $e' = 1 - e$, we obtain $h' \in \mathcal{P}^{p,p'}_{a',b,e',f}(L-1)$ with $h'_i = h_{i+1}$ for $0 \le i \le L - 1$.

Lemma 6.7. *Let $1 \le a, b < p'$ and $e, f \in \{0, 1\}$. For $L > 0$, let $h \in \mathcal{P}^{p,p'}_{a,b,e,f}(L)$ be such that $d(h) = e$ and $\pi(h) = 0$, and let $h' \in \mathcal{P}^{p,p'}_{a',b,e',f}(L')$ be obtained from h by the above process of path truncation. If $\Delta = a - a'$ then $\Delta = (-1)^{e'}$, $e = 1 - e'$, and*

- $L' = L - 1$;
- $m(h') = m(h)$;
- $\tilde{wt}(h') = \tilde{wt}(h) - \frac{1}{2}(L - m(h) + \Delta\beta(h))$.

In addition, $\alpha^{p,p'}_{a',b} = \alpha^{p,p'}_{a,b} + \Delta$ and $\beta^{p,p'}_{a',b,e',f} = \beta^{p,p'}_{a,b,e,f} + \Delta$.

Proof: That $\Delta = (-1)^{e'}$ and $e = 1 - e'$ and $L' = L - 1$ are immediate from the definition. Let h have striking sequence $\begin{pmatrix} a_1 & a_2 & a_3 & \cdots & a_l \\ b_1 & b_2 & b_3 & \cdots & b_l \end{pmatrix}^{(e,f,d)}$. If $a_1 + b_1 \ge 2$ and the first vertex of h is non-scoring, then h' has striking sequence $\begin{pmatrix} a_1 - 1 & a_2 & a_3 & \cdots & a_l \\ b_1 & b_2 & b_3 & \cdots & b_l \end{pmatrix}^{(1-e,f,d)}$ and $\pi(h') = 0$. If $a_1 + b_1 \ge 2$ and the first vertex of h is scoring, then h' has striking sequence $\begin{pmatrix} a_1 & a_2 & a_3 & \cdots & a_l \\ b_1 - 1 & b_2 & b_3 & \cdots & b_l \end{pmatrix}^{(1-e,f,d)}$ and $\pi(h') = 1$. If $a_1 + b_1 = 1$, then h' has striking sequence $\begin{pmatrix} a_2 & a_3 & \cdots & a_l \\ b_2 & b_3 & \cdots & b_l \end{pmatrix}^{(1-e,f,1-d)}$ and $\pi(h') = 0$. In each case, Lemma 2.2 gives $\tilde{wt}(h') = \tilde{wt}(h) - (b_2 + b_4 + \cdots)$, whereupon, since $\beta(h) = -\Delta((b_1 + b_3 + \cdots) - (b_2 + b_4 + \cdots))$, we obtain $\tilde{wt}(h') = \tilde{wt}(h) - \frac{1}{2}(L(h) - m(h) + \Delta\beta(h))$. Also when $L > 1$, we immediately see that $m(h') = m(h)$ in each case. For $L = 1$, it is readily verified that if $e \neq f$ then $m(h) = m(h') = 0$, and if $e = f$ then $m(h) = m(h') = 1$.

That $\alpha^{p,p'}_{a',b} = \alpha^{p,p'}_{a,b} + \Delta$ is immediate. Since $\pi(h) = 0$, we have $\lfloor a'p/p' \rfloor = \lfloor ap/p' \rfloor$. Then $\beta^{p,p'}_{a',b,e',f} = \lfloor bp/p' \rfloor - \lfloor a'p/p' \rfloor + f - e' = \lfloor bp/p' \rfloor - \lfloor ap/p' \rfloor + f - e + (1 - 2e') = \beta^{p,p'}_{a,b,e,f} + \Delta$, as required. □

Lemma 6.8. *Let $1 \le a, a', b < p'$ and $e, f \in \{0,1\}$ with $a - a' = -(-1)^e$ and $\delta^{p,p'}_{a,e} = 0$. Set $\Delta = a - a'$ and $e' = 1 - e$. If $d^L > 0$, let $\mu^*_{d^L} \ne a'$ and $\mu_{d^L} = a + \Delta$. If $d^L = 0$, restrict to $a \in \{1, p'-1\}$. Then $\boldsymbol{\mu}, \boldsymbol{\mu}^*, \boldsymbol{\nu}, \boldsymbol{\nu}^*$ are a mazy-four sandwiching (a', b), and if $L \ge 0$ then:*

$$\tilde{\chi}^{p,p'}_{a',b,e',f}(L,m)\left\{\begin{matrix}\boldsymbol{\mu}\ ;\boldsymbol{\nu}\\ \boldsymbol{\mu}^*;\boldsymbol{\nu}^*\end{matrix}\right\} = q^{-\frac{1}{2}(L+1-m+\Delta\beta)}\tilde{\chi}^{p,p'}_{a,b,e,f}(L+1,m)\left\{\begin{matrix}\boldsymbol{\mu}\ ;\boldsymbol{\nu}\\ \boldsymbol{\mu}^*;\boldsymbol{\nu}^*\end{matrix}\right\},$$

where $\beta = \beta^{p,p'}_{a,b,e,f}$.

In addition, $\alpha^{p,p'}_{a',b} = \alpha^{p,p'}_{a,b} + \Delta$, $\beta^{p,p'}_{a',b,e',f} = \beta^{p,p'}_{a,b,e,f} + \Delta$ and $\lfloor a'p/p' \rfloor = \lfloor ap/p' \rfloor$.

Proof: If $d^L > 0$ then since a is strictly between μ_{d^L} and $\mu^*_{d^L}$, and $a' \ne \mu_{d^L}$ and $a' \ne \mu^*_{d^L}$, it follows that a' is strictly between μ_{d^L} and $\mu^*_{d^L}$. Thus $\boldsymbol{\mu}, \boldsymbol{\mu}^*, \boldsymbol{\nu}, \boldsymbol{\nu}^*$ are a mazy-four sandwiching (a', b).

Let $h \in \mathcal{P}^{p,p'}_{a,b,e,f}(L+1,m)\left\{\begin{smallmatrix}\boldsymbol{\mu}\ ;\boldsymbol{\nu}\\ \boldsymbol{\mu}^*;\boldsymbol{\nu}^*\end{smallmatrix}\right\}$. First consider $d^L > 0$. Since $\mu_{d^L} = a + \Delta$ and $\Delta = \pm 1$, and since h attains $\mu^*_{d^L}$ before any μ_{d^L}, then necessarily $h_1 = a - \Delta = a'$. Then, on removing the first segment of h, we obtain a path h' for which, via Lemma 6.7, $h' \in \mathcal{P}^{p,p'}_{a',b,e',f}(L,m)\left\{\begin{smallmatrix}\boldsymbol{\mu}\ ;\boldsymbol{\nu}\\ \boldsymbol{\mu}^*;\boldsymbol{\nu}^*\end{smallmatrix}\right\}$. Clearly, each h' in this latter set arises this way. If $d^L = 0$, the same argument may be used after replacing μ_{d^L} by 0 (resp. p') when $a = 1$ (resp. $p' - 1$) and necessarily $\Delta = -1$ (resp. $+1$). The lemma then follows on using the expression for $\tilde{wt}(h')$ given in Lemma 6.7, and $\beta(h) = \beta^{p,p'}_{a,b,e,f}$ from Lemma 2.3.

The final statement follows from Lemma 6.7, and noting that $\delta^{p,p'}_{a,e} = 0$ and $a' = a + (-1)^e$ implies that $\lfloor a'p/p' \rfloor = \lfloor ap/p' \rfloor$. □

7. Generating the fermionic expressions

In this section, we prove the core result of this work: namely, for certain vectors $\boldsymbol{u}^L$ and $\boldsymbol{u}^R$, we precisely specify a subset of the set of paths $\mathcal{P}^{p,p'}_{a,b,c}(m_0)$ for which $F(\boldsymbol{u}^L, \boldsymbol{u}^R, m_0)$ is the generating function. Throughout this section, we fix a pair p, p' of coprime integers with $1 \leq p < p'$, and employ the notation of Sections 1.3, 1.5, 1.9, 1.11 and 1.12. We assume that $t > 0$ so that (only) the case $(p, p') = (1, 2)$ is not considered hereafter.

7.1. Runs and the core result

For $d \geq 1$, we refer to a set $\{\tau_j, \sigma_j, \Delta_j\}_{j=1}^d$ as a *run* if $\tau_1 = t+1$, each $\Delta_j = \pm 1$,

$$0 \leq \sigma_d < \tau_d < \sigma_{d-1} \leq \tau_{d-1} < \cdots < \sigma_2 \leq \tau_2 < \sigma_1 < t, \tag{7.1}$$

and

$$\tau_j = \sigma_j \quad \Longrightarrow \quad \Delta_j = \Delta_{j+1} \text{ and } 2 \leq j < d. \tag{7.2}$$

For $0 \leq i \leq t$, define $\eta(i)$ to be such that $\tau_{\eta(i)+1} \leq i < \tau_{\eta(i)}$, where we set $\tau_{d+1} = 0$. In addition, we define

$$d_0 = \begin{cases} 0 & \text{if } \sigma_1 < t_n; \\ 1 & \text{if } \sigma_1 \geq t_n. \end{cases} \tag{7.3}$$

In what follows, superscripts L or R (to designate *left* or *right*) may be appended to the quantities defined above.

Given a run $\mathcal{X}^L = \{\tau_j^L, \sigma_j^L, \Delta_j^L\}_{j=1}^{d^L}$, we define values a, e, t-dimensional vectors $\boldsymbol{u}^L$, $\boldsymbol{\Delta}^L$, and $(d^L - d_0^L)$-dimensional vectors $\boldsymbol{\mu}, \boldsymbol{\mu}^*$.

Define:

$$a = \sum_{m=2}^{d^L} \Delta_m^L (\kappa_{\tau_m^L} - \kappa_{\sigma_m^L}) + \begin{cases} \kappa_{\sigma_1^L} & \text{if } \Delta_1^L = -1; \\ p' - \kappa_{\sigma_1^L} & \text{if } \Delta_1^L = +1. \end{cases}$$

Define $e \in \{0, 1\}$ to be such that:

$$\Delta_{d^L}^L = \begin{cases} -(-1)^e & \text{if } \sigma_{d^L}^L = 0; \\ (-1)^{e+k^L} & \text{if } \sigma_{d^L}^L > 0, \end{cases}$$

where $k^L = \zeta(\sigma_{d^L}^L)$.

Define t-dimensional vectors $\boldsymbol{u}^L$ and $\boldsymbol{\Delta}^L$ as follows (c.f. (1.21) and (1.24)):

$$\boldsymbol{u}^L = \sum_{m=1}^{d^L} \boldsymbol{u}_{\sigma_m^L, \tau_m^L} + \begin{cases} 0 & \text{if } \Delta_1^L = -1; \\ \boldsymbol{e}_t & \text{if } \Delta_1^L = +1; \end{cases}$$

$$\boldsymbol{\Delta}^L = \sum_{m=1}^{d^L} \Delta_m^L \boldsymbol{u}_{\sigma_m^L, \tau_m^L} - \begin{cases} 0 & \text{if } \Delta_1^L = -1; \\ \boldsymbol{e}_t & \text{if } \Delta_1^L = +1. \end{cases}$$

Note that Δ_m^L are not the components of $\boldsymbol{\Delta}^L$.

Define values $\{\mu_j, \mu_j^*\}_{j=0}^{d^L-1}$ as follows:

$$\mu_0^* = \begin{cases} \kappa_{t_n} & \text{if } \Delta_1^L = -1; \\ p' - \kappa_{t_n} & \text{if } \Delta_1^L = +1; \end{cases}$$

$$\mu_0 = \begin{cases} 0 & \text{if } \Delta_1^L = -1; \\ p' & \text{if } \Delta_1^L = +1; \end{cases}$$

$$\mu_j^* = \sum_{m=2}^{j} \Delta_m^L (\kappa_{\tau_m^L} - \kappa_{\sigma_m^L}) + \begin{cases} \kappa_{\sigma_1^L} & \text{if } \Delta_1^L = -1; \\ p' - \kappa_{\sigma_1^L} & \text{if } \Delta_1^L = +1; \end{cases}$$

$$\mu_j = \mu_j^* + \Delta_{j+1}^L \kappa_{\tau_{j+1}^L},$$

for $1 \le j < d^L$. The $(d^L - d_0^L)$-dimensional vectors $\boldsymbol{\mu}$ and $\boldsymbol{\mu}^*$ are defined by $\boldsymbol{\mu} = (\mu_{d_0^L}, \ldots, \mu_{d^L-1})$ and $\boldsymbol{\mu}^* = (\mu_{d_0^L}^*, \ldots, \mu_{d^L-1}^*)$.

In a totally analogous way, a run $\mathcal{X}^R = \{\tau_j^R, \sigma_j^R, \Delta_j^R\}_{j=1}^{d^R}$, is used to define values b, f, $\{\nu_j, \nu_j^*\}_{j=0}^{d^R-1}$, t-dimensional vectors $\boldsymbol{u}^R$, $\boldsymbol{\Delta}^R$, and $(d^R - d_0^R)$-dimensional vectors $\boldsymbol{\nu}, \boldsymbol{\nu}^*$.

In view of (7.1), we see that $\boldsymbol{\mu}, \boldsymbol{\mu}^*, \boldsymbol{\nu}, \boldsymbol{\nu}^*$ satisfy the three criteria for being a mazy-four in the (p, p')-model sandwiching (a, b). That they are in fact interfacial will be established later Their neighbouring odd bands will be seen to be specified by the values $\{\tilde{\mu}_j, \tilde{\mu}_j^*\}_{j=0}^{d^L-1}$ and $\{\tilde{\nu}_j, \tilde{\nu}_j^*\}_{j=0}^{d^R-1}$, where we define:

$$\tilde{\mu}_0^* = \begin{cases} \tilde{\kappa}_{t_n} & \text{if } \Delta_1^L = -1; \\ p - \tilde{\kappa}_{t_n} & \text{if } \Delta_1^L = +1; \end{cases}$$

$$\tilde{\mu}_0 = \begin{cases} 0 & \text{if } \Delta_1^L = -1; \\ p & \text{if } \Delta_1^L = +1; \end{cases}$$

$$\tilde{\mu}_j^* = \sum_{m=2}^{j} \Delta_m^L (\tilde{\kappa}_{\tau_m^L} - \tilde{\kappa}_{\sigma_m^L}) + \begin{cases} \tilde{\kappa}_{\sigma_1^L} & \text{if } \Delta_1^L = -1; \\ p - \tilde{\kappa}_{\sigma_1^L} & \text{if } \Delta_1^L = +1; \end{cases}$$

$$\tilde{\mu}_j = \tilde{\mu}_j^* + \Delta_{j+1}^L \tilde{\kappa}_{\tau_{j+1}^L},$$

for $1 \le j < d^L$, with $\{\tilde{\nu}_j, \tilde{\nu}_j^*\}_{j=0}^{d^R-1}$ defined analogously.

Finally, note that $\boldsymbol{\Delta}^L = \boldsymbol{\Delta}(\mathcal{X}^L)$ and $\boldsymbol{\Delta}^R = \boldsymbol{\Delta}(\mathcal{X}^R)$, and use the procedure given in Section 1.10 to calculate γ' and well as $\{\alpha_j, \alpha_j'', \beta_j, \beta_j'', \gamma_j, \gamma_j', \gamma_j''\}_{j=0}^{t}$.

A run $\{\tau_j, \sigma_j, \Delta_j\}_{j=1}^d$ is said to be *naive* when if either $1 < j < d$, or $1 = j < d$ and $\sigma_1 \le t_n$, then

$$\begin{array}{lcl} \Delta_{j+1} = \Delta_j & \Longrightarrow & \tau_{j+1} = t_{\zeta(\sigma_j - 1)}, \\ \Delta_{j+1} \neq \Delta_j & \Longrightarrow & \tau_{j+1} = t_{\zeta(\sigma_j)}; \end{array}$$

and if $d > 1$ and $\sigma_1 > t_n$ then

$$\begin{array}{lcl} \Delta_2 = \Delta_1 & \Longrightarrow & \tau_2 = t_{n-1}, \\ \Delta_2 \neq \Delta_1 \text{ and } c_{n-1} > 1 & \Longrightarrow & \tau_2 = t_n - 1, \\ \Delta_2 \neq \Delta_1 \text{ and } c_{n-1} = 1 & \Longrightarrow & \tau_2 = t_{n-2}. \end{array}$$

Note that if $\{\tau_j, \sigma_j, \Delta_j\}_{j=1}^d$ is a naive run then for $1 \le j < d$, there exists (at least one value) k such that $\tau_{j+1} \le t_k < \sigma_j$.

Later, in Section 8, we show that the values $\{\tau_j, \sigma_j, \Delta_j\}_{j=1}^d$ that are obtained from each leaf node of a Takahashi tree as described in Section 1.7, provide a naive run.

The core result is the following:

Theorem 7.1. *Let $p' > 2p$, and let both $\mathcal{X}^L = \{\tau_j^L, \sigma_j^L, \Delta_j^L\}_{j=1}^{d^L}$ and $\mathcal{X}^R = \{\tau_j^R, \sigma_j^R, \Delta_j^R\}_{j=1}^{d^R}$ be naive runs. If $m_0 \equiv b - a \ (mod 2)$ then:*

$$\tilde{\chi}_{a,b,e,f}^{p,p'}(m_0) \left\{ \begin{matrix} \boldsymbol{\mu} \, ; \boldsymbol{\nu} \\ \boldsymbol{\mu}^* ; \boldsymbol{\nu}^* \end{matrix} \right\} = q^{\frac{1}{4}(\gamma_0 - \gamma')} F(\boldsymbol{u}^L, \boldsymbol{u}^R, m_0).$$

We prove this result in the following sections. The strategy for the proof is simple: we begin with the trivial generating functions for the $(1,3)$-model stated in Lemma 2.6,[1] and for $i = t-1, t-2, \ldots, 1$, apply a $\mathcal{B}$-transform if $i \neq t_k$ for all k, and apply a $\mathcal{BD}$-transform if $i = t_k$ for some k. If $(\boldsymbol{u}^L)_i = -1$ (resp. $+1$), we follow the transform with path extension (resp. truncation) on the left. If $(\boldsymbol{u}^R)_i = -1$ (resp. $+1$), we follow with path extension (resp. truncation) on the right. This process is carried out in Lemma 7.11 to yield, in Corollary 7.13, an expression for $\tilde{\chi}_{a,b,e,f}^{p,p'}(m_0, m_1) \left\{ \begin{smallmatrix} \boldsymbol{\mu} \, ; \boldsymbol{\nu} \\ \boldsymbol{\mu}^* ; \boldsymbol{\nu}^* \end{smallmatrix} \right\}$, for certain values of m_1. Then, at the end of Section 7.3, a final sum over m_1 is performed to prove Theorem 7.1.

In Section 7.4, we transfer our result across to generating functions in terms of the original weighting function of (2.5). In this way, for a specific c, we will have identified a subset of $\mathcal{P}_{a,b,c}^{p,p'}(L)$ for which the generating function is precisely $F(\boldsymbol{u}^L, \boldsymbol{u}^R, L)$.

We actually first prove an analogue of Theorem 7.1 involving *reduced* runs instead of naive runs, where a run $\{\tau_j, \sigma_j, \Delta_j\}_{j=1}^d$ is said to be reduced if:

$$0 \le \sigma_d < \tau_d < \sigma_{d-1} < \tau_{d-1} < \cdots < \sigma_2 < \tau_2 < \sigma_1 < t, \tag{7.4}$$

[1]It is actually possible to start with $\tilde{\chi}_{1,1,0,0}^{1,2}(0,0)$: an application of a single $\mathcal{B}$-transform yields the results of Lemma 2.6. We don't do this in order to avoid a further increase in the notational complexities in the main induction proof.

and if $d > 1$ then

1. $\sigma_1 > t_n$ and $\Delta_2 = \Delta_1 \implies \tau_2 = t_k$ for $1 \le k \le n-1$;
2. $\sigma_1 > t_n$ and $\Delta_2 \ne \Delta_1$ and $c_{n-1} > 1$
$\implies \tau_2 = t_n - 1$ or $\tau_2 = t_k$ for $1 \le k \le n-1$;
3. $\sigma_1 > t_n$ and $\Delta_2 \ne \Delta_1$ and $c_{n-1} = 1$
$\implies \tau_2 = t_k$ for $1 \le k \le n-2$;
4. $\sigma_1 \le t_n$ and $\Delta_2 = \Delta_1 \implies \tau_2 = t_k$ for $1 \le k \le \zeta(\sigma_1 - 1)$;
5. $\sigma_1 \le t_n$ and $\Delta_2 \ne \Delta_1 \implies \tau_2 = t_k$ for $1 \le k \le \zeta(\sigma_1)$,

and if $1 < j < d$ then

6. $\Delta_{j+1} = \Delta_j \implies \tau_{j+1} = t_k$ for $1 \le k \le \zeta(\sigma_j - 1)$;
7. $\Delta_{j+1} \ne \Delta_j \implies \tau_{j+1} = t_k$ for $1 \le k \le \zeta(\sigma_j)$.

The following result is immediately obtained:

Lemma 7.2. *Let $\{\tau_j, \sigma_j, \Delta_j\}_{j=1}^d$ be a run, and let the run $\{\tau'_j, \sigma'_j, \Delta'_j\}_{j=1}^{d'}$ be obtained from it by removing each triple $\{\tau_j, \sigma_j, \Delta_j\}$ whenever $\tau_j = \sigma_j$. Naturally $d' \le d$. If $\{\tau_j, \sigma_j, \Delta_j\}_{j=1}^d$ is a naive run then $\{\tau'_j, \sigma'_j, \Delta'_j\}_{j=1}^{d'}$ is a reduced run.*

Proof: A routine verification. □

The reduced run $\{\tau'_j, \sigma'_j, \Delta'_j\}_{j=1}^{d'}$ identified in the above lemma will be referred to as the reduced run *corresponding* to the run $\{\tau_j, \sigma_j, \Delta_j\}_{j=1}^d$.

Lemma 7.3. *Let $\{\tau_j, \sigma_j, \Delta_j\}_{j=1}^d$ be a naive run and $\{\tau'_j, \sigma'_j, \Delta'_j\}_{j=1}^{d'}$ the corresponding reduced run. Let $d_0 < j < d$. If $\tau'_{j+1} = t_k$ for some k and $c_k = 1$ then:*

1. $\sigma'_j = t_{k+1} \implies \Delta'_{j+1} \ne \Delta'_j$;
2. $\sigma'_j = t_{k+1} + 1 \implies \Delta'_{j+1} = \Delta'_j$;
3. $\sigma'_j > t_{k+1} + 1 \implies \sigma'_j > t_{k+2}$.

Proof: We have $t_{k+1} = t_k + 1$ and $\tau'_{j+1} < \sigma'_j < \tau'_j$. Let J and J' be such that $\tau_{J+1} = \tau'_{j+1}$ and $\sigma_{J'} = \sigma'_j$. Then $J' \le J$ and $\sigma_i = \tau_i$ for $J' < i \le J$. Moreover, $\Delta_{J'+1} = \Delta_{J'+2} = \cdots = \Delta_J = \Delta_{J+1}$. First note that $\sigma_J = \tau_J = t_{k+1}$ is forbidden here since $\Delta_J = \Delta_{J+1}$ would then dictate that $\tau_{J+1} = t_{\zeta(\sigma_J - 1)} < t_k$ thereby contradicting the specified value of $\tau_{J+1} = \tau'_{j+1} = t_k$.

Now consider the case $\sigma'_j = t_{k+1}$. Necessarily $J' = J$. Then $\Delta'_{j+1} = \Delta'_j$ is excluded since this would also imply that $\tau_{J+1} = t_{\zeta(\sigma_J - 1)} < t_k$.

Now consider the case $\sigma'_j = t_{k+1} + 1$. Since $\sigma_J = \tau_J = t_{k+1}$ is forbidden then we again have $J' = J$. Here we exclude $\Delta'_{j+1} \ne \Delta'_j$ since this would imply that $\Delta_{J+1} \ne \Delta_J$ and then $\tau_{J+1} = t_{\zeta(\sigma_J)} = t_{\zeta(\sigma'_j)} = t_{k+1}$.

Finally consider the case $\sigma'_j > t_{k+1} + 1$. Here $J' < J$ because if $J' = J$ then $\tau_{J+1} = t_{\zeta(\sigma'_j)} \ge t_{k+1}$ or $\tau_{J+1} = t_{\zeta(\sigma'_j - 1)} \ge t_{k+1}$. Therefore $\sigma_J = \tau_J$ and $\Delta_J = \Delta_{J+1}$. If $\sigma_J > t_{k+1} + 1$ then $\tau'_{j+1} = \tau_{J+1} = t_{\zeta(\sigma_J - 1)} \ge t_{k+1}$. Therefore, since $\sigma_J = t_{k+1}$ is forbidden, we conclude that $\sigma_J = t_{k+1} + 1$. So $\tau_J = t_{k+1} + 1$. But either $\tau_J = t_{k'}$ for some k', or $\tau_J = t_n - 1$, $J = 2$ and $\sigma_1 > t_n$. In the former case, necessarily $k' = k + 2$ (and thus $c_{k+1} = 1$), whereupon $\sigma'_j = \sigma_{J'} > \sigma_J = \tau_J = t_{k+2}$, as required. In the latter case, necessarily $n \ge k + 2$ and $J' = j = 1$. Then $\sigma'_1 = \sigma_1 > t_n \ge t_{k+2}$ as required. □

Now note that the quantities a, e, $\boldsymbol{u}^L$, and $\boldsymbol{\Delta}^L$ (resp. b, f ,$\boldsymbol{u}^R$ and $\boldsymbol{\Delta}^R$) obtained from a run $\{\tau_j^L,\sigma_j^L,\Delta_j^L\}_{j=1}^{d^L}$ (resp. $\{\tau_j^R,\sigma_j^R,\Delta_j^R\}_{j=1}^{d^R}$) are equal to those obtained from the corresponding reduced run $\{\tau_j^{L\prime},\sigma_j^{L\prime},\Delta_j^{L\prime}\}_{j=1}^{d^{L\prime}}$ (resp. $\{\tau_j^{R\prime},\sigma_j^{R\prime},\Delta_j^{R\prime}\}_{j=1}^{d^{R\prime}}$).

The following lemma shows that the generating function is unchanged on replacing naive runs with their corresponding reduced runs.

Lemma 7.4. *Let* $\{\tau_j^L,\sigma_j^L,\Delta_j^L\}_{j=1}^{d^L}$ *and* $\{\tau_j^R,\sigma_j^R,\Delta_j^R\}_{j=1}^{d^R}$ *be naive runs, and let* $\{\mu_j,\mu_j^*\}_{j=0}^{d^L-1}$ *and* $\{\nu_j,\nu_j^*\}_{j=0}^{d^R-1}$ *be obtained from these as above.*

Then let $\{\tau_j^{L\prime},\sigma_j^{L\prime},\Delta_j^{L\prime}\}_{j=1}^{d^{L\prime}}$ *and* $\{\tau_j^{R\prime},\sigma_j^{R\prime},\Delta_j^{R\prime}\}_{j=1}^{d^{R\prime}}$ *be the corresponding reduced runs, and let* $\{\mu_j',\mu_j^{*\prime}\}_{j=0}^{d^{L\prime}-1}$ *and* $\{\nu_j',\nu_j^{*\prime}\}_{j=0}^{d^{R\prime}-1}$ *be obtained from these as above. Then:*

$$\tilde{\chi}^{p,p'}_{a,b,e,f}(m_0)\left\{\begin{matrix}\mu_{d_0^L},\mu_1,\ldots,\mu_{d^L-1};\nu_{d_0^R},\nu_1,\ldots,\nu_{d^R-1}\\ \mu^*_{d_0^L},\mu^*_1,\ldots,\mu^*_{d^L-1};\nu^*_{d_0^R},\nu^*_1,\ldots,\nu^*_{d^R-1}\end{matrix}\right\}$$
$$=\tilde{\chi}^{p,p'}_{a,b,e,f}(m_0)\left\{\begin{matrix}\mu'_{d_0^L},\mu'_1,\ldots,\mu'_{d^{L\prime}-1};\nu'_{d_0^R},\nu'_1,\ldots,\nu'_{d^{R\prime}-1}\\ \mu'^*_{d_0^L},\mu'^*_1,\ldots,\mu'^*_{d^{L\prime}-1};\nu'^*_{d_0^R},\nu'^*_1,\ldots,\nu'^*_{d^{R\prime}-1}\end{matrix}\right\}.$$

Proof: Let j be the smallest value for which $\tau_j^L=\sigma_j^L$. If such a value exists, then necessarily $2\le j<d^L$ and $\Delta_j^L=\Delta_{j+1}^L$. From the definitions, we see that $\mu_j^*=\mu_{j-1}^*$. Thus, by Lemma 5.6,

$$\tilde{\chi}^{p,p'}_{a,b,e,f}(m_0)\left\{\begin{matrix}\ldots,\mu_{j-2},\mu_{j-1},\mu_j,\ldots;\nu_{d_0^R},\nu_1,\ldots,\nu_{d^R-1}\\ \ldots,\mu^*_{j-2},\mu^*_{j-1},\mu^*_j,\ldots;\nu^*_{d_0^R},\nu^*_1,\ldots,\nu^*_{d^R-1}\end{matrix}\right\}$$
$$=\tilde{\chi}^{p,p'}_{a,b,e,f}(m_0)\left\{\begin{matrix}\ldots,\mu_{j-2},\mu_j,\ldots;\nu_{d_0^R},\nu_1,\ldots,\nu_{d^R-1}\\ \ldots,\mu^*_{j-2},\mu^*_j,\ldots;\nu^*_{d_0^R},\nu^*_1,\ldots,\nu^*_{d^R-1}\end{matrix}\right\}.$$

By recursively applying this process, and doing likewise for any j for which $\tau_j^R=\sigma_j^R$, we obtain the required result. □

Thus, to prove Theorem 7.1 for a pair of arbitrary naive runs, it is only necessary to prove it for the corresponding reduced runs. This is what will be done.

7.2. Induction parameters

In this and the following Section 7.3, we assume that $p'>2p$. We also fix a pair $\{\tau_j^L,\sigma_j^L,\Delta_j^L\}_{j=1}^{d^L}$, $\{\tau_j^R,\sigma_j^R,\Delta_j^R\}_{j=1}^{d^R}$ of reduced runs. In our main induction (Lemma 7.11), we will for $i=t-1,t-2,\ldots,0$, step through a sequence of $(p^{(i)},p^{(i)\prime})$-models, constructing the generating function for a certain set of paths at each stage. In this section, we define various sets of parameters that pertain to those sets of paths, and identify some basic relationships between them.

Firstly, for $0\le i<t$, let $k(i)$ be such that $t_{k(i)}\le i<t_{k(i)+1}$ (i.e. $k(i)=\zeta(i+1)$), and define $p^{(i)}$ and $p^{(i)\prime}$ to be the positive coprime integers for which $p^{(i)\prime}/p^{(i)}$ has continued fraction $[t_{k(i)+1}+1-i,c_{k(i)+1},\ldots,c_n]$. Thus $p^{(i)\prime}/p^{(i)}$ has rank $t^{(i)}$ where we set $t^{(i)}=t-i$. As in Section 1.5, we obtain Takahashi lengths $\{\kappa_j^{(i)}\}_{j=0}^{t^{(i)}}$ and truncated Takahashi lengths $\{\tilde{\kappa}_j^{(i)}\}_{j=0}^{t^{(i)}}$ for $p^{(i)\prime}/p^{(i)}$.

Lemma 7.5. *Let $1 \le i < t$. If $i \neq t_{k(i)}$ then:*

$$\begin{aligned} p^{(i-1)\prime} &= p^{(i)\prime} + p^{(i)}; \\ p^{(i-1)} &= p^{(i)}; \\ \kappa_j^{(i-1)} &= \kappa_{j-1}^{(i)} + \tilde{\kappa}_{j-1}^{(i)} && (1 \le j \le t^{(i-1)}); \\ \tilde{\kappa}_j^{(i-1)} &= \tilde{\kappa}_{j-1}^{(i)} && (1 \le j \le t^{(i-1)}). \end{aligned}$$

If $i = t_{k(i)}$ then:

$$\begin{aligned} p^{(i-1)\prime} &= 2p^{(i)\prime} - p^{(i)}; \\ p^{(i-1)} &= p^{(i)\prime} - p^{(i)}; \\ \kappa_j^{(i-1)} &= 2\kappa_{j-1}^{(i)} - \tilde{\kappa}_{j-1}^{(i)} && (2 \le j \le t^{(i-1)}); \\ \tilde{\kappa}_j^{(i-1)} &= \tilde{\kappa}_{j-1}^{(i)} - \tilde{\kappa}_{j-1}^{(i)} && (2 \le j \le t^{(i-1)}). \end{aligned}$$

Proof: If $i \neq t_{k(i)}$ then $k(i-1) = k(i)$. Then $p^{(i)\prime}/p^{(i)}$ and $p^{(i-1)\prime}/p^{(i-1)}$ have continued fractions $[t_{k(i)} + 1 - i, c_{k(i)+1}, \ldots, c_n]$ and $[t_{k(i)} + 2 - i, c_{k(i)+1}, \ldots, c_n]$ respectively. That $p^{(i-1)\prime} = p^{(i)\prime} + p^{(i)}$ and $p^{(i-1)} = p^{(i)}$ follows immediately. The expressions for $\kappa_j^{(i-1)}$ and $\tilde{\kappa}_j^{(i-1)}$ then follow from Lemma E.1.

If $i = t_{k(i)}$ then $k(i-1) = k(i) - 1$. Then $p^{(i)\prime}/p^{(i)}$ and $p^{(i-1)\prime}/p^{(i-1)}$ have continued fractions $[c_{k(i)} + 1, c_{k(i)+1}, \ldots, c_n]$ and $[2, c_{k(i)}, c_{k(i)+1}, \ldots, c_n]$ respectively. That $p^{(i-1)\prime} = 2p^{(i)\prime} - p^{(i)}$ and $p^{(i-1)} = p^{(i)\prime} - p^{(i)}$ follows immediately. The expressions for $\kappa_j^{(i-1)}$ and $\tilde{\kappa}_j^{(i-1)}$ then follow from combining Lemma E.2 with Lemma E.1. □

For $0 \le i < t$, we now define $e^{(i)}, f^{(i)} \in \{0, 1\}$. With $k^L(i)$ such that $t_{k^L(i)} < \sigma^L_{\eta^L(i)} \le t_{k^L(i)+1}$, define $e^{(i)}$ such that:

$$\Delta^L_{\eta^L(i)} = \begin{cases} -(-1)^{e^{(i)}} & \text{if } i \ge \sigma^L_{\eta^L(i)}; \\ (-1)^{e^{(i)} + k(i) - k^L(i)} & \text{if } i < \sigma^L_{\eta^L(i)}, \end{cases} \tag{7.5}$$

and with $k^R(i)$ such that $t_{k^R(i)} < \sigma^R_{\eta^R(i)} \le t_{k^R(i)+1}$, define $f^{(i)}$ such that:

$$\Delta^R_{\eta^R(i)} = \begin{cases} -(-1)^{f^{(i)}} & \text{if } i \ge \sigma^R_{\eta^R(i)}; \\ (-1)^{f^{(i)} + k(i) - k^R(i)} & \text{if } i < \sigma^R_{\eta^R(i)}. \end{cases} \tag{7.6}$$

We now define what will be the starting and ending points of the paths that we consider at the ith induction step. For $0 \le i < t$, define:

$$a^{(i)} = \begin{cases} -\Delta^L_{\eta^L(i)} + \sum_{m=2}^{\eta^L(i)} \Delta^L_m \kappa^{(i)}_{\tau^L_m - i} - \sum_{m=1}^{\eta^L(i)-1} \Delta^L_m \kappa^{(i)}_{\sigma^L_m - i} & \text{if } \Delta^L_1 = -1 \text{ and } i \ge \sigma^L_{\eta^L(i)}; \\ \sum_{m=2}^{\eta^L(i)} \Delta^L_m \kappa^{(i)}_{\tau^L_m - i} - \sum_{m=1}^{\eta^L(i)} \Delta^L_m \kappa^{(i)}_{\sigma^L_m - i} & \text{if } \Delta^L_1 = -1 \text{ and } i \le \sigma^L_{\eta^L(i)}; \\ p^{(i)\prime} - \Delta^L_{\eta^L(i)} + \sum_{m=2}^{\eta^L(i)} \Delta^L_m \kappa^{(i)}_{\tau^L_m - i} - \sum_{m=1}^{\eta^L(i)-1} \Delta^L_m \kappa^{(i)}_{\sigma^L_m - i} & \text{if } \Delta^L_1 = +1 \text{ and } i \ge \sigma^L_{\eta^L(i)}; \\ p^{(i)\prime} + \sum_{m=2}^{\eta^L(i)} \Delta^L_m \kappa^{(i)}_{\tau^L_m - i} - \sum_{m=1}^{\eta^L(i)} \Delta^L_m \kappa^{(i)}_{\sigma^L_m - i} & \text{if } \Delta^L_1 = +1 \text{ and } i \le \sigma^L_{\eta^L(i)}. \end{cases}$$

Similarly, for $0 \le i < t$ define:

$$b^{(i)} = \begin{cases} -\Delta^R_{\eta^R(i)} + \sum_{m=2}^{\eta^R(i)} \Delta^R_m \kappa^{(i)}_{\tau^R_m - i} - \sum_{m=1}^{\eta^R(i)-1} \Delta^R_m \kappa^{(i)}_{\sigma^R_m - i} & \text{if } \Delta^R_1 = -1 \text{ and } i \ge \sigma^R_{\eta^R(i)}; \\ \sum_{m=2}^{\eta^R(i)} \Delta^R_m \kappa^{(i)}_{\tau^R_m - i} - \sum_{m=1}^{\eta^R(i)} \Delta^R_m \kappa^{(i)}_{\sigma^R_m - i} & \text{if } \Delta^R_1 = -1 \text{ and } i \le \sigma^R_{\eta^R(i)}; \\ p^{(i)\prime} - \Delta^R_{\eta^R(i)} + \sum_{m=2}^{\eta^R(i)} \Delta^R_m \kappa^{(i)}_{\tau^R_m - i} - \sum_{m=1}^{\eta^R(i)-1} \Delta^R_m \kappa^{(i)}_{\sigma^R_m - i} & \text{if } \Delta^R_1 = +1 \text{ and } i \ge \sigma^R_{\eta^R(i)}; \\ p^{(i)\prime} + \sum_{m=2}^{\eta^R(i)} \Delta^R_m \kappa^{(i)}_{\tau^R_m - i} - \sum_{m=1}^{\eta^R(i)} \Delta^R_m \kappa^{(i)}_{\sigma^R_m - i} & \text{if } \Delta^R_1 = +1 \text{ and } i \le \sigma^R_{\eta^R(i)}. \end{cases}$$

We now define what will be the starting and ending points of certain intermediate paths that we consider at the ith induction step. For $1 \leq i < t$, define:

$$a^{(i)\prime} = \begin{cases} -\Delta^L_{\eta^L(i)}(1+\delta_{i,t_{k(i)}}) + \sum_{m=2}^{\eta^L(i)} \Delta^L_m \kappa^{(i-1)}_{\tau^L_m - i+1} - \sum_{m=1}^{\eta^L(i)-1} \Delta^L_m \kappa^{(i-1)}_{\sigma^L_m - i+1} & \text{if } \Delta^L_1 = -1 \text{ and } i \geq \sigma^L_{\eta^L(i)}; \\ \sum_{m=2}^{\eta^L(i)} \Delta^L_m \kappa^{(i-1)}_{\tau^L_m - i+1} - \sum_{m=1}^{\eta^L(i)} \Delta^L_m \kappa^{(i-1)}_{\sigma^L_m - i+1} & \text{if } \Delta^L_1 = -1 \text{ and } i < \sigma^L_{\eta^L(i)}; \\ p^{(i-1)\prime} - \Delta^L_{\eta^L(i)}(1+\delta_{i,t_{k(i)}}) + \sum_{m=2}^{\eta^L(i)} \Delta^L_m \kappa^{(i-1)}_{\tau^L_m - i+1} - \sum_{m=1}^{\eta^L(i)-1} \Delta^L_m \kappa^{(i-1)}_{\sigma^L_m - i+1} & \text{if } \Delta^L_1 = +1 \text{ and } i \geq \sigma^L_{\eta^L(i)}; \\ p^{(i-1)\prime} + \sum_{m=2}^{\eta^L(i)} \Delta^L_m \kappa^{(i-1)}_{\tau^L_m - i+1} - \sum_{m=1}^{\eta^L(i)} \Delta^L_m \kappa^{(i-1)}_{\sigma^L_m - i+1} & \text{if } \Delta^L_1 = +1 \text{ and } i < \sigma^L_{\eta^L(i)}. \end{cases}$$

Similarly, for $1 \leq i < t$ define:

$$b^{(i)\prime} = \begin{cases} -\Delta^R_{\eta^R(i)}(1+\delta_{i,t_{k(i)}}) + \sum_{m=2}^{\eta^R(i)} \Delta^R_m \kappa^{(i-1)}_{\tau^R_m - i+1} - \sum_{m=1}^{\eta^R(i)-1} \Delta^R_m \kappa^{(i-1)}_{\sigma^R_m - i+1} & \text{if } \Delta^R_1 = -1 \text{ and } i \geq \sigma^R_{\eta^R(i)}; \\ \sum_{m=2}^{\eta^R(i)} \Delta^R_m \kappa^{(i-1)}_{\tau^R_m - i+1} - \sum_{m=1}^{\eta^R(i)} \Delta^R_m \kappa^{(i-1)}_{\sigma^R_m - i+1} & \text{if } \Delta^R_1 = -1 \text{ and } i < \sigma^R_{\eta^R(i)}; \\ p^{(i-1)\prime} - \Delta^R_{\eta^R(i)}(1+\delta_{i,t_{k(i)}}) + \sum_{m=2}^{\eta^R(i)} \Delta^R_m \kappa^{(i-1)}_{\tau^R_m - i+1} - \sum_{m=1}^{\eta^R(i)-1} \Delta^R_m \kappa^{(i-1)}_{\sigma^R_m - i+1} & \text{if } \Delta^R_1 = +1 \text{ and } i \geq \sigma^R_{\eta^R(i)}; \\ p^{(i-1)\prime} + \sum_{m=2}^{\eta^R(i)} \Delta^R_m \kappa^{(i-1)}_{\tau^R_m - i+1} - \sum_{m=1}^{\eta^R(i)} \Delta^R_m \kappa^{(i-1)}_{\sigma^R_m - i+1} & \text{if } \Delta^R_1 = +1 \text{ and } i < \sigma^R_{\eta^R(i)}. \end{cases}$$

We now define values that will turn out to be $\lfloor a^{(i)}p^{(i)}/p^{(i)\prime} \rfloor$ and $\lfloor b^{(i)}p^{(i)}/p^{(i)\prime} \rfloor$. For $0 \le i < t$, define:

$$\tilde{a}^{(i)} = \begin{cases} -e^{(i)} + \displaystyle\sum_{m=2}^{\eta^L(i)} \Delta_m^L \tilde{\kappa}^{(i)}_{\tau_m^L - i} - \sum_{m=1}^{\eta^L(i)-1} \Delta_m^L \tilde{\kappa}^{(i)}_{\sigma_m^L - i} & \text{if } \Delta_1^L = -1 \text{ and } i \ge \sigma^L_{\eta^L(i)}; \\ -e^{(i)} + \displaystyle\sum_{m=2}^{\eta^L(i)} \Delta_m^L \tilde{\kappa}^{(i)}_{\tau_m^L - i} - \sum_{m=1}^{\eta^L(i)} \Delta_m^L \tilde{\kappa}^{(i)}_{\sigma_m^L - i} & \text{if } \Delta_1^L = -1 \text{ and } i < \sigma^L_{\eta^L(i)}; \\ p^{(i)} - e^{(i)} + \displaystyle\sum_{m=2}^{\eta^L(i)} \Delta_m^L \tilde{\kappa}^{(i)}_{\tau_m^L - i} - \sum_{m=1}^{\eta^L(i)-1} \Delta_m^L \tilde{\kappa}^{(i)}_{\sigma_m^L - i} & \text{if } \Delta_1^L = +1 \text{ and } i \ge \sigma^L_{\eta^L(i)}; \\ p^{(i)} - e^{(i)} + \displaystyle\sum_{m=2}^{\eta^L(i)} \Delta_m^L \tilde{\kappa}^{(i)}_{\tau_m^L - i} - \sum_{m=1}^{\eta^L(i)} \Delta_m^L \tilde{\kappa}^{(i)}_{\sigma_m^L - i} & \text{if } \Delta_1^L = +1 \text{ and } i < \sigma^L_{\eta^L(i)}. \end{cases}$$

Similarly, for $0 \le i < t$ define:

$$\tilde{b}^{(i)} = \begin{cases} -f^{(i)} + \displaystyle\sum_{m=2}^{\eta^R(i)} \Delta_m^R \tilde{\kappa}^{(i)}_{\tau_m^R - i} - \sum_{m=1}^{\eta^R(i)-1} \Delta_m^R \tilde{\kappa}^{(i)}_{\sigma_m^R - i} & \text{if } \Delta_1^R = -1 \text{ and } i \ge \sigma^R_{\eta^R(i)}; \\ -f^{(i)} + \displaystyle\sum_{m=2}^{\eta^R(i)} \Delta_m^R \tilde{\kappa}^{(i)}_{\tau_m^R - i} - \sum_{m=1}^{\eta^R(i)} \Delta_m^R \tilde{\kappa}^{(i)}_{\sigma_m^R - i} & \text{if } \Delta_1^R = -1 \text{ and } i < \sigma^R_{\eta^R(i)}; \\ p^{(i)} - f^{(i)} + \displaystyle\sum_{m=2}^{\eta^R(i)} \Delta_m^R \tilde{\kappa}^{(i)}_{\tau_m^R - i} - \sum_{m=1}^{\eta^R(i)-1} \Delta_m^R \tilde{\kappa}^{(i)}_{\sigma_m^R - i} & \text{if } \Delta_1^R = +1 \text{ and } i \ge \sigma^R_{\eta^R(i)}; \\ p^{(i)} - f^{(i)} + \displaystyle\sum_{m=2}^{\eta^R(i)} \Delta_m^R \tilde{\kappa}^{(i)}_{\tau_m^R - i} - \sum_{m=1}^{\eta^R(i)} \Delta_m^R \tilde{\kappa}^{(i)}_{\sigma_m^R - i} & \text{if } \Delta_1^R = +1 \text{ and } i < \sigma^R_{\eta^R(i)}. \end{cases}$$

Lemma 7.6. *Let $1 \le i < t$. Then:*

$$a^{(i)\prime} = \begin{cases} a^{(i)} + \tilde{a}^{(i)} + e^{(i)} & \text{if } i \ne t_{k(i)}; \\ 2a^{(i)} - \tilde{a}^{(i)} - e^{(i)} & \text{if } i = t_{k(i)}, \end{cases}$$

and

$$b^{(i)\prime} = \begin{cases} b^{(i)} + \tilde{b}^{(i)} + f^{(i)} & \text{if } i \ne t_{k(i)}; \\ 2b^{(i)} - \tilde{b}^{(i)} - f^{(i)} & \text{if } i = t_{k(i)}. \end{cases}$$

Proof: These results follow immediately from Lemma 7.5. □

Lemma 7.7. *Let* $1 \le i < t$, $\eta^L = \eta^L(i)$ *and* $\eta^R = \eta^R(i)$. *Then:*

$$a^{(i-1)} = \begin{cases} a^{(i)\prime} + \Delta^L_{\eta^L+1} & \text{if } i = \tau^L_{\eta^L+1}; \\ a^{(i)\prime} + \Delta^L_{\eta^L} & \text{if } i > \sigma^L_{\eta^L} \text{ and } i = t_{k(i)}; \\ a^{(i)\prime} - \Delta^L_{\eta^L} & \text{if } i = \sigma^L_{\eta^L} \text{ and } i \ne t_{k(i)}; \\ a^{(i)\prime} & \text{otherwise,} \end{cases}$$

and

$$b^{(i-1)} = \begin{cases} b^{(i)\prime} + \Delta^R_{\eta^R+1} & \text{if } i = \tau^R_{\eta^R+1}; \\ b^{(i)\prime} + \Delta^R_{\eta^R} & \text{if } i > \sigma^R_{\eta^R} \text{ and } i = t_{k(i)}; \\ b^{(i)\prime} - \Delta^R_{\eta^R} & \text{if } i = \sigma^R_{\eta^R} \text{ and } i \ne t_{k(i)}; \\ b^{(i)\prime} & \text{otherwise.} \end{cases}$$

Proof: In the case $i = \tau^L_{\eta^L+1}$, we have $i < \sigma^L_{\eta^L}$, $\sigma^L_{\eta^L+1} \le i-1 < \tau^L_{\eta^L+1}$ and $\eta^L(i-1) = \eta^L + 1$. Then, since $\kappa_1^{(i-1)} = 2$, we obtain $a^{(i-1)} = a^{(i)\prime} + \Delta^L_{\eta^L+1}$.

For $\sigma^L_{\eta^L} < i < \tau^L_{\eta^L}$, then manifestly $a^{(i-1)} = a^{(i)\prime} + \Delta^L_{\eta^L}\delta_{i,t_{k(i)}}$.

In the case $i = \sigma^L_{\eta^L}$, we have $\tau^L_{\eta^L+1} \le i-1 < \sigma^L_{\eta^L}$. Since $\kappa_1^{(i-1)} = 2$, we immediately obtain $a^{(i-1)} = a^{(i)\prime} + \Delta^L_{\eta^L}(-1 + \delta_{i,t_{k(i)}})$.

Finally, for $\tau^L_{\eta^L+1} < i < \sigma^L_{\eta^L}$, we immediately see that $a^{(i-1)} = a^{(i)\prime}$.

The expressions for $b^{(i-1)}$ are obtained in precisely the same way. □

Lemma 7.8. *Let* $1 \le i < t$, $\eta^L = \eta^L(i)$ *and* $\eta^R = \eta^R(i)$. *Then:*

$$\tilde{a}^{(i-1)} = \begin{cases} \tilde{a}^{(i)} + \Delta^L_{\eta^L+1} & \text{if } i \ne t_{k(i)}, i = \tau^L_{\eta^L+1} \text{ and } e^{(i-1)} = e^{(i)}; \\ \tilde{a}^{(i)} & \text{if } i \ne t_{k(i)} \text{ otherwise;} \\ a^{(i)} - 1 - \tilde{a}^{(i)} + \Delta^L_{\eta^L+1} & \text{if } i = t_{k(i)}, i = \tau^L_{\eta^L+1} \text{ and } e^{(i-1)} \ne e^{(i)}; \\ a^{(i)} - 1 - \tilde{a}^{(i)} & \text{if } i = t_{k(i)} \text{ otherwise,} \end{cases}$$

and

$$\tilde{b}^{(i-1)} = \begin{cases} \tilde{b}^{(i)} + \Delta^R_{\eta^R+1} & \text{if } i \ne t_{k(i)}, i = \tau^R_{\eta^R+1} \text{ and } f^{(i-1)} = f^{(i)}; \\ \tilde{b}^{(i)} & \text{if } i \ne t_{k(i)} \text{ otherwise;} \\ b^{(i)} - 1 - \tilde{b}^{(i)} + \Delta^R_{\eta^R+1} & \text{if } i = t_{k(i)}, i = \tau^R_{\eta^R+1} \text{ and } f^{(i-1)} \ne f^{(i)}; \\ b^{(i)} - 1 - \tilde{b}^{(i)} & \text{if } i = t_{k(i)} \text{ otherwise.} \end{cases}$$

Proof: First consider $i \ne t_{k(i)}$, so that $\tilde{\kappa}_j^{(i-1)} = \tilde{\kappa}_{j-1}^{(i)}$ for $1 \le j \le t^{(i-1)}$ by Lemma 7.5.

In the subcase $i = \tau^L_{\eta^L+1}$, we have $i < \sigma^L_{\eta^L}$ and $\sigma^L_{\eta^L+1} \le i-1 < \tau^L_{\eta^L+1}$ and $\eta^L(i-1) = \eta^L + 1$. Then, since $\tilde{\kappa}_1^{(i-1)} = 1$, we have $\tilde{a}^{(i-1)} - \tilde{a}^{(i)} = -e^{(i-1)} + \Delta^L_{\eta^L+1} + e^{(i)}$. The $e^{(i)} = e^{(i-1)}$ case follows immediately and the $e^{(i)} = 1 - e^{(i-1)}$ case follows on using $\Delta^L_{\eta^L+1} = 2e^{(i-1)} - 1$, which is verified immediately from the definition of $e^{(i-1)}$.

In the subcase $i = \sigma^L_{\eta^L}$, since $\tilde{\kappa}^{(i-1)}_1 = 1$, we have $\tilde{a}^{(i-1)} - \tilde{a}^{(i)} = -e^{(i-1)} - \Delta^L_{\eta^L} + e^{(i)} = 0$ because $e^{(i-1)} = 1 - c^{(i)}$ and $\Delta^L_{\eta^L} = 2e^{(i)} - 1$.

For $\sigma^L_{\eta^L} < i < \tau^L_{\eta^L}$ and $\tau^L_{\eta^L+1} < i < \sigma^L_{\eta^L}$, we see that $e^{(i-1)} = e^{(i)}$, whence $\tilde{a}^{(i-1)} = \tilde{a}^{(i)}$.

Now when $i = t_{k(i)}$, Lemma 7.5 gives $\tilde{\kappa}^{(i-1)}_j = \kappa^{(i)}_{j-1} - \tilde{\kappa}^{(i)}_{j-1}$ for $2 \leq j \leq t^{(i-1)}$.

In the subcase $i = \tau^L_{\eta^L+1}$, since $\tilde{\kappa}^{(i-1)}_1 = 1$, we have $\tilde{a}^{(i-1)} - a^{(i)} + 1 + \tilde{a}^{(i)} = -e^{(i-1)} + 1 + \Delta^L_{\eta^L+1} - e^{(i)}$. The $e^{(i)} = 1 - e^{(i-1)}$ case follows immediately and the $e^{(i)} = e^{(i-1)}$ case follows on using $\Delta^L_{\eta^L+1} = 2e^{(i-1)} - 1$.

In the subcase $i = \sigma^L_{\eta^L}$, since $\tilde{\kappa}^{(i-1)}_1 = 1$, we have $\tilde{a}^{(i-1)} - a^{(i)} + 1 + \tilde{a}^{(i)} = -e^{(i-1)} + \Delta^L_{\eta^L} + 1 - e^{(i)} - \Delta^L_{\eta^L} = 0$ because $e^{(i-1)} = 1 - e^{(i)}$.

For $\tau^L_{\eta^L+1} < i < \sigma^L_{\eta^L}$, we obtain $\tilde{a}^{(i-1)} - a^{(i)} + 1 + \tilde{a}^{(i)} = -e^{(i-1)} - e^{(i)} + 1 = 0$, since here $k(i-1) = k(i) - 1$ implies that $e^{(i-1)} = 1 - e^{(i)}$.

For $\sigma^L_{\eta^L} < i < \tau^L_{\eta^L}$, we obtain $\tilde{a}^{(i-1)} - a^{(i)} + 1 + \tilde{a}^{(i)} = -e^{(i-1)} + \Delta^L_{\eta^L} - e^{(i)} + 1 = 0$, since here $e^{(i-1)} = e^{(i)}$ and $\Delta^L_{\eta^L} = 2e^{(i)} - 1$.

The expressions for $\tilde{b}^{(i-1)}$ are obtained in precisely the same way. □

For $1 \leq j < d^L$ and $0 \leq i < \tau^L_{j+1}$ (so that $j < \eta^L(i)$), define:

$$\mu^{(i)*}_j = \sum_{m=2}^{j} \Delta^L_m(\kappa^{(i)}_{\tau^L_m - i} - \kappa^{(i)}_{\sigma^L_m - i}) + \begin{cases} \kappa^{(i)}_{\sigma^L_1 - i} & \text{if } \Delta^L_1 = -1; \\ p^{(i)\prime} - \kappa^{(i)}_{\sigma^L_1 - i} & \text{if } \Delta^L_1 = +1; \end{cases}$$

$$\mu^{(i)}_j = \mu^{(i)*}_j + \Delta^L_{j+1}\kappa^{(i)}_{\tau^L_{j+1} - i},$$

and for $0 \leq i < t_n$, define:

$$\mu^{(i)*}_0 = \begin{cases} \kappa^{(i)}_{t_n - i} & \text{if } \Delta^L_1 = -1; \\ p^{(i)\prime} - \kappa^{(i)}_{t_n - i} & \text{if } \Delta^L_1 = +1, \end{cases}$$

$$\mu^{(i)}_0 = \begin{cases} 0 & \text{if } \Delta^L_1 = -1; \\ p^{(i)\prime} & \text{if } \Delta^L_1 = +1. \end{cases}$$

We also set $d^L_0(i) = 0$ if both $i < t_n$ and $\sigma^L_1 < t_n$, and $d^L_0(i) = 1$ otherwise. For $0 \leq i < t$, we define the $(\eta^L(i) - d^L_0(i))$-dimensional vectors $\boldsymbol{\mu}^{(i)}$ and $\boldsymbol{\mu}^{(i)*}$, and the $(\eta^L(i+1) - d^L_0(i+1))$-dimensional vectors $\boldsymbol{\mu}^{(i)}_\uparrow$ and $\boldsymbol{\mu}^{(i)*}_\uparrow$ by:

$$\boldsymbol{\mu}^{(i)} = (\mu^{(i)}_{d^L_0(i)}, \ldots, \mu^{(i)}_{\eta^L(i)-1}), \qquad \boldsymbol{\mu}^{(i)*} = (\mu^{(i)*}_{d^L_0(i)}, \ldots, \mu^{(i)*}_{\eta^L(i)-1}),$$

$$\boldsymbol{\mu}^{(i)}_\uparrow = (\mu^{(i)}_{d^L_0(i+1)}, \ldots, \mu^{(i)}_{\eta^L(i+1)-1}), \qquad \boldsymbol{\mu}^{(i)*}_\uparrow = (\mu^{(i)*}_{d^L_0(i+1)}, \ldots, \mu^{(i)*}_{\eta^L(i+1)-1}).$$

For $1 \leq j < d^R$ and $0 \leq i < \tau^R_{j+1}$ (so that $j < \eta^R(i)$), define:

$$\nu^{(i)*}_j = \sum_{m=2}^{j} \Delta^R_m(\kappa^{(i)}_{\tau^R_m - i} - \kappa^{(i)}_{\sigma^R_m - i}) + \begin{cases} \kappa^{(i)}_{\sigma^R_1 - i} & \text{if } \Delta^R_1 = -1; \\ p^{(i)\prime} - \kappa^{(i)}_{\sigma^R_1 - i} & \text{if } \Delta^R_1 = +1; \end{cases}$$

$$\nu^{(i)}_j = \nu^{(i)*}_j + \Delta^R_{j+1}\kappa^{(i)}_{\tau^R_{j+1} - i},$$

and for $0 \leq i < t_n$, define:

$$\nu_0^{(i)*} = \begin{cases} \kappa_{t_n - i}^{(i)} & \text{if } \Delta_1^R = -1; \\ p^{(i)\prime} - \kappa_{t_n-i}^{(i)} & \text{if } \Delta_1^R = +1, \end{cases}$$

$$\nu_0^{(i)} = \begin{cases} 0 & \text{if } \Delta_1^R = -1; \\ p^{(i)\prime} & \text{if } \Delta_1^R = +1. \end{cases}$$

We also set $d_0^R(i) = 0$ if both $i < t_n$ and $\sigma_1^R < t_n$, and $d_0^R(i) = 1$ otherwise. For $0 \leq i < t$, we define the $(\eta^R(i) - d_0^R(i))$-dimensional vectors $\boldsymbol{\nu}^{(i)}$ and $\boldsymbol{\nu}^{(i)*}$, and the $(\eta^R(i+1) - d_0^R(i+1))$-dimensional vectors $\boldsymbol{\nu}_\uparrow^{(i)}$ and $\boldsymbol{\nu}_\uparrow^{(i)*}$ by:

$$\boldsymbol{\nu}^{(i)} = (\nu_{d_0^R(i)}^{(i)}, \ldots, \nu_{\eta^R(i)-1}^{(i)}), \qquad \boldsymbol{\nu}^{(i)*} = (\nu_{d_0^R(i)}^{(i)*}, \ldots, \nu_{\eta^R(i)-1}^{(i)*}),$$
$$\boldsymbol{\nu}_\uparrow^{(i)} = (\nu_{d_0^R(i+1)}^{(i)}, \ldots, \nu_{\eta^R(i+1)-1}^{(i)}), \qquad \boldsymbol{\nu}_\uparrow^{(i)*} = (\nu_{d_0^R(i+1)}^{(i)*}, \ldots, \nu_{\eta^R(i+1)-1}^{(i)*}).$$

(The parentheses that delimit $\boldsymbol{\mu}^{(i)}$, $\boldsymbol{\mu}^{(i)*}$, $\boldsymbol{\mu}_\uparrow^{(i)}$, $\boldsymbol{\mu}_\uparrow^{(i)*}$ $\boldsymbol{\nu}^{(i)}$, $\boldsymbol{\nu}^{(i)*}$, $\boldsymbol{\nu}_\uparrow^{(i)}$, $\boldsymbol{\nu}_\uparrow^{(i)*}$ will be dropped when these symbols are incorporated into the notation for the path generating functions.)

Lemma 7.9. *Let $1 \leq j < d^L$. Then:*

$$\begin{aligned}
\mu_j^{(i-1)*} &= a^{(i-1)} - \Delta_{j+1}^L && \text{if } i = \tau_{j+1}^L; \\
\mu_j^{(i-1)*} &= a^{(i-1)} - \Delta_{j+1}^L(\kappa_{\tau_{j+1}^L - i+1}^{(i-1)} - 1) && \text{if } \sigma_{j+1}^L < i \leq \tau_{j+1}^L; \\
\mu_j^{(i-1)*} &= a^{(i-1)} - \Delta_{j+1}^L(\kappa_{\tau_{j+1}^L - i+1}^{(i-1)} - 2) && \text{if } i = \sigma_{j+1}^L; \\
\mu_j^{(i-1)} &= a^{(i-1)} + \Delta_{j+1}^L && \text{if } \sigma_{j+1}^L < i \leq \tau_{j+1}^L; \\
\mu_j^{(i-1)} &= a^{(i)\prime} + (1 + \delta_{i,t_{k(i)}})\Delta_{j+1}^L && \text{if } i = \sigma_{j+1}^L.
\end{aligned}$$

If $\sigma_1^L < t_n$ then:

$$\begin{aligned}
\mu_0^{(i-1)*} &= a^{(i-1)} - \Delta_1^L && \text{if } i = t_n; \\
\mu_0^{(i-1)*} &= a^{(i-1)} - \Delta_1^L(\kappa_{t_n - i+1}^{(i-1)} - 1) && \text{if } \sigma_1^L < i \leq t_n; \\
\mu_0^{(i-1)*} &= a^{(i-1)} - \Delta_1^L(\kappa_{t_n - i+1}^{(i-1)} - 2) && \text{if } i = \sigma_1^L; \\
\mu_0^{(i-1)} &= a^{(i-1)} + \Delta_1^L && \text{if } \sigma_1^L < i \leq t_n; \\
\mu_0^{(i-1)} &= a^{(i)\prime} + (1 + \delta_{i,t_{k(i)}})\Delta_1^L && \text{if } i = \sigma_1^L.
\end{aligned}$$

Now let $1 \leq j < d^R$. Then:

$$\begin{aligned}
\nu_j^{(i-1)*} &= b^{(i-1)} - \Delta_{j+1}^R && \text{if } i = \tau_{j+1}^R; \\
\nu_j^{(i-1)*} &= b^{(i-1)} - \Delta_{j+1}^R(\kappa_{\tau_{j+1}^R - i+1}^{(i-1)} - 1) && \text{if } \sigma_{j+1}^R < i \leq \tau_{j+1}^R; \\
\nu_j^{(i-1)*} &= b^{(i-1)} - \Delta_{j+1}^R(\kappa_{\tau_{j+1}^R - i+1}^{(i-1)} - 2) && \text{if } i = \sigma_{j+1}^R; \\
\nu_j^{(i-1)} &= b^{(i-1)} + \Delta_{j+1}^R && \text{if } \sigma_{j+1}^R < i \leq \tau_{j+1}^R; \\
\nu_j^{(i-1)} &= b^{(i)\prime} + (1 + \delta_{i,t_{k(i)}})\Delta_{j+1}^R && \text{if } i = \sigma_{j+1}^R.
\end{aligned}$$

If $\sigma_1^R < t_n$ then:

$$\begin{aligned}
\nu_0^{(i-1)*} &= b^{(i-1)} - \Delta_1^R && \textit{if } i = t_n;\\
\nu_0^{(i-1)*} &= b^{(i-1)} - \Delta_1^R(\kappa_{t_n-i+1}^{(i-1)} - 1) && \textit{if } \sigma_1^R < i \le t_n;\\
\nu_0^{(i-1)*} &= b^{(i-1)} - \Delta_1^R(\kappa_{t_n-i+1}^{(i-1)} - 2) && \textit{if } i = \sigma_1^R;\\
\nu_0^{(i-1)} &= b^{(i-1)} + \Delta_1^R && \textit{if } \sigma_1^R < i \le t_n;\\
\nu_0^{(i-1)} &= b^{(i)\prime} + (1 + \delta_{i,t_{k(i)}})\Delta_1^R && \textit{if } i = \sigma_1^R.
\end{aligned}$$

Proof: The first four expressions follow immediately from the definitions after noting that $\sigma_{j+1}^L \le i \le \tau_{j+1}^L$ implies that $\eta^L(i-1) = j+1$ (and using $\kappa_1^{(i-1)} = 2$). The fifth is immediate. The next five follow from the definitions after noting that $\sigma_1^L \le i \le t_n$ implies that $\eta^L(i-1) = 1$. The other expressions follow similarly. □

Using (7.4), it is easily verified that $\boldsymbol{\mu}^{(i)}, \boldsymbol{\mu}^{(i)*}, \boldsymbol{\nu}^{(i)}, \boldsymbol{\nu}^{(i)*}$ satisfy the first three criteria for being a mazy-four in the $(p^{(i)}, p^{(i)\prime})$-model sandwiching $(a^{(i)}, b^{(i)})$. That they satisfy the fourth criterion is most readily determined during the main induction. However, the burden will be lessened by, for $1 \le j < d^L$ and $i < \tau_{j+1}^L$, defining:

$$\begin{aligned}
\tilde{\mu}_j^{(i)*} &= \sum_{m=2}^{j} \Delta_m^L(\tilde{\kappa}_{\tau_m^L-i}^{(i)} - \tilde{\kappa}_{\sigma_m^L-i}^{(i)}) + \begin{cases} \tilde{\kappa}_{\sigma_1^L-i}^{(i)} & \text{if } \Delta_1^L = -1;\\ p^{(i)} - \tilde{\kappa}_{\sigma_1^L-i}^{(i)} & \text{if } \Delta_1^L = +1; \end{cases}\\
\tilde{\mu}_j^{(i)} &= \tilde{\mu}_j^{(i)*} + \Delta_{j+1}^L \tilde{\kappa}_{\tau_{j+1}^L-i}^{(i)},
\end{aligned}$$

for $1 \le j < d^R$ and $i < \tau_{j+1}^R$, defining:

$$\begin{aligned}
\tilde{\nu}_j^{(i)*} &= \sum_{m=2}^{j} \Delta_m^R(\tilde{\kappa}_{\tau_m^R-i}^{(i)} - \tilde{\kappa}_{\sigma_m^R-i}^{(i)}) + \begin{cases} \tilde{\kappa}_{\sigma_1^R-i}^{(i)} & \text{if } \Delta_1^R = -1;\\ p^{(i)} - \tilde{\kappa}_{\sigma_1^R-i}^{(i)} & \text{if } \Delta_1^R = +1; \end{cases}\\
\tilde{\nu}_j^{(i)} &= \tilde{\nu}_j^{(i)*} + \Delta_{j+1}^R \tilde{\kappa}_{\tau_{j+1}^R-i}^{(i)}.
\end{aligned}$$

and for $0 \le i < t_n$, defining:

$$\begin{aligned}
\tilde{\mu}_0^{(i)*} &= \begin{cases} \tilde{\kappa}_{t_n-i}^{(i)} & \text{if } \Delta_1^L = -1;\\ p^{(i)} - \tilde{\kappa}_{t_n-i}^{(i)} & \text{if } \Delta_1^L = +1; \end{cases}\\
\tilde{\mu}_0^{(i)} &= \begin{cases} 0 & \text{if } \Delta_1^L = -1;\\ p^{(i)} & \text{if } \Delta_1^L = +1, \end{cases}\\
\tilde{\nu}_0^{(i)*} &= \begin{cases} \tilde{\kappa}_{t_n-i}^{(i)} & \text{if } \Delta_1^R = -1;\\ p^{(i)} - \tilde{\kappa}_{t_n-i}^{(i)} & \text{if } \Delta_1^R = +1; \end{cases}\\
\tilde{\nu}_0^{(i)} &= \begin{cases} 0 & \text{if } \Delta_1^R = -1;\\ p^{(i)} & \text{if } \Delta_1^R = +1, \end{cases}
\end{aligned}$$

Lemma 7.10. *Let $0 \le j < d^L$ with $0 < i < \tau^L_{j+1}$ if $j > 0$, and $0 < i < t_n$ if $j = 0$. If $i \ne t_{k(i)}$ then:*

$$\begin{aligned} \mu_j^{(i-1)} &= \mu_j^{(i)} + \tilde{\mu}_j^{(i)}; & \tilde{\mu}_j^{(i-1)} &= \tilde{\mu}_j^{(i)};\\ \mu_j^{(i-1)*} &= \mu_j^{(i)*} + \tilde{\mu}_j^{(i)*}; & \tilde{\mu}_j^{(i-1)*} &= \tilde{\mu}_j^{(i)*}, \end{aligned}$$

and if $i = t_{k(i)}$ then:

$$\begin{aligned} \mu_j^{(i-1)} &= 2\mu_j^{(i)} - \tilde{\mu}_j^{(i)}; & \tilde{\mu}_j^{(i-1)} &= \mu_j^{(i)} - \tilde{\mu}_j^{(i)};\\ \mu_j^{(i-1)*} &= 2\mu_j^{(i)*} - \tilde{\mu}_j^{(i)*}; & \tilde{\mu}_j^{(i-1)*} &= \mu_j^{(i)*} - \tilde{\mu}_j^{(i)*}. \end{aligned}$$

Let $0 \le j < d^R$ with $0 < i < \tau^R_{j+1}$ if $j > 0$, and $0 < i < t_n$ if $j = 0$. If $i \ne t_{k(i)}$ then:

$$\begin{aligned} \nu_j^{(i-1)} &= \nu_j^{(i)} + \tilde{\nu}_j^{(i)}; & \tilde{\nu}_j^{(i-1)} &= \tilde{\nu}_j^{(i)};\\ \nu_j^{(i-1)*} &= \nu_j^{(i)*} + \tilde{\nu}_j^{(i)*}; & \tilde{\nu}_j^{(i-1)*} &= \tilde{\nu}_j^{(i)*}, \end{aligned}$$

and if $i = t_{k(i)}$ then:

$$\begin{aligned} \nu_j^{(i-1)} &= 2\nu_j^{(i)} - \tilde{\nu}_j^{(i)}; & \tilde{\nu}_j^{(i-1)} &= \nu_j^{(i)} - \tilde{\nu}_j^{(i)};\\ \nu_j^{(i-1)*} &= 2\nu_j^{(i)*} - \tilde{\nu}_j^{(i)*}; & \tilde{\nu}_j^{(i-1)*} &= \nu_j^{(i)*} - \tilde{\nu}_j^{(i)*}. \end{aligned}$$

Proof: These results follow readily from Lemma 7.5. □

For each t-dimensional vector $\boldsymbol{u} = (u_1, u_2, \ldots, u_t)$, define the $(t-1)$-dimensional vector $\boldsymbol{u}^{(\flat,k)} = (u_1^{(\flat,k)}, u_2^{(\flat,k)}, \ldots, u_{t-1}^{(\flat,k)})$ by

$$u_i^{(\flat,k)} = \begin{cases} 0 & \text{if } t_{k'} < i \le t_{k'+1},\ k' \equiv k \pmod 2;\\ u_i & \text{if } t_{k'} < i \le t_{k'+1},\ k' \not\equiv k \pmod 2, \end{cases} \tag{7.7}$$

and the $(t-1)$-dimensional vector $\boldsymbol{u}^{(\sharp,k)} = (u_1^{(\sharp,k)}, u_2^{(\sharp,k)}, \ldots, u_{t-1}^{(\sharp,k)})$ by

$$u_i^{(\sharp,k)} = \begin{cases} u_i & \text{if } t_{k'} < i \le t_{k'+1},\ k' \equiv k \pmod 2;\\ 0 & \text{if } t_{k'} < i \le t_{k'+1},\ k' \not\equiv k \pmod 2, \end{cases} \tag{7.8}$$

For convenience, we sometimes write $\boldsymbol{u}_{(\flat,k)}$ instead of $\boldsymbol{u}^{(\flat,k)}$, and $\boldsymbol{u}_{(\sharp,k)}$ instead of $\boldsymbol{u}^{(\sharp,k)}$.

Now for $0 \le i \le t-3$, define:

$$F^{(i)}(\boldsymbol{u}^L, \boldsymbol{u}^R, m_i, m_{i+1}; q) =$$

$$\sum q^{\frac14 \hat{\boldsymbol{m}}^{(i+1)T} \boldsymbol{C} \hat{\boldsymbol{m}}^{(i+1)} + \frac14 m_i^2 - \frac12 m_i m_{i+1} - \frac12 (\boldsymbol{u}^L_{(\flat,k(i))} + \boldsymbol{u}^R_{(\sharp,k(i))}) \cdot \boldsymbol{m}^{(i)} + \frac14 \gamma_i''} \times \prod_{j=i+1}^{t-1} \begin{bmatrix} m_j - \frac12 (\boldsymbol{C}^* \hat{\boldsymbol{m}}^{(i)} - \boldsymbol{u}^L - \boldsymbol{u}^R)_j \\ m_j \end{bmatrix}_q . \tag{7.9}$$

With $(Q_1, Q_2, \ldots, Q_{t-1}) = \boldsymbol{Q}(\boldsymbol{u}^L + \boldsymbol{u}^R)$ as defined in Section 1.12, the sum here is to be taken over all $(m_{i+2}, m_{i+3}, \ldots, m_{t-1}) \equiv (Q_{i+2}, Q_{i+3}, \ldots, Q_{t-1})$. The $(t-1)$-dimensional $\boldsymbol{m}^{(i)} = (0, \ldots, 0, m_{i+1}, m_{i+2}, m_{i+3}, \ldots, m_{t-1})$ has its first i components equal to zero. The t-dimensional $\hat{\boldsymbol{m}}^{(i)} = (0, \ldots, 0, m_i, m_{i+1}, m_{i+2}, \ldots, m_{t-1})$ has its first i components equal to zero. The matrix $\boldsymbol{C}$ is as defined in Section

1.11. We define $F^{(t-2)}(\boldsymbol{u}^L, \boldsymbol{u}^R, m_{t-2}, m_{t-1}; q)$ using (7.9), but with the summation omitted from the right side.

We also define:

$$F^{(t-1)}(\boldsymbol{u}^L, \boldsymbol{u}^R, m_{t-1}, m_t; q) = q^{\frac{1}{4}m_{t-1}^2 + \frac{1}{4}\gamma''_{t-1}} \delta_{m_t,0}. \tag{7.10}$$

For convenience, we also set $Q_t = 0$.

By Lemma 2.8, $\begin{bmatrix} m+n \\ m \end{bmatrix}_{q^{-1}} = q^{-mn} \begin{bmatrix} m+n \\ m \end{bmatrix}_q$. Then since $C_{i+1,i} = -1$, it follows that for $0 \leq i \leq t-2$:

$$F^{(i)}(\boldsymbol{u}^L, \boldsymbol{u}^R, m_i, m_{i+1}; q^{-1}) =$$

$$\sum q^{\frac{1}{4}\hat{\boldsymbol{m}}^{(i+1)T}\boldsymbol{C}\hat{\boldsymbol{m}}^{(i+1)} - \frac{1}{4}m_i^2 - \frac{1}{2}(\boldsymbol{u}^L_{(\flat,k(i)-1)} + \boldsymbol{u}^R_{(\sharp,k(i)-1)})\cdot \boldsymbol{m}^{(i)} - \frac{1}{4}\gamma''_i} \times \prod_{j=i+1}^{t-1} \begin{bmatrix} m_j - \frac{1}{2}(\boldsymbol{C}^*\hat{\boldsymbol{m}}^{(i)} - \boldsymbol{u}^L - \boldsymbol{u}^R)_j \\ m_j \end{bmatrix}_q, \tag{7.11}$$

where, as above, the sum is taken over all $(m_{i+2}, m_{i+3}, \ldots, m_{t-1}) \equiv (Q_{i+2}, Q_{i+3}, \ldots, Q_{t-1})$, except in the $i = t-2$ case where the summation is omitted. Of course, we also have:

$$F^{(t-1)}(\boldsymbol{u}^L, \boldsymbol{u}^R, m_{t-1}, m_t; q^{-1}) = q^{-\frac{1}{4}m_{t-1}^2 - \frac{1}{4}\gamma''_{t-1}} \delta_{m_t,0}. \tag{7.12}$$

7.3. The induction

In this section, we bring together the results on the $\mathcal{B}$-transform (Corollary 5.8), the $\mathcal{D}$-transform (Corollary 5.10), path extension (Lemmas 6.4 and 6.2), and path truncation (Lemmas 6.8 and 6.6), to obtain Theorem 7.1 by means of a huge induction proof.

Throughout this section, we restrict consideration to the cases for which $p' > 2p$. We fix runs $\mathcal{X}^L = \{\tau_j^L, \sigma_j^L, \Delta_j^L\}_{j=1}^{d^L}$ and $\mathcal{X}^R = \{\tau_j^R, \sigma_j^R, \Delta_j^R\}_{j=1}^{d^R}$ and make use of the definitions of Section 7.1. Except in the final Corollary 7.14, these runs will be assumed to be reduced and the definitions of Section 7.2 will be used. We also make use of the results of Appendices C and D. For typographical convenience, we abbreviate $\rho^{p^{(i)},p^{(i)\prime}}$ (defined in Section 2.2) to $\rho^{(i)}$.

Lemma 7.11. *Let $0 \le i < t$. If $m_i \equiv Q_i$ and $m_{i+1} \equiv Q_{i+1}$ then:*

$$\tilde{\chi}^{p^{(i)},p^{(i)\prime}}_{a^{(i)},b^{(i)},e^{(i)},f^{(i)}}(m_i, m_{i+1} + 2\delta^{p^{(i)},p^{(i)\prime}}_{a^{(i)},e^{(i)}})\left\{\begin{matrix}\boldsymbol{\mu}^{(i)} \ ; \boldsymbol{\nu}^{(i)} \\ \boldsymbol{\mu}^{(i)*}; \boldsymbol{\nu}^{(i)*}\end{matrix}\right\} = F^{(i)}(\boldsymbol{u}^L, \boldsymbol{u}^R, m_i, m_{i+1}).$$

In addition,

- $\alpha^{p^{(i)},p^{(i)\prime}}_{a^{(i)},b^{(i)}} = \alpha_i''$;
- $\beta^{p^{(i)},p^{(i)\prime}}_{a^{(i)},b^{(i)},e^{(i)},f^{(i)}} = \beta_i'$;
- $\lfloor a^{(i)}p^{(i)}/p^{(i)\prime}\rfloor = \tilde{a}^{(i)}$;
- $\lfloor b^{(i)}p^{(i)}/p^{(i)\prime}\rfloor = \tilde{b}^{(i)}$;
- *if* $\delta^{p^{(i)},p^{(i)\prime}}_{a^{(i)},e^{(i)}} = 1$ *then* $i+1 = \tau_j^L$ *for some* j *and* $\mu_{j-1}^{(i)} = a^{(i)} - (-1)^{e^{(i)}}$;
- *if* $\delta^{p^{(i)},p^{(i)\prime}}_{b^{(i)},f^{(i)}} = 1$ *then* $i+1 = \tau_j^R$ *for some* j *and* $\nu_{j-1}^{(i)} = b^{(i)} - (-1)^{f^{(i)}}$;
- *if there exists* j *and* k *for which* $t_k \le i$ *and* $\tau_{j+1}^L \le i < t_{k+1} < \sigma_j^L$ *then* $a^{(i)}$ *is interfacial in the* $(p^{(i)}, p^{(i)\prime})$*-model.*
  If $t_k \le i$ *and* $\tau_{j+1}^L \le i < t_{k+1} \le \sigma_j^L$ *and* $a^{(i)}$ *is not interfacial in the* $(p^{(i)}, p^{(i)\prime})$*-model then* $\tau_j^L = \sigma_j^L + 1$;
- *if there exists* j *and* k *for which* $t_k \le i$ *and* $\tau_{j+1}^R \le i < t_{k+1} < \sigma_j^R$ *then* $b^{(i)}$ *is interfacial in the* $(p^{(i)}, p^{(i)\prime})$*-model.*
  If $t_k \le i$ *and* $\tau_{j+1}^R \le i < t_{k+1} \le \sigma_j^R$ *and* $b^{(i)}$ *is not interfacial in the* $(p^{(i)}, p^{(i)\prime})$*-model then* $\tau_j^R = \sigma_j^R + 1$;
- $\boldsymbol{\mu}^{(i)}$, $\boldsymbol{\mu}^{(i)*}, \boldsymbol{\nu}^{(i)}$, $\boldsymbol{\nu}^{(i)*}$ *are an interfacial mazy-four in the* $(p^{(i)}, p^{(i)\prime})$*-model sandwiching* $(a^{(i)}, b^{(i)})$;
- *If* $d_0^L(i) \le j < \eta^L(i)$ *then* $\rho^{(i)}(\mu_j^{(i)*}) = \tilde{\mu}_j^{(i)*}$ *and* $\rho^{(i)}(\mu_j^{(i)}) = \tilde{\mu}_j^{(i)}$;
  if $d_0^R(i) \le j < \eta^R(i)$ *then* $\rho^{(i)}(\nu_j^{(i)*}) = \tilde{\nu}_j^{(i)*}$ *and* $\rho^{(i)}(\nu_j^{(i)}) = \tilde{\nu}_j^{(i)}$.

Proof: This is proved by downward induction. For $i = t-1$, we have $i \ge t_n$, $i \ge \sigma_1^L$ and $i \ge \sigma_1^R$, and consequently, $d_0^L(i) = d_0^R(i) = 1$ and $\eta^L(i) = \eta^R(i) = 1$. The definitions of Section 7.2 then yield $p^{(i)\prime} = 3$, $p^{(i)} = 1$, and if $\Delta_1^L = -1$ then $a^{(i)} = 1$, $e^{(i)} = 0$ and $(\boldsymbol{\Delta}^L)_t = 0$; and if $\Delta_1^L = +1$ then $a^{(i)} = 2$, $e^{(i)} = 1$ and $(\boldsymbol{\Delta}^L)_t = -1$. Similarly, if $\Delta_1^R = -1$ then $b^{(i)} = 1$, $f^{(i)} = 0$ and $(\boldsymbol{\Delta}^R)_t = 0$; and if $\Delta_1^R = +1$ then $b^{(i)} = 2$, $f^{(i)} = 1$ and $(\boldsymbol{\Delta}^R)_t = -1$. In particular, we now immediately obtain

nbining all the above, and using the expression for γ''_{i-1} given by (1.25) and
yields:

$${}^{,p^{(i-1)\prime}}_{,b^{(i-1)},e^{(i-1)},f^{(i-1)}}(m_{i-1}, m_i + 2\delta^{p^{(i-1)},p^{(i-1)\prime}}_{a^{(i-1)},e^{(i-1)}}) \begin{Bmatrix} \boldsymbol{\mu}^{(i-1)} ; \boldsymbol{\nu}^{(i-1)} \\ \boldsymbol{\mu}^{(i-1)*}; \boldsymbol{\nu}^{(i-1)*}\end{Bmatrix}$$

$$= \sum q^{\frac14 \hat{\boldsymbol{m}}^{(i)T} \boldsymbol{C} \hat{\boldsymbol{m}}^{(i)} + \frac14 m_{i-1}^2 - \frac12 m_{i-1}m_i - \frac12(\boldsymbol{u}^L_{(\flat,k)} + \boldsymbol{u}^R_{(\sharp,k)})\cdot \boldsymbol{m}^{(i-1)} + \frac14\gamma''_{i-1}}$$

$$\times \prod_{j=i}^{t-1} \begin{bmatrix} m_j - \frac12(\boldsymbol{C}^*\hat{\boldsymbol{m}}^{(i-1)} - \boldsymbol{u}^L - \boldsymbol{u}^R)_j \\ m_j\end{bmatrix}_q$$

$$= F^{(i-1)}(\boldsymbol{u}^L, \boldsymbol{u}^R, m_{i-1}, m_i),$$

the required result when $i\neq t_k$, since $k=k(i)=k(i-1)$.
is $i \neq t_k$ case, making use of (1.25), (1.26), and Lemma 7.8, we also
tely obtain:

$$\begin{aligned}
\alpha^{p^{(i-1)},p^{(i-1)\prime}}_{a^{(i-1)},b^{(i-1)}} &= \alpha''_{i-1};\\
\beta^{p^{(i-1)},p^{(i-1)\prime}}_{a^{(i-1)},b^{(i-1)},e^{(i-1)},f^{(i-1)}} &= \beta'_{i-1};\\
\left\lfloor \frac{a^{(i-1)}p^{(i-1)}}{p^{(i-1)\prime}}\right\rfloor &= \tilde a^{(i-1)};\\
\left\lfloor \frac{b^{(i-1)}p^{(i-1)}}{p^{(i-1)\prime}}\right\rfloor &= \tilde b^{(i-1)}.
\end{aligned}$$

consider the case for which $i = t_k$. Equation (1.27) gives $\alpha_i = \alpha''_i$, $\beta_i =$
nd $\gamma_i = -\alpha_i^2 - \gamma''_i$. Then Lemma 4.2 gives $\alpha^{p^{(i)\prime}-p^{(i)},p^{(i)\prime}}_{a^{(i)},b^{(i)}} = \alpha''_i = \alpha_i$
${}^{-p^{(i)},p^{(i)\prime}}_{b^{(i)},1-e^{(i)},1-f^{(i)}} = \alpha''_i - \beta'_i = \beta_i$. Let $m_{i-1}\equiv Q_{i-1}$. On setting $M =$
${}^L + u_i^R$, equations (1.29), (1.32) and (1.33) imply that $M - m_i \equiv Q_{i+1}$.
of the induction hypothesis and Lemmas 5.10, 7.5, 7.6 and 7.10 yields:

$${}^{,p^{(i-1)\prime}}_{(i)\prime,1-e^{(i)},1-f^{(i)}}(M, m_i; q) \begin{Bmatrix}\boldsymbol{\mu}_\uparrow^{(i-1)};\boldsymbol{\nu}_\uparrow^{(i-1)} \\ \boldsymbol{\mu}_\uparrow^{(i-1)*};\boldsymbol{\nu}_\uparrow^{(i-1)*}\end{Bmatrix}$$

$$= \sum_{m_{i+1}\equiv Q_{i+1}} q^{\frac14(m_i^2 + (M-m_i)^2 - \alpha_i^2 - \beta_i^2)} \begin{bmatrix}\frac12(M+m_i-m_{i+1})\\ m_i\end{bmatrix}_q$$

$$\times\quad F^{(i)}(\boldsymbol{u}^L,\boldsymbol{u}^R,m_i,m_{i+1};q^{-1}).$$

0 also gives $\alpha^{p^{(i-1)},p^{(i-1)\prime}}_{a^{(i)\prime},b^{(i)\prime}} = 2\alpha''_i - \beta'_i = \alpha_i+\beta_i$, and $\beta^{p^{(i-1)},p^{(i-1)\prime}}_{a^{(i)\prime},b^{(i)\prime},1-e^{(i)},1-f^{(i)}} =$
$_i$, and Lemmas 7.6 and C.3 imply that $\lfloor a^{(i)\prime}p^{(i-1)}/p^{(i-1)\prime}\rfloor = a^{(i)} - 1 - \tilde a^{(i)}$
${}^{i-1)}/p^{(i-1)\prime}\rfloor = b^{(i)} - 1 - \tilde b^{(i)}$. Lemma C.3 also implies that $\delta^{p^{(i-1)},p^{(i-1)\prime}}_{a^{(i)\prime},1-e^{(i)}} =$
$= 0$. Lemma 5.10 also implies that $\boldsymbol{\mu}_\uparrow^{(i-1)}$, $\boldsymbol{\mu}_\uparrow^{(i-1)*}$, $\boldsymbol{\nu}_\uparrow^{(i-1)}$, $\boldsymbol{\nu}_\uparrow^{(i-1)*}$
rfacial mazy-four in the $(p^{(i-1)},p^{(i-1)\prime})$-model sandwiching $(a^{(i)\prime},b^{(i)\prime})$.
ith Lemma C.4, Lemma 7.10 also shows that $\rho^{(i-1)}(\mu_j^{(i-1)*}) = \tilde\mu_j^{(i-1)*}$
$\mu_j^{(i-1)}) = \tilde\mu_j^{(i-1)}$ for $d_0^L(i)\le j<\eta^L(i)$, and that $\rho^{(i-1)}(\nu_j^{(i-1)*}) = \tilde\nu_j^{(i-1)*}$
$\nu_j^{(i-1)}) = \tilde\nu_j^{(i-1)}$ for $d_0^R(i)\le j<\eta^R(i)$.

$\delta^{p^{(i)},p^{(i)\prime}}_{a^{(i)},e^{(i)}} = 0$. Via Section 1.10, we obtain $\alpha''_{t-1} = \beta'_{t-1} = (\boldsymbol{\Delta}^L)_t - (\boldsymbol{\Delta}^R)_t$ and $\gamma''_{t-1} = -((\boldsymbol{\Delta}^L)_t - (\boldsymbol{\Delta}^R)_t)^2$. For $i = t-1$, the first statement of our induction proposition is now seen to hold via Lemma 2.6. Each of the bulleted items follows readily.

Now assume the result holds for a particular i with $1\le i<t$. Let $k=k(i)$ so that $t_k\le i<t_{k+1}$, and let $\eta^L=\eta^L(i)$ and $\eta^R=\eta^R(i)$ so that $\tau_{\eta^L+1}\le i<\tau_{\eta^L}$ and $\tau_{\eta^R+1}\le i<\tau_{\eta^R}$. We also set $\eta^{L\prime}=\eta^L(i-1)$ and $\eta^{R\prime}=\eta^R(i-1)$. Then $\eta^{L\prime}=\eta^L$ unless $i=\tau^L_{\eta^L+1}$ in which case $\eta^{L\prime}=\eta^L+1$; and $\eta^{R\prime}=\eta^R$ unless $i=\tau^R_{\eta^R+1}$ in which case $\eta^{R\prime}=\eta^R+1$.

First consider the case $i\neq t_k$. Equation (1.27) gives $\alpha_i=\alpha''_i$, $\beta_i=\beta'_i$ and $\gamma_i=\gamma''_i$. Let $m_{i-1}\equiv Q_{i-1}$. On setting $M=m_{i-1}+u_i^L+u_i^R$, equations (1.30), (1.32) and (1.33) imply that $M\equiv Q_{i+1}$. Then, use of the induction hypothesis and Lemmas 5.8, 7.5, 7.6 and 7.10 yields:

$$\tilde\chi^{p^{(i-1)},p^{(i-1)\prime}}_{a^{(i)\prime},b^{(i)\prime},e^{(i)},f^{(i)}}(M,m_i)\begin{Bmatrix}\boldsymbol{\mu}_\uparrow^{(i-1)};\boldsymbol{\nu}_\uparrow^{(i-1)}\\ \boldsymbol{\mu}_\uparrow^{(i-1)*};\boldsymbol{\nu}_\uparrow^{(i-1)*}\end{Bmatrix}$$

$$=\sum_{m_{i+1}\equiv Q_{i+1}} q^{\frac14(M-m_i)^2-\frac14\beta_i^2}\begin{bmatrix}\frac12(M+m_{i+1})\\ m_i\end{bmatrix}_q F^{(i)}(\boldsymbol{u}^L,\boldsymbol{u}^R,m_i,m_{i+1}).$$

Lemma 5.8 also gives $\alpha^{p^{(i-1)},p^{(i-1)\prime}}_{a^{(i)\prime},b^{(i)\prime}}=\alpha_i+\beta_i$, and $\beta^{p^{(i-1)},p^{(i-1)\prime}}_{a^{(i)\prime},b^{(i)\prime},e^{(i)},f^{(i)}}=\beta_i$. Lemma C.1 implies, via Lemma 7.6, that $\lfloor a^{(i)\prime}p^{(i-1)}/p^{(i-1)\prime}\rfloor=\tilde a^{(i)}$ and $\lfloor b^{(i)\prime}p^{(i-1)}/p^{(i-1)\prime}\rfloor=\tilde b^{(i)}$. Lemma C.1 also implies that $\delta^{p^{(i-1)},p^{(i-1)\prime}}_{a^{(i)\prime},e^{(i)}}=\delta^{p^{(i-1)},p^{(i-1)\prime}}_{b^{(i)\prime},f^{(i)}}=0$. Lemma 5.8 also shows that $\boldsymbol{\mu}_\uparrow^{(i-1)}$, $\boldsymbol{\mu}_\uparrow^{(i-1)*}$, $\boldsymbol{\nu}_\uparrow^{(i-1)}$, $\boldsymbol{\nu}_\uparrow^{(i-1)*}$ are an interfacial mazy-four in the $(p^{(i-1)},p^{(i-1)\prime})$-model sandwiching $(a^{(i)\prime},b^{(i)\prime})$. Together with Lemma C.4(1), Lemma 7.10 also shows that $\rho^{(i-1)}(\mu_j^{(i-1)*})=\tilde\mu_j^{(i-1)*}$ and $\rho^{(i-1)}(\mu_j^{(i-1)})=\tilde\mu_j^{(i-1)}$ for $d_0^L(i)\le j<\eta^L(i)$, and that $\rho^{(i-1)}(\nu_j^{(i-1)*})=\tilde\nu_j^{(i-1)*}$ and $\rho^{(i-1)}(\nu_j^{(i-1)})=\tilde\nu_j^{(i-1)}$ for $d_0^R(i)\le j<\eta^R(i)$.

Since $M=m_{i-1}+u_i^L+u_i^R$, on noting that $t_k<i<t_{k+1}$, we have:

$$M+m_{i+1}=2m_i-(\boldsymbol{C}^*\hat{\boldsymbol{m}}^{(i-1)}-\boldsymbol{u}^L-\boldsymbol{u}^R)_i,$$

and

$$\begin{aligned}
\hat{\boldsymbol{m}}^{(i+1)T}\boldsymbol{C}\hat{\boldsymbol{m}}^{(i+1)}&+m_i^2-2m_im_{i+1}+(M-m_i)^2\\
&=\hat{\boldsymbol{m}}^{(i)T}\boldsymbol{C}\hat{\boldsymbol{m}}^{(i)}+M^2-2Mm_i\\
&=\hat{\boldsymbol{m}}^{(i)T}\boldsymbol{C}\hat{\boldsymbol{m}}^{(i)}+m_{i-1}^2-2m_im_{i-1}+2(m_{i-1}-m_i)(u_i^L+u_i^R)+(u_i^L+u_i^R)^2.
\end{aligned}$$

(In the case $i=t-1$, we require this expression after substituting $m_t=0$.) Thence,

$$\tilde\chi^{p^{(i-1)},p^{(i-1)\prime}}_{a^{(i)\prime},b^{(i)\prime},e^{(i)},f^{(i)}}(m_{i-1}+u_i^L+u_i^R,m_i)\begin{Bmatrix}\boldsymbol{\mu}_\uparrow^{(i-1)};\boldsymbol{\nu}_\uparrow^{(i-1)}\\ \boldsymbol{\mu}_\uparrow^{(i-1)*};\boldsymbol{\nu}_\uparrow^{(i-1)*}\end{Bmatrix}$$

$$=\sum q^{\frac14\hat{\boldsymbol{m}}^{(i)T}\boldsymbol{C}\hat{\boldsymbol{m}}^{(i)}+\frac14m_{i-1}^2-\frac12m_im_{i-1}+\frac12(m_{i-1}-m_i)(u_i^L+u_i^R)+\frac14(u_i^L+u_i^R)^2+\frac14\gamma_i-\frac14\beta_i^2}$$

$$\times\quad q^{-\frac12(\boldsymbol{u}^L_{(\flat,k)}+\boldsymbol{u}^R_{(\sharp,k)})\cdot\boldsymbol{m}^{(i)}}\prod_{j=i}^{t-1}\begin{bmatrix}m_j-\frac12(\boldsymbol{C}^*\hat{\boldsymbol{m}}^{(i-1)}-\boldsymbol{u}^L-\boldsymbol{u}^R)_j\\ m_j\end{bmatrix}_q,$$

where the sum is over all $(m_{i+1},m_{i+2},\ldots,m_{t-1})\equiv(Q_{i+1},Q_{i+2},\ldots,Q_{t-1})$.

We now check that we can apply Lemma 6.6 (for path truncation on the right) if $i=\sigma^R_{\eta^R}$, and Lemma 6.2 (for path extension on the right) if $i=\tau^R_{\eta^R+1}$.

If $i=\sigma^R_{\eta^R}$ then $u_i^R=1$, and we see that $\Delta^R_{\eta^R}=-(-1)^{f^{(i)}}=(-1)^{f^{(i-1)}}$ from (7.6), and $b^{(i-1)}=b^{(i)\prime}-\Delta^R_{\eta^R}$ from Lemma 7.7. Also note that $\eta^{R\prime}=\eta^R$. If $\eta^R>1$ or both $\eta^R=1$ and $\sigma^R_1<t_n$ then Lemma 7.9 implies that $\nu^{(i-1)}_{\eta^R-1}=b^{(i)\prime}+\Delta^R_{\eta^R}$ and $\nu^{(i-1)*}_{\eta^R-1}\neq b^{(i-1)}$. Otherwise, when $i=\sigma^R_1>t_n$ then $b^{(i)\prime}\in\{1,p^{(i-1)\prime}-1\}$ from the definition.

If $i=\tau^R_{\eta^R+1}$ (when necessarily $\eta^R<d^R$) then $u_i^R=-1$ and $\eta^{R\prime}=\eta^R+1$. We see that $\Delta^R_{\eta^{R\prime}}=-(-1)^{f(i-1)}$ from (7.6) and $b^{(i-1)}=b^{(i)\prime}+\Delta^R_{\eta^{R\prime}}$ via Lemma 7.7. In addition, Lemma 7.9 gives $\nu^{(i-1)*}_{\eta^R}=b^{(i-1)}-\Delta^R_{\eta^{R\prime}}$ and $\nu^{(i-1)}_{\eta^R}=b^{(i-1)}+\Delta^R_{\eta^{R\prime}}$. The definition of a reduced run implies that this case, $i=\tau^R_{\eta^R+1}$ and $i\neq t_k$, only occurs when $t_{n-1}\le t_n-1=\tau^R_2=i<t_n<\sigma^R_1$. The induction hypothesis then implies that $b^{(i)}$ is interfacial in the $(p^{(i)},p^{(i)\prime})$-model and $\delta^{p^{(i)},p^{(i)\prime}}_{b^{(i)},f^{(i)}}=0$. Thereupon, by Lemma C.1, $b^{(i)\prime}$ is interfacial in the $(p^{(i-1)},p^{(i-1)\prime})$-model. Also note that $d^R_0(i)=1$ and so $\boldsymbol{\nu}^{(i-1)}_{\uparrow}=\boldsymbol{\nu}^{(i-1)*}_{\uparrow}=()$ here.

Then, using Lemma 6.6 if $i=\sigma^R_{\eta^R}$, or Lemma 6.2 if $i=\tau^R_{\eta^R+1}$, yields:

$$\tilde{\chi}^{p^{(i-1)},p^{(i-1)\prime}}_{a^{(i)\prime},b^{(i-1)},e^{(i)},f^{(i-1)}}(m_{i-1}+u_i^L,m_i)\left\{\begin{matrix}\boldsymbol{\mu}^{(i-1)}_{\uparrow};\boldsymbol{\nu}^{(i-1)}\\ \boldsymbol{\mu}^{(i-1)*}_{\uparrow};\boldsymbol{\nu}^{(i-1)*}\end{matrix}\right\}$$

$$=q^{-\frac{1}{2}u_i^R(m_{i-1}+u_i^L+u_i^R-\Delta^R_{\eta^{R\prime}}(\alpha_i+\beta_i))}$$

$$\times\quad\tilde{\chi}^{p^{(i-1)},p^{(i-1)\prime}}_{a^{(i)\prime},b^{(i)\prime},e^{(i)},f^{(i)}}(m_{i-1}+u_i^L+u_i^R,m_i)\left\{\begin{matrix}\boldsymbol{\mu}^{(i-1)}_{\uparrow};\boldsymbol{\nu}^{(i-1)}_{\uparrow}\\ \boldsymbol{\mu}^{(i-1)*}_{\uparrow};\boldsymbol{\nu}^{(i-1)*}_{\uparrow}\end{matrix}\right\},$$

noting that $d^R_0(i)=d^R_0(i-1)$ because $i\neq t_n$. If neither $i=\sigma^R_{\eta^R}$ nor $i=\tau^R_{\eta^R+1}$ then (noting that $i\neq t_k$) $u_i^R=0$, $f^{(i-1)}=f^{(i)}$ from (7.6) and, via Lemma 7.7, $b^{(i-1)}=b^{(i)\prime}$. The preceding expression thus also holds (trivially) in this case.

Note that $\delta^{p^{(i-1)},p^{(i-1)\prime}}_{b^{(i-1)},f^{(i-1)}}=1$ only in the case $i=\tau^R_{\eta^R+1}$, when $\nu^{(i-1)}_{\eta^{R\prime}-1}=b^{(i-1)}-(-1)^{f^{(i-1)}}$.

Lemmas 6.6 and Lemma 6.2 also imply that:

$$\begin{aligned}
\alpha^{p^{(i-1)},p^{(i-1)\prime}}_{a^{(i)\prime},b^{(i-1)}}&=\alpha_i+\beta_i-u_i^R\Delta^R_{\eta^{R\prime}};\\
\beta^{p^{(i-1)},p^{(i-1)\prime}}_{a^{(i)\prime},b^{(i-1)},e^{(i)},f^{(i-1)}}&=\beta_i-u_i^R\Delta^R_{\eta^{R\prime}};\\
\left\lfloor\frac{b^{(i-1)}p^{(i-1)}}{p^{(i-1)\prime}}\right\rfloor&=\begin{cases}\tilde{b}^{(i)}+\Delta^R_{\eta^{R\prime}} & \text{if } i=\tau^R_{\eta^R+1}\text{ and } f^{(i)}=f^{(i-1)};\\ \tilde{b}^{(i)} & \text{otherwise.}\end{cases}
\end{aligned}$$

They also imply that $\boldsymbol{\mu}^{(i-1)}_{\uparrow}$, $\boldsymbol{\mu}^{(i-1)*}_{\uparrow}$, $\boldsymbol{\nu}^{(i-1)}$, $\boldsymbol{\nu}^{(i-1)*}$ are a mazy-four sandwiching $(a^{(i)\prime},b^{(i-1)})$. That they are actually interfacial follows if in the $i=\tau^R_{\eta^R+1}$ case, we show that $\nu^{(i-1)*}_{\eta^R}$ and $\nu^{(i-1)}_{\eta^R}$ are interfacial. From above, $\nu^{(i-1)*}_{\eta^R}=b^{(i)\prime}$ is interfacial. That $\nu^{(i-1)}_{\eta^R}$ is interfacial will be established below.

We now check that we can apply Lemma 6.8 (f
if $i=\sigma^L_{\eta^L}$, and Lemma 6.4 (for path extension on t

If $i=\sigma^L_{\eta^L}$ then $u_i^L=1$, and we see that $\Delta^L_{\eta^L}$
(7.5), and $a^{(i-1)}=a^{(i)\prime}-\Delta^L_{\eta^L}$ from Lemma 7.7. Al
or both $\eta^L=1$ and $\sigma^L_1<t_n$ then Lemma 7.9 impli
$\mu^{(i-1)*}_{\eta^L-1}\neq a^{(i-1)}$. Otherwise, when $i=\sigma^L_1>t_n$, t
the definition.

If $i=\tau^L_{\eta^L+1}$ (when necessarily $\eta^L<d^L$) then u
that $\Delta^L_{\eta^L(i-1)}=-(-1)^{e(i-1)}$ from (7.5) and, via I
In addition, Lemma 7.9 gives $\mu^{(i-1)*}_{\eta^L}=a^{(i-1)}-$
The definition of a reduced run implies that this
occurs when $t_{n-1}\le t_n-1=\tau^L_2=i<t_n<\sigma^L_1$
implies that $a^{(i)}$ is interfacial in the $(p^{(i)},p^{(i)\prime})$-mo
by Lemma C.1, $a^{(i)\prime}$ is interfacial in the $(p^{(i-1}$
$d^L_0(i)=1$ and so $\boldsymbol{\mu}^{(i-1)}_{\uparrow}=\boldsymbol{\mu}^{(i-1)*}_{\uparrow}=()$ here.

Then, on using Lemma 6.8 if $i=\sigma^L_{\eta^L}$, or Le

noting that $d^L_0(i)=d^L_0(i-1)$ because $i\neq t_n$
then (noting that $i\neq t_k$) $u_i^L=0$, $e^{(i-1)}=$
$a^{(i-1)}=a^{(i)\prime}$. The preceding expression thus

Note that $\delta^{p^{(i-1)},p^{(i-1)\prime}}_{a^{(i-1)},e^{(i-1)}}=1$ only in the
$a^{(i-1)}-(-1)^{e^{(i-1)}}$.

Lemmas 6.8 and Lemma 6.4 also imply t

They also imply that $\boldsymbol{\mu}^{(i-1)}$, $\boldsymbol{\mu}^{(i-1)*}$, $\boldsymbol{\nu}^{(i-1)}$
$(a^{(i-1)},b^{(i-1)})$. That they are actually inte
we show that $\mu^{(i-1)*}_{\eta^L}$ and $\mu^{(i-1)}_{\eta^L}$ are inter
interfacial. That $\mu^{(i-1)}_{\eta^L}$ is interfacial will be

Since $M = m_{i-1} + u_i^L + u_i^R$, on noting that $i = t_k$, we have:

$$M + m_i - m_{i+1} = 2m_i - (\boldsymbol{C}^* \hat{\boldsymbol{m}}^{(i-1)} - \boldsymbol{u}^L - \boldsymbol{u}^R)_i$$

(in the case $i = t-1$, we require this expression after substituting $m_t = 0$), and

$$\begin{aligned}\hat{\boldsymbol{m}}^{(i+1)T} \boldsymbol{C} \hat{\boldsymbol{m}}^{(i+1)} &+ (M - m_i)^2 \\ &= \hat{\boldsymbol{m}}^{(i)T} \boldsymbol{C} \hat{\boldsymbol{m}}^{(i)} + M^2 - 2Mm_i \\ &= \hat{\boldsymbol{m}}^{(i)T} \boldsymbol{C} \hat{\boldsymbol{m}}^{(i)} + m_{i-1}^2 - 2m_i m_{i-1} + 2(m_{i-1} - m_i)(u_i^L + u_i^R) + (u_i^L + u_i^R)^2.\end{aligned}$$

Use of expression (7.11) or (7.12) for $F_{a,b}^{(i)}(\boldsymbol{u}^L, \boldsymbol{u}^R, m_i, m_{i+1}; q^{-1})$ then gives:

$$\begin{aligned}&\tilde{\chi}^{p^{(i-1)}, p^{(i-1)\prime}}_{a^{(i)\prime}, b^{(i)\prime}, 1-e^{(i)}, 1-f^{(i)}}(m_{i-1} + u_i^L + u_i^R, m_i) \left\{ \begin{matrix} \boldsymbol{\mu}_\uparrow^{(i-1)} & ; \boldsymbol{\nu}_\uparrow^{(i-1)} \\ \boldsymbol{\mu}_\uparrow^{(i-1)*} & ; \boldsymbol{\nu}_\uparrow^{(i-1)*} \end{matrix} \right\} \\ &= \sum q^{\frac{1}{4} \hat{\boldsymbol{m}}^{(i)T} \boldsymbol{C} \hat{\boldsymbol{m}}^{(i)} + \frac{1}{4} m_{i-1}^2 - \frac{1}{2} m_i m_{i-1} + \frac{1}{2}(m_{i-1} - m_i)(u_i^L + u_i^R) + \frac{1}{4}(u_i^L + u_i^R)^2 + \frac{1}{4}\gamma_i - \frac{1}{4}\beta_i^2} \\ &\qquad \times \quad q^{-\frac{1}{2}(\boldsymbol{u}^L_{(\flat, k-1)} + \boldsymbol{u}^R_{(\sharp, k-1)}) \cdot \boldsymbol{m}^{(i)}} \prod_{j=i}^{t-1} \begin{bmatrix} m_j - \frac{1}{2}(\boldsymbol{C}^* \hat{\boldsymbol{m}}^{(i-1)} - \boldsymbol{u}^L - \boldsymbol{u}^R)_j \\ m_j \end{bmatrix}_q,\end{aligned}$$

where the sum is over all $(m_{i+1}, m_{i+2}, \ldots, m_{t-1}) \equiv (Q_{i+1}, Q_{i+2}, \ldots, Q_{t-1})$.

Since $i = t_k$, it follows that $u_i^R = -1$ if $i = \tau^R_{\eta^R+1}$ or $\sigma^R_{\eta^R} < i < \tau^R_{\eta^R}$. We will apply Lemma 6.2 (for path extension on the right) in these cases.

In the case $i = \tau^R_{\eta^R+1}$, we have $i + 1 \neq \tau^R_{\eta^R}$ which implies via the induction hypothesis that $\delta^{p^{(i)}, p^{(i)\prime}}_{b^{(i)}, f^{(i)}} = 0$. Then Lemma C.3 implies that $b^{(i)\prime}$ is interfacial in the $(p^{(i-1)}, p^{(i-1)\prime})$-model. Note that if $\eta^R = 2$ and $\sigma_1^R \geq t_n$ then $\boldsymbol{\nu}_\uparrow^{(i-1)} = \boldsymbol{\nu}_\uparrow^{(i-1)*} = ()$; if $\eta^R = 2$ and $\sigma_1^R < t_n$ (necessarily $k < n$) then $\boldsymbol{\nu}_\uparrow^{(i-1)} = (\nu_0^{(i-1)})$ and $\boldsymbol{\nu}_\uparrow^{(i-1)*} = (\nu_0^{(i-1)*})$, with $\nu_0^{(i-1)} \neq b^{(i-1)} \neq \nu_0^{(i-1)*}$ direct from the definitions after noting that $i < \sigma_1^R < t_n$; and if $\eta^R = j+1$ for $j > 1$ then $\boldsymbol{\nu}_\uparrow^{(i-1)} = (\ldots, \nu_{j-1}^{(i-1)})$ and $\boldsymbol{\nu}_\uparrow^{(i-1)*} = (\ldots, \nu_{j-1}^{(i-1)*})$, with $\nu_{j-1}^{(i-1)} \neq b^{(i-1)} \neq \nu_{j-1}^{(i-1)*}$ direct from the definitions after noting that $i < \sigma_j^R < \tau_j^R$. We now proceed via Lemma 6.2 as in the $i \neq t_k$ case, to obtain:

$$\begin{aligned}&\tilde{\chi}^{p^{(i-1)}, p^{(i-1)\prime}}_{a^{(i)\prime}, b^{(i-1)}, 1-e^{(i)}, f^{(i-1)}}(m_{i-1} + u_i^L, m_i) \left\{ \begin{matrix} \boldsymbol{\mu}_\uparrow^{(i-1)} & ; \boldsymbol{\nu}^{(i-1)} \\ \boldsymbol{\mu}_\uparrow^{(i-1)*} & ; \boldsymbol{\nu}^{(i-1)*} \end{matrix} \right\} \\ &\qquad = q^{-\frac{1}{2} u_i^R (m_{i-1} + u_i^L + u_i^R - \Delta^R_{\eta^{R\prime}}(\alpha_i + \beta_i))} \\ &\qquad\qquad \times \quad \tilde{\chi}^{p^{(i-1)}, p^{(i-1)\prime}}_{a^{(i)\prime}, b^{(i)\prime}, 1-e^{(i)}, 1-f^{(i)}}(m_{i-1} + u_i^L + u_i^R, m_i) \left\{ \begin{matrix} \boldsymbol{\mu}_\uparrow^{(i-1)} & ; \boldsymbol{\nu}_\uparrow^{(i-1)} \\ \boldsymbol{\mu}_\uparrow^{(i-1)*} & ; \boldsymbol{\nu}_\uparrow^{(i-1)*} \end{matrix} \right\},\end{aligned}$$

having noted that $d_0^R(i) = d_0^R(i-1)$. As in the $i \neq t_k$ case, we obtain that $\boldsymbol{\mu}_\uparrow^{(i-1)}$, $\boldsymbol{\mu}_\uparrow^{(i-1)*}$, $\boldsymbol{\nu}^{(i-1)}$, $\boldsymbol{\nu}^{(i-1)*}$ are a mazy-four sandwiching $(a^{(i)\prime}, b^{(i-1)})$, and that they are actually interfacial once it is established that $\nu_{\eta^R}^{(i-1)}$ is interfacial. This is done below.

In the case $\sigma^R_{\eta^R} < i < \tau^R_{\eta^R}$, we have $\eta^{R\prime} = \eta^R$ and $\Delta^R_{\eta^R} = -(-1)^{f^{(i-1)}} = (-1)^{1-f^{(i)}}$ from (7.6) and, via Lemma 7.7, $b^{(i-1)} = b^{(i)\prime} + \Delta^R_{\eta^R}$. Note that if $k = n$

(necessarily $\eta^R = 1$) then $\boldsymbol{\nu}_\uparrow^{(i-1)} = \boldsymbol{\nu}_\uparrow^{(i-1)*} = ()$; if $k < n$ and $\eta^R = 1$ then $\boldsymbol{\nu}_\uparrow^{(i-1)} = (\nu_0^{(i-1)})$ and $\boldsymbol{\nu}_\uparrow^{(i-1)*} = (\nu_0^{(i-1)*})$, with $\nu_0^{(i-1)} \neq b^{(i-1)} \neq \nu_0^{(i-1)*}$ direct from the definitions after noting that $i < t_n$; and if $k < n$ and $\eta^R = j$ for $j > 1$ then $\boldsymbol{\nu}_\uparrow^{(i-1)} = (\ldots, \nu_{j-1}^{(i-1)})$ and $\boldsymbol{\nu}_\uparrow^{(i-1)*} = (\ldots, \nu_{j-1}^{(i-1)*})$, with $\nu_{j-1}^{(i-1)} \neq b^{(i-1)} \neq \nu_{j-1}^{(i-1)*}$ direct from the definitions after noting that $i < \tau_j^R$. Then Lemma 6.2 yields:

$$\tilde{\chi}^{p^{(i-1)},p^{(i-1)\prime}}_{a^{(i)\prime},b^{(i-1)},1-e^{(i)},f^{(i-1)}}(m_{i-1}+u_i^L, m_i)\left\{\begin{matrix}\boldsymbol{\mu}_\uparrow^{(i-1)} & ;\boldsymbol{\nu}_\uparrow^{(i-1)} & ,b^{(i-1)}+\Delta^R_{\eta^R}\\ \boldsymbol{\mu}_\uparrow^{(i-1)*} & ;\boldsymbol{\nu}_\uparrow^{(i-1)*} & ,b^{(i-1)}-\Delta^R_{\eta^R}\end{matrix}\right\}$$

$$= q^{-\frac{1}{2}u_i^R(m_{i-1}+u_i^L+u_i^R-\Delta^R_{\eta^{R\prime}}(\alpha_i+\beta_i))}$$

$$\times \quad \tilde{\chi}^{p^{(i-1)},p^{(i-1)\prime}}_{a^{(i)\prime},b^{(i)\prime},1-e^{(i)},1-f^{(i)}}(m_{i-1}+u_i^L+u_i^R, m_i)\left\{\begin{matrix}\boldsymbol{\mu}_\uparrow^{(i-1)} & ;\boldsymbol{\nu}_\uparrow^{(i-1)}\\ \boldsymbol{\mu}_\uparrow^{(i-1)*} & ;\boldsymbol{\nu}_\uparrow^{(i-1)*}\end{matrix}\right\}.$$

Here, if $i = t_n$ and $\sigma_1^R < t_n$, we have $d_0^R(i) = 1$ and $d_0^R(i-1) = 0$. Also note that $\eta^{R\prime} = \eta^R = 1$. Then $\nu_0^{(i-1)} = b^{(i-1)} + \Delta_1^R$ and $\nu_0^{(i-1)*} = b^{(i-1)} - \Delta_1^R$ by Lemma 7.9, whereupon we obtain precisely the expression obtained above in the $i = \tau^R_{\eta^R+1}$ case. Also note that $i+1 \neq \tau_1^R$ again implies, via $\delta^{p^{(i)},p^{(i)\prime}}_{b^{(i)},f^{(i)}} = 0$ and Lemma C.3, that $b^{(i)\prime}$ is interfacial in the $(p^{(i-1)}, p^{(i-1)\prime})$-model. Lemma 6.2 shows that $\boldsymbol{\mu}_\uparrow^{(i-1)}$, $\boldsymbol{\mu}_\uparrow^{(i-1)*}$, $\boldsymbol{\nu}^{(i-1)}$, $\boldsymbol{\nu}^{(i-1)*}$ are a mazy-four sandwiching $(a^{(i)\prime}, b^{(i-1)})$. Since $\nu_0^{(i-1)*} = b^{(i)\prime}$ is interfacial, to show that $\boldsymbol{\mu}_\uparrow^{(i-1)}$, $\boldsymbol{\mu}_\uparrow^{(i-1)*}$, $\boldsymbol{\nu}^{(i-1)}$, $\boldsymbol{\nu}^{(i-1)*}$ are an interfacial mazy-four requires only that it be established that $\nu_0^{(i-1)}$ is interfacial. This is done below.

Otherwise for $\sigma^R_{\eta^R} < i < \tau^R_{\eta^R}$, we have $\nu^{(i-1)}_{\eta^R-1} = b^{(i-1)} + \Delta^R_{\eta^R}$ by Lemma 7.9, and $\eta^{R\prime} = \eta^R$, whence use of Lemma 5.5 again yields precisely the expression obtained above in the $i = \tau^R_{\eta^R+1}$ case.

In addition, the same expression clearly also holds in the case $\tau^R_{\eta^R+1} < i \leq \sigma^R_{\eta^R}$, for which $u_i^R = 0$ and $\eta^{R\prime} = \eta^R$ (in the $i = \sigma^R_{\eta^R}$ case, note that $k(i-1) = k-1 = k^R(i-1)$ and consequently $f^{(i-1)} = 1 - f^{(i)}$ by (7.6)).

Note that in the case $\sigma^R_{\eta^R} < i < \tau^R_{\eta^R}$, we have $f^{(i-1)} = f^{(i)}$ from (7.6). Lemma 6.2 then implies that $\delta^{p^{(i-1)},p^{(i-1)\prime}}_{b^{(i-1)},f^{(i-1)}} = 0$ in this case. Consequently, $\delta^{p^{(i-1)},p^{(i-1)\prime}}_{b^{(i-1)},f^{(i-1)}} = 1$ only in the case $i = \tau^R_{\eta^R+1}$, and then only if $\nu^{(i-1)}_{\eta^R-1} = b^{(i-1)} - (-1)^{f^{(i-1)}}$.

Lemma 6.2 also implies that:

$$\alpha^{p^{(i-1)},p^{(i-1)\prime}}_{a^{(i)\prime},b^{(i-1)}} = \alpha_i + \beta_i - u_i^R\Delta^R_{\eta^{R\prime}};$$

$$\beta^{p^{(i-1)},p^{(i-1)\prime}}_{a^{(i)\prime},b^{(i-1)},e^{(i)},f^{(i-1)}} = \beta_i - u_i^R\Delta^R_{\eta^{R\prime}};$$

$$\left\lfloor\frac{b^{(i-1)}p^{(i-1)}}{p^{(i-1)\prime}}\right\rfloor = \begin{cases} b^{(i)} - 1 - \tilde{b}^{(i)} + \Delta^R_{\eta^{R\prime}} & \text{if } i = \tau^R_{\eta^R+1} \text{ and } f^{(i)} \neq f^{(i-1)};\\ b^{(i)} - 1 - \tilde{b}^{(i)} & \text{otherwise.}\end{cases}$$

Since $i = t_k$, it follows that $u_i^L = -1$ if $i = \tau^L_{\eta^L+1}$ or $\sigma^L_{\eta^L} < i < \tau^L_{\eta^L}$. We will apply Lemma 6.4 (for path extension on the left) in these cases.

In the case $i = \tau^L_{\eta^L+1}$, we have $i+1 \neq \tau^L_{\eta^L}$ which implies via the induction hypothesis that $\delta^{p^{(i)},p^{(i)\prime}}_{a^{(i)},e^{(i)}} = 0$. Then Lemma C.3 implies that $a^{(i)\prime}$ is interfacial in the $(p^{(i-1)},p^{(i-1)\prime})$-model. Note that if $\eta^L = 2$ and $\sigma^L_1 \geq t_n$ then $\boldsymbol{\mu}^{(i-1)}_{\uparrow} = \boldsymbol{\mu}^{(i-1)*}_{\uparrow} = ()$; if $\eta^L = 2$ and $\sigma^L_1 < t_n$ (necessarily $k < n$) then $\boldsymbol{\mu}^{(i-1)}_{\uparrow} = (\mu^{(i-1)}_0)$ and $\boldsymbol{\mu}^{(i-1)*}_{\uparrow} = (\mu^{(i-1)*}_0)$, with $\mu^{(i-1)}_0 \neq a^{(i-1)} \neq \mu^{(i-1)*}_0$ direct from the definitions after noting that $i < \sigma^L_1 < t_n$; and if $\eta^L = j+1$ for $j > 1$ then $\boldsymbol{\mu}^{(i-1)}_{\uparrow} = (\ldots,\mu^{(i-1)}_{j-1})$ and $\boldsymbol{\mu}^{(i-1)*}_{\uparrow} = (\ldots,\mu^{(i-1)*}_{j-1})$, with $\mu^{(i-1)}_{j-1} \neq a^{(i-1)} \neq \mu^{(i-1)*}_{j-1}$ direct from the definitions after noting that $i < \sigma^L_j < \tau^L_j$. We now proceed via Lemma 6.4 as in the $i \neq t_k$ case, to obtain:

$$\tilde{\chi}^{p^{(i-1)},p^{(i-1)\prime}}_{a^{(i-1)},b^{(i-1)},e^{(i-1)},f^{(i-1)}}(m_{i-1}, m_i + 2\delta^{p^{(i-1)},p^{(i-1)\prime}}_{a^{(i-1)},e^{(i-1)}}) \begin{Bmatrix} \boldsymbol{\mu}^{(i-1)} & ; \boldsymbol{\nu}^{(i-1)} \\ \boldsymbol{\mu}^{(i-1)*} & ; \boldsymbol{\nu}^{(i-1)*} \end{Bmatrix}$$

$$= q^{-\frac{1}{2}u^L_i(m_{i-1}-m_i+u^L_i+\Delta^L_{\eta^{L\prime}}(\beta_i - u^R_i \Delta^R_{\eta^{R\prime}}))}$$

$$\times \quad \tilde{\chi}^{p^{(i-1)},p^{(i-1)\prime}}_{a^{(i)\prime},b^{(i-1)},1-e^{(i)},f^{(i-1)}}(m_{i-1}+u^L_i, m_i) \begin{Bmatrix} \boldsymbol{\mu}^{(i-1)}_{\uparrow} & ; \boldsymbol{\nu}^{(i-1)} \\ \boldsymbol{\mu}^{(i-1)*}_{\uparrow} & ; \boldsymbol{\nu}^{(i-1)*} \end{Bmatrix},$$

after noting that $d^L_0(i) = d^L_0(i-1)$. As in the $i \neq t_k$ case, we obtain that $\boldsymbol{\mu}^{(i-1)}$, $\boldsymbol{\mu}^{(i-1)*}$, $\boldsymbol{\nu}^{(i-1)}$, $\boldsymbol{\nu}^{(i-1)*}$ are an interfacial mazy-four sandwiching $(a^{(i-1)}, b^{(i-1)})$, and that they are actually interfacial once it is established that $\mu^{(i-1)}_{\eta^L}$ is interfacial. This is done below.

In the case $\sigma^L_{\eta^L} < i < \tau^L_{\eta^L}$, we have $\eta^{L\prime} = \eta^L$ and $\Delta^L_{\eta^L} = -(-1)^{e^{(i-1)}} = (-1)^{1-e^{(i)}}$ from (7.5), and via Lemma 7.7, $a^{(i-1)} = a^{(i)\prime} + \Delta^L_{\eta^L}$. Note that if $k = n$ (necessarily $\eta^L = 1$) then $\boldsymbol{\mu}^{(i-1)}_{\uparrow} = \boldsymbol{\mu}^{(i-1)*}_{\uparrow} = ()$; if $k < n$ and $\eta^L = 1$ then $\boldsymbol{\mu}^{(i-1)}_{\uparrow} = (\mu^{(i-1)}_0)$ and $\boldsymbol{\mu}^{(i-1)*}_{\uparrow} = (\mu^{(i-1)*}_0)$, with $\mu^{(i-1)}_0 \neq a^{(i-1)} \neq \mu^{(i-1)*}_0$ direct from the definitions after noting that $i < t_n$; and if $k < n$ and $\eta^L = j$ for $j > 1$ then $\boldsymbol{\mu}^{(i-1)}_{\uparrow} = (\ldots,\mu^{(i-1)}_{j-1})$ and $\boldsymbol{\mu}^{(i-1)*}_{\uparrow} = (\ldots,\mu^{(i-1)*}_{j-1})$, with $\mu^{(i-1)}_{j-1} \neq a^{(i-1)} \neq \mu^{(i-1)*}_{j-1}$ direct from the definitions after noting that $i < \tau^L_j$. Then Lemma 6.4 yields:

$$\tilde{\chi}^{p^{(i-1)},p^{(i-1)\prime}}_{a^{(i-1)},b^{(i-1)},e^{(i-1)},f^{(i-1)}}(m_{i-1}, m_i + 2\delta^{p^{(i-1)},p^{(i-1)\prime}}_{a^{(i-1)},e^{(i-1)}}) \begin{Bmatrix} \boldsymbol{\mu}^{(i-1)}_{\uparrow}, a^{(i-1)} + \Delta^L_{\eta^L} & ; \boldsymbol{\nu}^{(i-1)} \\ \boldsymbol{\mu}^{(i-1)*}_{\uparrow}, a^{(i-1)} - \Delta^L_{\eta^L} & ; \boldsymbol{\nu}^{(i-1)*} \end{Bmatrix}$$

$$= q^{-\frac{1}{2}u^L_i(m_{i-1}-m_i+u^L_i+\Delta^L_{\eta^{L\prime}}(\beta_i - u^R_i \Delta^R_{\eta^{R\prime}}))}$$

$$\times \quad \tilde{\chi}^{p^{(i-1)},p^{(i-1)\prime}}_{a^{(i)},b^{(i-1)},1-e^{(i)},f^{(i-1)}}(m_{i-1}+u^L_i, m_i) \begin{Bmatrix} \boldsymbol{\mu}^{(i-1)}_{\uparrow} & ; \boldsymbol{\nu}^{(i-1)} \\ \boldsymbol{\mu}^{(i-1)*}_{\uparrow} & ; \boldsymbol{\nu}^{(i-1)*} \end{Bmatrix}.$$

Here, if $i = t_n$ and $\sigma^L_1 < t_n$, we have $d^L_0(i) = 1$ and $d^L_0(i-1) = 0$. Also note that $\eta^{L\prime} = \eta^L = 1$. Then $\mu^{(i-1)}_0 = a^{(i-1)} + \Delta^L_1$ and $\mu^{(i-1)*}_0 = a^{(i-1)} - \Delta^L_1$ by Lemma 7.9, whereupon we obtain precisely the expression obtained above in the $i = \tau^L_{\eta^L+1}$ case. Also note that $i+1 \neq \tau^L_1$ again implies, via $\delta^{p^{(i)},p^{(i)\prime}}_{a^{(i)},e^{(i)}} = 0$ and Lemma C.3, that $a^{(i)\prime}$ is interfacial in the $(p^{(i-1)},p^{(i-1)\prime})$-model. Lemma 6.4 shows that $\boldsymbol{\mu}^{(i-1)}$, $\boldsymbol{\mu}^{(i-1)*}$, $\boldsymbol{\nu}^{(i-1)}$, $\boldsymbol{\nu}^{(i-1)*}$ are a mazy-four sandwiching $(a^{(i-1)}, b^{(i-1)})$. Since $\mu^{(i-1)*}_0 = a^{(i)\prime}$ is interfacial, to show that $\boldsymbol{\mu}^{(i-1)}$, $\boldsymbol{\mu}^{(i-1)*}$, $\boldsymbol{\nu}^{(i-1)}$, $\boldsymbol{\nu}^{(i-1)*}$ are an interfacial mazy-four requires only that it be established that $\mu_0^{(i-1)}$ is interfacial. This is done below.

Otherwise for $\sigma^L_{\eta^L} < i < \tau^L_{\eta^L}$, we have $\mu^{(i-1)}_{\eta^L-1} = a^{(i-1)} + \Delta^L_{\eta^L}$ by Lemma 7.9, and $\eta^{L\prime} = \eta^L$, whence use of Lemma 5.5 again yields precisely the expression obtained above in the $i = \tau^L_{\eta^L+1}$ case.

In addition, the same expression clearly also holds in the case $\tau^L_{\eta^L+1} < i \le \sigma^L_{\eta^L}$, for which $u^L_i = 0$ and $\eta^{L\prime} = \eta^L$ (in the $i = \sigma^L_{\eta^L}$ case, note that $k(i-1) = k-1 = k^L(i-1)$ and consequently $e^{(i-1)} = 1 - e^{(i)}$ by (7.5)).

Note that in the case $\sigma^L_{\eta^L} < i < \tau^L_{\eta^L}$, we have $e^{(i-1)} = e^{(i)}$ from (7.5). Lemma 6.4 then implies that $\delta^{p^{(i-1)},p^{(i-1)\prime}}_{a^{(i-1)},e^{(i-1)}} = 0$ in this case. Consequently, $\delta^{p^{(i-1)},p^{(i-1)\prime}}_{a^{(i-1)},e^{(i-1)}} = 1$ only in the case $i = \tau^L_{\eta^L+1}$, and then only if $\mu^{(i-1)}_{\eta^L-1} = a^{(i-1)} - (-1)^{e^{(i-1)}}$.

Lemma 6.4 also implies that:

$$\alpha^{p^{(i-1)},p^{(i-1)\prime}}_{a^{(i-1)},b^{(i-1)}} = \alpha_i + \beta_i - u^R_i\Delta^R_{\eta^{R\prime}} + u^L_i\Delta^L_{\eta^{L\prime}};$$

$$\beta^{p^{(i-1)},p^{(i-1)\prime}}_{a^{(i-1)},b^{(i-1)},e^{(i-1)},f^{(i-1)}} = \beta_i - u^R_i\Delta^R_{\eta^{R\prime}} + u^L_i\Delta^L_{\eta^{L\prime}};$$

$$\left\lfloor \frac{a^{(i-1)}p^{(i-1)}}{p^{(i-1)\prime}} \right\rfloor = \begin{cases} a^{(i)} - 1 - \tilde{a}^{(i)} + \Delta^L_{\eta^{L\prime}} & \text{if } i = \tau^L_{\eta^L+1} \text{ and } e^{(i)} \ne e^{(i-1)}; \\ a^{(i)} - 1 - \tilde{a}^{(i)} & \text{otherwise.} \end{cases}$$

Combining all the above cases for $i = t_k$ gives:

$$\tilde{\chi}^{p^{(i-1)},p^{(i-1)\prime}}_{a^{(i-1)},b^{(i-1)},e^{(i-1)},f^{(i-1)}}\left(m_{i-1}, m_i + 2\delta^{p^{(i-1)},p^{(i-1)\prime}}_{a^{(i-1)},e^{(i-1)}}\right)\begin{Bmatrix}\boldsymbol{\mu}^{(i-1)} & ;\boldsymbol{\nu}^{(i-1)} \\ \boldsymbol{\mu}^{(i-1)*} & ;\boldsymbol{\nu}^{(i-1)*}\end{Bmatrix}$$

$$= \sum q^{\frac14\hat{\boldsymbol{m}}^{(i)T}\boldsymbol{C}\hat{\boldsymbol{m}}^{(i)} + \frac14 m^2_{i-1} - \frac12 m_{i-1}m_i - \frac12(\boldsymbol{u}^L_{(\flat,k-1)} + \boldsymbol{u}^R_{(\sharp,k-1)})\cdot\boldsymbol{m}^{(i-1)} + \frac14\gamma''_{i-1}}$$

$$\times \prod_{j=i}^{t-1} \begin{bmatrix} m_j - \frac12(\boldsymbol{C}^*\hat{\boldsymbol{m}}^{(i-1)} - \boldsymbol{u}^L - \boldsymbol{u}^R)_j \\ m_j \end{bmatrix}_q$$

$$= F^{(i-1)}(\boldsymbol{u}^L, \boldsymbol{u}^R, m_{i-1}, m_i),$$

which is the required result when $i = t_k$, since $k(i-1) = k-1$.

In this $i = t_k$ case, making use of (1.25), (1.26), and Lemma 7.8, we also immediately obtain:

$$\alpha^{p^{(i-1)},p^{(i-1)\prime}}_{a^{(i-1)},b^{(i-1)}} = \alpha''_{i-1};$$

$$\beta^{p^{(i-1)},p^{(i-1)\prime}}_{a^{(i-1)},b^{(i-1)},e^{(i-1)},f^{(i-1)}} = \beta'_{i-1};$$

$$\left\lfloor \frac{a^{(i-1)}p^{(i-1)}}{p^{(i-1)\prime}} \right\rfloor = \tilde{a}^{(i-1)};$$

$$\left\lfloor \frac{b^{(i-1)}p^{(i-1)}}{p^{(i-1)\prime}} \right\rfloor = \tilde{b}^{(i-1)}.$$

It remains to consider the four final bulleted items of the induction statement. First consider $t_k \le i-1$ and $\tau^L_{j+1} \le i-1 < t_{k+1} < \sigma^L_j$. If $i < t_{k+1}$ then $a^{(i)}$ is interfacial in the $(p^{(i)}, p^{(i)\prime})$-model by the induction hypothesis. Then Lemma

C.1 implies that $a^{(i)\prime}$ is interfacial in the $(p^{(i-1)}, p^{(i-1)\prime})$-model. If $i = t_{k+1}$ first note that since $i \neq \tau_j^L$, the induction hypothesis yields $\delta^{p^{(i)},p^{(i)\prime}}_{a^{(i)},e^{(i)}} = 0$. Then Lemma C.3 implies that $a^{(i)\prime}$ is interfacial in the $(p^{(i-1)}, p^{(i-1)\prime})$-model. In both cases, $a^{(i-1)} = a^{(i)\prime}$ by Lemma 7.7 and thus $a^{(i-1)}$ is interfacial in the $(p^{(i-1)}, p^{(i-1)\prime})$-model as required.

Now consider $t_k \leq i-1$ and $\tau_{j+1}^L \leq i-1 < t_{k+1} = \sigma_j^L$ with $a^{(i-1)}$ not interfacial in the $(p^{(i-1)}, p^{(i-1)\prime})$-model. $a^{(i-1)} = a^{(i)\prime}$ by Lemma 7.7. In the case $i < t_{k+1}$ note that $i+1 \neq \tau_j^L$, whereupon the induction hypothesis yields $\delta^{p^{(i)},p^{(i)\prime}}_{a^{(i)},e^{(i)}} = 0$, and then Lemma C.1 implies that $a^{(i)}$ is not interfacial in the $(p^{(i)}, p^{(i)\prime})$-model. The induction hypothesis then implies that $\tau_j^L = \sigma_j^L + 1$ as required. In the case $i = t_{k+1}$ $(= \sigma_j^L)$, Lemma C.3 implies that $\delta^{p^{(i)},p^{(i)\prime}}_{a^{(i)},e^{(i)}} = 1$. The induction hypothesis then implies that $i+1 = \tau_j^L$ as required.

An entirely analogous argument yields the next bulleted item.

We now tackle the remaining cases of the final two bulleted items together (we only consider the cases with superscript 'L': those with superscript 'R' follow similarly). The remaining cases arise when $d_0^L(i-1) < d_0^L(i)$ and $\eta^L(i-1) > \eta^L(i)$. In the former of these cases, necessarily $\sigma_1^L < i = t_n$ when $d_0^L(i-1) = 0$ and $d_0^L(i) = 1$, and thus here we set $j = 0$. In the latter, necessarily $i = \tau^L_{\eta^L+1}$ when $\eta^L(i-1) = \eta^L + 1$ and $\eta^L(i) = \eta^L$, and thus here we set $j = \eta^L$ (>0). In either case, the above analysis has shown that $\mu_j^{(i-1)*} = a^{(i-1)} - \Delta_j^L = a^{(i)\prime}$ is interfacial. It remains to show that $\mu_j^{(i-1)} = a^{(i)\prime} + 2\Delta_{j+1}^L$ is interfacial and that $\rho^{(i-1)}(\mu_j^{(i-1)*}) = \tilde{\mu}_j^{(i-1)*}$ and $\rho^{(i-1)}(\mu_j^{(i-1)}) = \tilde{\mu}_j^{(i-1)}$.

In the $j = 0$ case (when $\sigma_1^L < i = t_n$), the definitions directly give $\mu_0^{(i-1)} = 0$ if $\Delta_1^L = -1$ and $\mu_0^{(i-1)} = p^{(i-1)\prime}$ if $\Delta_1^L = 1$, so that $\mu_0^{(i-1)}$ is certainly interfacial. We then immediately have $\rho^{(i-1)}(\mu_0^{(i-1)}) = 0 = \tilde{\mu}_0^{(i-1)}$ if $\Delta_1^L = -1$ and $\rho^{(i-1)}(\mu_0^{(i-1)}) = p^{(i-1)} = \tilde{\mu}_0^{(i-1)}$ if $\Delta_1^L = 1$, as required. It follows from Note D.4 that $\rho^{(i-1)}(\mu_0^{(i-1)*}) = 1 = \tilde{\mu}_0^{(i-1)*}$ if $\Delta_1^L = -1$ and $\rho^{(i-1)}(\mu_0^{(i-1)*}) = p^{(i-1)} - 1 = \tilde{\mu}_0^{(i-1)*}$ if $\Delta_1^L = 1$, as required.

For the $j > 0$ cases, if we can show that $\rho^{(i-1)}(\mu_j^{(i-1)*}) = \tilde{\mu}_j^{(i-1)*}$ and that $a^{(i)\prime} + 2\Delta_{j+1}^L$ is interfacial then, via Note D.4, it will follow that $\rho^{(i-1)}(\mu_j^{(i-1)}) = \rho^{(i-1)}(\mu_j^{(i-1)*}) + \Delta_{j+1}^L = \tilde{\mu}_j^{(i-1)*} + \Delta_{j+1}^L = \tilde{\mu}_j^{(i-1)}$, the final equality following from the definition of $\tilde{\mu}_j^{(i-1)}$. To show this, we consider various cases.

With $i = \tau_{j+1}^L$, that $\{\tau_J^L, \sigma_J^L, \Delta_J^L\}_{J=1}^{d^L}$ is a reduced run implies that $i = t_k$ unless $i = t_n - 1 \neq t_{n-1}$.

In the $i = t_n - 1 \neq t_{n-1}$ case, we have $j = 1$, $t_n < \sigma_1^L < t$, $\Delta_2^L \neq \Delta_1^L$ and $p^{(i-1)}/p^{(i-1)\prime} = 3 + 1/c_n$. For $1 \leq r < c_n$, the rth odd band in the $(p^{(i-1)}, p^{(i-1)\prime})$-model lies between heights $3r$ and $3r+1$. We readily obtain $\kappa_{r+2}^{(i-1)} = 3r+1$ and $\tilde{\kappa}_{r+2}^{(i-1)} = r$. Then, if $\Delta_1^L = -1$, we have $a^{(i)\prime} = \kappa^{(i-1)}_{\sigma_1^L - i + 1} = 3(\sigma_1^L - t_n) + 1$ and therefore $\rho^{(i-1)}(\mu_1^{(i-1)*}) = \sigma_1^L - t_n = \tilde{\kappa}^{(i-1)}_{\sigma_1^L - t_n + 2} = \tilde{\mu}_1^{(i-1)*}$. In addition, $\mu_1^{(i-1)} = a^{(i)\prime} + 2\Delta_2^L = a^{(i)\prime} + 2$ is clearly interfacial. If $\Delta_1^L = 1$, we have $a^{(i)\prime} = p^{(i-1)\prime} - \kappa^{(i-1)}_{\sigma_1^L - i + 1} = 3(c_n - \sigma_1^L + t_n)$ and therefore $\rho^{(i-1)}(\mu_1^{(i-1)*}) = c_n - \sigma_1^L + t_n =$

$p^{(i-1)} - \tilde{\kappa}^{(i-1)}_{\sigma_1^L - t_n + 2} = \tilde{\mu}_1^{(i-1)*}$. In addition, $\mu_1^{(i-1)} = a^{(i)\prime} + 2\Delta_2^L = a^{(i)\prime} - 2$ is clearly interfacial.

We now tackle the cases for which $i = \tau_{j+1}^L = t_k$. Since $i+1 \neq \tau_j^L$ we obtain $\delta^{p^{(i)},p^{(i)\prime}}_{a^{(i)},e^{(i)}} = 0$. Now set $\hat{a} = a^{(i)} + \Delta_{j+1}^L$, and set $\hat{e} = e^{(i)}$ if $\lfloor \hat{a}p^{(i)}/p^{(i)\prime} \rfloor = \lfloor a^{(i)}p^{(i)}/p^{(i)\prime} \rfloor$ and $\hat{e} = 1 - e^{(i)}$ otherwise. We claim that $\delta^{p^{(i)},p^{(i)\prime}}_{\hat{a},\hat{e}} = 0$ (this will be established below). Thereupon, invoking Lemma D.2 in the case $\lfloor \hat{a}p^{(i)}/p^{(i)\prime} \rfloor = \lfloor a^{(i)}p^{(i)}/p^{(i)\prime} \rfloor$ or Lemma D.1 in the case $\lfloor \hat{a}p^{(i)}/p^{(i)\prime} \rfloor \neq \lfloor a^{(i)}p^{(i)}/p^{(i)\prime} \rfloor$ shows that $a^{(i)\prime} + 2\Delta_{j+1}^L$ (and $a^{(i)\prime}$) is interfacial in the $(p^{(i-1)}, p^{(i-1)\prime})$-model. Using first the fact that $a^{(i)\prime}$ is interfacial in the $(p^{(i-1)}, p^{(i-1)\prime})$-model, then Lemma 7.8 noting that (7.5) implies $\Delta_{j+1}^L = 2e^{(i-1)} - 1$, then the definitions of $\tilde{a}^{(i-1)}$ and $\tilde{\mu}_j^{(i-1)*}$, yields:

$$\begin{aligned}
\rho^{(i-1)}(a^{(i)\prime}) &= \lfloor a^{(i)\prime}p^{(i-1)}/p^{(i-1)\prime} \rfloor + 1 - e^{(i)} \\
&= a^{(i)} - \tilde{a}^{(i)} - e^{(i)} = \tilde{a}^{(i-1)} + e^{(i-1)} - \Delta_{j+1}^L \\
&= \tilde{\mu}_j^{(i-1)*},
\end{aligned}$$

as required.

To establish our claim that $\delta^{p^{(i)},p^{(i)\prime}}_{\hat{a},\hat{e}} = 0$ requires the consideration of a number of cases. It is convenient to separately treat $c_k > 1$ and $c_k = 1$. Note that because $p^{(i)\prime}/p^{(i)}$ has continued fraction $[c_k + 1, c_{k+1}, \ldots, c_n]$, each neighbouring pair of odd bands in the $(p^{(i)}, p^{(i)\prime})$-model is separated by either c_k or $c_k + 1$ even bands.

For $c_k > 1$, first consider $\sigma_j^L > t_{k+1}$. Here, since $\tau_{j+1}^L = t_k = i < t_{k+1} < \sigma_j^L$, we have that $a^{(i)}$ is interfacial in the $(p^{(i)}, p^{(i)\prime})$-model. Furthermore, since $c_k > 1$, each pair of odd bands is separated by at least two even bands, whereupon $\delta^{p^{(i)},p^{(i)\prime}}_{\hat{a},\hat{e}} = 0$ immediately.

Now consider $c_k > 1$ and $\sigma_j^L \leq t_{k+1}$. Since $k(i) = k^L(i) = k$, we obtain $\Delta_j^L = (-1)^{e^{(i)}}$ from (7.5). We readily calculate $\mu_{j-1}^{(i)} - a^{(i)} = \Delta_j^L \kappa^{(i)}_{\sigma_j^L - i} = \Delta_j^L(\sigma_j^L - i + 1)$ where we have relied on the above continued fraction expansion to evaluate $\kappa^{(i)}_{\sigma_j^L - i}$. Thus $2 \leq |\mu_{j-1}^{(i)} - a^{(i)}| \leq c_k + 1$. Then $\delta^{p^{(i)},p^{(i)\prime}}_{\hat{a},\hat{e}} = 0$ follows because, on the one hand, $\mu_{j-1}^{(i)}$ is interfacial by the induction hypothesis, on the second hand, neighbouring odd bands in the $(p^{(i)}, p^{(i)\prime})$-model are separated by either c_k or $c_k + 1$ even bands, and on the other hand, if $\sigma_j^L = t_k + 1$ then $\Delta_{j+1}^L \neq \Delta_j^L$. Note that this reasoning applies even if $j = 1$ when necessarily $k \leq n - 1$, if we interpret both 0 and $p^{(i)\prime}$ as bordering odd bands.

For $c_k = 1$, first consider $\sigma_j^L = t_{k+1}$. Lemma 7.3 implies that $\Delta_{j+1}^L \neq \Delta_j^L$. Since $k(i) = k^L(i) = k$, we have $\Delta_{j+1}^L = -\Delta_j^L = -(-1)^{e^{(i)}}$ via (7.5). If $a^{(i)}$ is interfacial then $\delta^{p^{(i)},p^{(i)\prime}}_{a^{(i)},e^{(i)}} = 0$ implies that $\lfloor \hat{a}p^{(i)}/p^{(i)\prime} \rfloor \neq \lfloor a^{(i)}p^{(i)}/p^{(i)\prime} \rfloor$ and therefore $\delta^{p^{(i)},p^{(i)\prime}}_{\hat{a},\hat{e}} = 0$. If $a^{(i)}$ is not interfacial then $\lfloor \hat{a}p^{(i)}/p^{(i)\prime} \rfloor = \lfloor a^{(i)}p^{(i)}/p^{(i)\prime} \rfloor$, which since $\hat{a} = a^{(i)} + \Delta_{j+1}^L = a^{(i)} - (-1)^{\hat{e}}$ immediately implies that $\delta^{p^{(i)},p^{(i)\prime}}_{\hat{a},\hat{e}} = 0$.

Now consider $c_k = 1$ and $\sigma_j^L = t_{k+1} + 1$. Lemma 7.3 implies that $\Delta_{j+1}^L = \Delta_j^L$. Since $k^L(i) = k + 1$ and $k(i) = k$, we have $\Delta_{j+1}^L = \Delta_j^L = -(-1)^{e^{(i)}}$. Then $\delta^{p^{(i)},p^{(i)\prime}}_{\hat{a},\hat{e}} = 0$ as in the previous case.

Now consider $c_k = 1$ and $\sigma_j^L > t_{k+1} + 1$. Lemma 7.3 implies that in fact, $\sigma_j^L > t_{k+2}$. Here, since $\tau_{j+1}^L = t_k = i < t_{k+1} < \sigma_j^L$, the induction hypothesis implies that $a^{(i)}$ is interfacial in the $(p^{(i)}, p^{(i)\prime})$-model. Now, however, $\delta_{\hat{a},\hat{e}}^{p^{(i)},p^{(i)\prime}} = 0$ is immediate only in the $\lfloor \hat{a}p^{(i)}/p^{(i)\prime} \rfloor \neq \lfloor a^{(i)}p^{(i)}/p^{(i)\prime} \rfloor$ case. In the case $\lfloor \hat{a}p^{(i)}/p^{(i)\prime} \rfloor = \lfloor a^{(i)}p^{(i)}/p^{(i)\prime} \rfloor$, since $a^{(i)}$ is interfacial and $\delta_{a^{(i)},e^{(i)}}^{p^{(i)},p^{(i)\prime}} = 0$, it follows that $\hat{a} = a^{(i)} + (-1)^{e^{(i)}}$. Then $\Delta_{j+1}^L = (-1)^{e^{(i)}} = -(-1)^{e^{(i+1)}}$, the final equality arising from (7.5) because $i+1 = t_{k+1} < \sigma_j^L$ implies that $k(i+1) = k+1$ and $k^L(i+1) = k^L(i)$. Again via the induction hypothesis, $\tau_{j+1}^L < t_{k+1} = i+1 < t_{k+2} < \sigma_j^L$ implies that $a^{(i+1)}$ is interfacial in the $(p^{(i)}, p^{(i)\prime})$-model and $i+2 < \sigma_j^L$ implies that $\delta_{a^{(i+1)},e^{(i+1)}}^{p^{(i+1)},p^{(i+1)\prime}} = 0$, so that $\lfloor (a^{(i+1)} + \Delta_{j+1}^L)p^{(i+1)}/p^{(i+1)\prime} \rfloor \neq \lfloor a^{(i+1)}p^{(i+1)}/p^{(i+1)\prime} \rfloor$. Invoking Lemma D.1 with $p' = p^{(i+1)\prime}$, $p = p^{(i)}$, and $\{a, \hat{a}\} = \{a^{(i+1)}, a^{(i+1)} + \Delta_{j+1}^L\}$ now shows that $\lfloor a^{(i)}p^{(i)}/p^{(i)\prime} \rfloor = \lfloor (a^{(i)} + 2\Delta_{j+1}^L)p^{(i)}/p^{(i)\prime} \rfloor$, whereupon, certainly $\delta_{\hat{a},\hat{e}}^{p^{(i)},p^{(i)\prime}} = 0$.

This proves the ninth bulleted item for all i with $0 \le i < \tau_{j+1}^L$ when $j > 0$ and all i with $0 \le i < t_n$ when $j = 0$. An entirely analogous argument gives the remaining bulleted item.

We now see that our proposition holds for i replaced by $i-1$, and thus the lemma is proved by induction. □

Before taking the final step to proving Theorem 7.1, we need the following result concerning the quantities defined in Section 1.10, for the vector $\boldsymbol{u} = \boldsymbol{u}^L + \boldsymbol{u}^R$.

Lemma 7.12. *For* $0 \le j \le t$,

$$\alpha_j'' \equiv Q_j \ (mod 2);$$
$$\beta_j' \equiv Q_j - Q_{j+1} \ (mod 2).$$

Proof: Since $\alpha_t'' = 0$, $\beta_t' = 0$ and $Q_t = Q_{t+1} = 0$, this result is manifest for $j = t$.

We now proceed by downward induction. Thus assume the result holds for a particular $j > 0$. When $j \neq t_{k(j)}$, equations (1.25) and (1.27) imply that $\beta_{j-1}' \equiv \beta_j' + (\boldsymbol{u}^L)_j - (\boldsymbol{u}^R)_j$. Equations (1.33), (1.32) and (1.30) imply that $Q_{j-1} \equiv Q_{j+1} - (\boldsymbol{u}^L)_j - (\boldsymbol{u}^R)_j$. Thus the induction hypothesis immediately gives $\beta_{j-1}' \equiv Q_{j-1} - Q_j$ in this case.

When $j = t_{k(j)}$, equations (1.25) and (1.27) imply that $\beta_{j-1}' \equiv \alpha_j'' - \beta_j' + (\boldsymbol{u}^L)_j - (\boldsymbol{u}^R)_j$. Equations (1.33), (1.32) and (1.29) imply that $Q_{j-1} \equiv Q_j + Q_{j+1} - (\boldsymbol{u}^L)_j - (\boldsymbol{u}^R)_j$. Thus the induction hypothesis also gives $\beta_{j-1}' \equiv Q_{j-1} - Q_j$ in this case.

In both cases, equations (1.26) and (1.27) give $\alpha_{j-1}'' = \alpha_j'' + \beta_{j-1}'$, whence the induction hypothesis together with the above result immediately gives $\alpha_{j-1}'' \equiv Q_{j-1}$ as required. □

In order to isolate the $i = 0$ case of Lemma 7.11, we note that comparison of the definitions of Sections 7.1 and 7.2, gives $a = a^{(0)}$, $b = b^{(0)}$, $e = e^{(0)}$, $f = f^{(0)}$, $p = p^{(0)}$, $p' = p^{(0)\prime}$, $\boldsymbol{\mu} = \boldsymbol{\mu}^{(0)}$, $\boldsymbol{\mu}^* = \boldsymbol{\mu}^{(0)*}$, $\boldsymbol{\nu} = \boldsymbol{\nu}^{(0)}$, $\boldsymbol{\nu}^* = \boldsymbol{\nu}^{(0)*}$, and since $k(0) = 0$, $\boldsymbol{u}_{(\flat,k(0))}^L = \boldsymbol{u}_\flat^L$ and $\boldsymbol{u}_{(\sharp,k(0))}^R = \boldsymbol{u}_\sharp^R$. We now readily obtain:

Corollary 7.13. *Let $p' > 2p$, $m_0 \equiv \alpha^{p,p'}_{a,b}$ (mod2) and $m_1 \equiv \alpha^{p,p'}_{a,b} + \beta^{p,p'}_{a,b,e,f}$ (mod2). Then:*

$$\tilde{\chi}^{p,p'}_{a,b,e,f}(m_0, m_1 + 2\delta^{p,p'}_{a,e}) \left\{ \begin{matrix} \boldsymbol{\mu} \,; \boldsymbol{\nu} \\ \boldsymbol{\mu}^*; \boldsymbol{\nu}^* \end{matrix} \right\}$$

$$= \sum q^{\frac{1}{4}\hat{\boldsymbol{m}}^T \boldsymbol{C}\hat{\boldsymbol{m}} - \frac{1}{4}m_0^2 - \frac{1}{2}(\boldsymbol{u}^L_\flat + \boldsymbol{u}^R_\sharp)\cdot \boldsymbol{m} + \frac{1}{4}\gamma_0} \prod_{j=1}^{t-1} \begin{bmatrix} m_j - \frac{1}{2}(\boldsymbol{C}^*\hat{\boldsymbol{m}} + \boldsymbol{u}^L + \boldsymbol{u}^R)_j \\ m_j \end{bmatrix}_q,$$

with the sum to be taken over all $(m_2, m_3, \ldots, m_{t-1}) \equiv (Q_2, Q_3, \ldots, Q_{t-1})$, with $\boldsymbol{m} = (m_1, m_2, m_3, \ldots, m_{t-1})$ and $\hat{\boldsymbol{m}} = (m_0, m_1, m_2, m_3, \ldots, m_{t-1})$. For the case $t = 2$, the summation in the above expression is omitted. For $t = 1$, we have:

$$\tilde{\chi}^{p,p'}_{a,b,e,f}(m_0, m_1) \left\{ \begin{matrix} \boldsymbol{\mu} \,; \boldsymbol{\nu} \\ \boldsymbol{\mu}^*; \boldsymbol{\nu}^* \end{matrix} \right\} = q^{\frac{1}{4}m_0^2 + \frac{1}{4}\gamma_0} \delta_{m_1,0}.$$

(In this $t = 1$ case, necessarily $\delta^{p,p'}_{a,e} = 0$.)

In addition, $\boldsymbol{\mu}, \boldsymbol{\mu}^*, \boldsymbol{\nu}, \boldsymbol{\nu}^*$ are an interfacial mazy-four in the (p,p')-model sandwiching (a,b), with $\rho^{p,p'}(\mu_j) = \tilde{\mu}_j$ and $\rho^{p,p'}(\mu^*_j) = \tilde{\mu}^*_j$ for $d^L_0 \le j < d^L$, and $\rho^{p,p'}(\nu_j) = \tilde{\nu}_j$ and $\rho^{p,p'}(\nu^*_j) = \tilde{\nu}^*_j$ for $d^R_0 \le j < d^R$.

Also, $\delta^{p,p'}_{b,f} = 1$ only if $\tau^R_{d^R} = 1$ and $\nu_{d^R-1} = b - (-1)^f$. If $\sigma^R_{d^R} > t_1$ then b is interfacial in the (p,p')-model. If $\sigma^R_{d^R} = t_1$ and b is not interfacial in the (p,p')-model then $\tau^R_{d^R} = t_1 + 1$.

If b is interfacial and $\sigma^R_{d^R} > 0$ then

$$\rho^{p,p'}(b) = \sum_{m=2}^{d^R} \Delta^R_m (\tilde{\kappa}_{\tau^R_m} - \tilde{\kappa}_{\sigma^R_m}) + \begin{cases} \tilde{\kappa}_{\sigma^R_1} & \text{if } \Delta^R_1 = -1; \\ p - \tilde{\kappa}_{\sigma^R_1} & \text{if } \Delta^R_1 = +1. \end{cases}$$

If b is not interfacial and $\sigma^R_{d^R} > 0$ then

$$\left\lfloor \frac{bp}{p'} \right\rfloor = \sum_{m=2}^{d^R} \Delta^R_m (\tilde{\kappa}_{\tau^R_m} - \tilde{\kappa}_{\sigma^R_m}) - \frac{1}{2}(1 - \Delta^R_{d^R}) + \begin{cases} \tilde{\kappa}_{\sigma^R_1} & \text{if } \Delta^R_1 = -1; \\ p - \tilde{\kappa}_{\sigma^R_1} & \text{if } \Delta^R_1 = +1. \end{cases}$$

Proof: Lemma 7.11 gives $\alpha^{p,p'}_{a,b} = \alpha''_0 = \alpha_0$ and $\beta^{p,p'}_{a,b,e,f} = \beta'_0 = \beta_0$, whereupon, the first statement follows from Lemma 7.12. The statements in the second and third paragraphs follow from the $i = 0$ case of Lemma 7.11 after noting that $t_0 < 0 = \tau^R_{d^R+1} < t_1$. For the final paragraph, the $i = 0$ case of Lemma 7.11 implies that if $\sigma^R_{d^R} > 0$ then

$$\left\lfloor \frac{bp}{p'} \right\rfloor = \tilde{b}^{(0)} = -f + \sum_{m=2}^{d^R} \Delta^R_m (\tilde{\kappa}_{\tau^R_m} - \tilde{\kappa}_{\sigma^R_m}) + \begin{cases} \tilde{\kappa}_{\sigma^R_1} & \text{if } \Delta^R_1 = -1; \\ p - \tilde{\kappa}_{\sigma^R_1} & \text{if } \Delta^R_1 = +1. \end{cases}$$

Note that $\sigma^R_{d^R} > 0$ implies that $\tau^R_{d^R} > 1$ which, from above, implies that $\delta^{p,p'}_{b,f} = 0$. Then if b is interfacial, $\rho^{p,p'}(b) = \lfloor bp/p' \rfloor + f$ giving the desired result in this case. If b is not interfacial then, from above, $\sigma^R_{d^R} \le t_1$ and therefore $k^R(0) = 0$. Since $k(0) = 0$, (7.6) yields $\Delta^R_{d^R} = (-1)^f = 1 - 2f$, from which the desired result follows. □

Proof of Theorem 7.1: For reduced runs, Theorem 7.1 now follows from Lemma 5.4 and Corollary 7.13 since $\alpha^{p,p'}_{a,b} = b - a$ and, via Lemma 7.12, $\alpha^{p,p'}_{a,b} + \beta^{p,p'}_{a,b,e,f} \equiv$

$Q_1 \pmod 2$. That it holds in the case of arbitrary naive runs then follows from Lemma 7.4. □

For later convenience, we show here that the values $\{\mu_j^*\}_{j=1}^{d^L-1}$ and $\{\nu_j^*\}_{j=1}^{d^R-1}$ that are obtained from *naive* runs are interfacial. It is not possible using the results of this section to show that all $\{\mu_j\}_{j=1}^{d^L-1}$ and $\{\nu_j\}_{j=1}^{d^R-1}$ are also interfacial. However, that they actually are will be deduced in Section 9.2.

Corollary 7.14. *Let $\{\tau_j^L, \sigma_j^L, \Delta_j^L\}_{j=1}^{d^L}$ and $\{\tau_j^R, \sigma_j^R, \Delta_j^R\}_{j=1}^{d^R}$ be naive runs.*

If $1 \le j < d^L$ then μ_j^* is interfacial with $\rho^{p,p'}(\mu_j^*) = \tilde{\mu}_j^*$. In addition, if $1 \le j < d^L$, and $\tau_{j+1}^L \ne \sigma_{j+1}^L$ then μ_j is interfacial with $\rho^{p,p'}(\mu_j) = \tilde{\mu}_j$.

If $1 \le j < d^R$ then ν_j^* is interfacial with $\rho^{p,p'}(\nu_j^*) = \tilde{\nu}_j^*$. In addition, if $1 \le j < d^R$, and $\tau_{j+1}^R \ne \sigma_{j+1}^R$ then ν_j is interfacial with $\rho^{p,p'}(\nu_j) = \tilde{\nu}_j$.

Proof: That this holds for reduced runs is immediate from Corollary 7.13. If $\tau_j^L = \sigma_j^L$ then $\mu_j^* = \mu_{j-1}^*$ and $\tilde{\mu}_j^* = \tilde{\mu}_{j-1}^*$. Since $\tau_j^L = \sigma_j^L$ cannot occur for $j = 1$, the required result follows in the μ_j^* case. With $\tau_{j+1}^L \ne \sigma_{j+1}^L$, the result for the μ_j cases follows immediately from that for reduced runs.

The final paragraph follows in the same way. □

7.4. Transferring to the original weighting

In this and the following sections, we fix naive runs $\mathcal{X}^L = \{\tau_j^L, \sigma_j^L, \Delta_j^L\}_{j=1}^{d^L}$ and $\mathcal{X}^R = \{\tau_j^R, \sigma_j^R, \Delta_j^R\}_{j=1}^{d^R}$, and make use of the definitions of Section 7.1. The set $\mathcal{P}_{a,b,c}^{p,p'}(L)\left\{\begin{smallmatrix}\boldsymbol{\mu}\ ;\ \boldsymbol{\nu}\\ \boldsymbol{\mu}^*;\ \boldsymbol{\nu}^*\end{smallmatrix}\right\}$ and its generating function $\chi_{a,b,c}^{p,p'}(L)\left\{\begin{smallmatrix}\boldsymbol{\mu}\ ;\boldsymbol{\nu}\\ \boldsymbol{\mu}^*;\boldsymbol{\nu}^*\end{smallmatrix}\right\}$ are defined in Section 5.4.

The following lemma relates this generating function, defined in terms of the original weight function (2.1), to that considered in Theorem 7.1 (the latter generating function was defined in (5.1) in terms of the modified weight function (2.5)).

Lemma 7.15. *Let $p' > 2p$. Set $c = b + \Delta_{d^R}^R$.*

If $\sigma_{d^R}^R = 0$ and either $\lfloor pb/p' \rfloor = \lfloor pc/p' \rfloor$ or $c \in \{0, p'\}$ then:

$$\chi_{a,b,c}^{p,p'}(L)\left\{\begin{matrix}\boldsymbol{\mu}\ ;\boldsymbol{\nu}\\ \boldsymbol{\mu}^*;\boldsymbol{\nu}^*\end{matrix}\right\} = q^{-\frac{1}{2}(L+\Delta_{d^R}^R(a-b))}\tilde{\chi}_{a,b,e,f}^{p,p'}(L)\left\{\begin{matrix}\boldsymbol{\mu}\ ;\boldsymbol{\nu}\\ \boldsymbol{\mu}^*;\boldsymbol{\nu}^*\end{matrix}\right\}.$$

Otherwise:

$$\chi_{a,b,c}^{p,p'}(L)\left\{\begin{matrix}\boldsymbol{\mu}\ ;\boldsymbol{\nu}\\ \boldsymbol{\mu}^*;\boldsymbol{\nu}^*\end{matrix}\right\} = \tilde{\chi}_{a,b,e,f}^{p,p'}(L)\left\{\begin{matrix}\boldsymbol{\mu}\ ;\boldsymbol{\nu}\\ \boldsymbol{\mu}^*;\boldsymbol{\nu}^*\end{matrix}\right\}.$$

Proof: Let $\tilde{h} \in \mathcal{P}_{a,b,e,f}^{p,p'}(L)\left\{\begin{smallmatrix}\boldsymbol{\mu}\ ;\boldsymbol{\nu}\\ \boldsymbol{\mu}^*;\boldsymbol{\nu}^*\end{smallmatrix}\right\}$, and let $h \in \mathcal{P}_{a,b,c}^{p,p'}(L)\left\{\begin{smallmatrix}\boldsymbol{\mu}\ ;\boldsymbol{\nu}\\ \boldsymbol{\mu}^*;\boldsymbol{\nu}^*\end{smallmatrix}\right\}$ have the same sequence of heights: $h_i = \tilde{h}_i$ for $0 \le i \le L$.

First consider $\sigma_{d^R}^R = 0$. The $i = 1$ case of Lemma 7.9 implies that $\nu_{d^R-1} = b + \Delta_{d^R}^R$ and hence $\tilde{h}_{L-1} = b - \Delta_{d^R}^R$. The definition of f (in Section 7.1) gives $\Delta_{d^R}^R = -(-1)^f$. Then, by (2.4), the Lth vertex of $\tilde{h}$ is scoring with weight $\frac{1}{2}(L + \Delta_{d^R}^R(a-b))$. On the other hand, with $c = b + \Delta_{d^R}^R$, if $\lfloor pb/p' \rfloor = \lfloor pc/p' \rfloor$ and $0 < c < p'$ then Table 2.1 shows that the Lth vertex of h is non-scoring. This is also the case if $c \in \{0, p'\}$ because the 0th and $(p'-1)$th bands are defined to be even when $p' > 2p$.

Therefore $wt(h) = \tilde{wt}(\tilde{h}) - \frac{1}{2}(L + \Delta^R_{dR}(a-b))$. In the case where $0 < c < p'$ and $\lfloor pb/p' \rfloor \neq \lfloor pc/p' \rfloor$, the Lth vertex is scoring in both $\tilde{h}$ and h, and therefore $wt(h) = \tilde{wt}(\tilde{h})$. The required identities between generating functions then follow whenever $\sigma^R_{dR} = 0$.

For $\sigma^R_{dR} > 0$, note that $\tau^R_{dR} > \sigma^R_{dR}$ implies that $\tau^R_{dR} > 1$ and hence, via Corollary 7.13, that $\delta^{p,p'}_{b,f} = 0$.

In the case $0 < \sigma^R_{dR} \leq t_1$, we have $k^R = 0$ and therefore $\Delta^R_{dR} = (-1)^f$. From $\delta^{p,p'}_{b,f} = 0$ then follows that $wt(h) = \tilde{wt}(\tilde{h})$, and hence the required identity between the generating functions results.

If $\sigma^R_{dR} > t_1$, then b is interfacial by Corollary 7.13. Again $\delta^{p,p'}_{b,f} = 0$ implies that $wt(h) = \tilde{wt}(\tilde{h})$ whereupon the lemma follows. □

The following theorem specifies the subset of $\mathcal{P}^{p,p'}_{a,b,c}(L)$ (weighted as in (2.2)) for which the generating function is $F(\boldsymbol{u}^L, \boldsymbol{u}^R, L)$. It deals with both the cases $p' > 2p$ and $p' < 2p$.

Theorem 7.16. *If b is interfacial, let $c \in \{b \pm 1\}$. Otherwise if b is not interfacial, set $c = b + \Delta^R_{dR}$. If $L \equiv b - a \pmod 2$ then:*

$$\chi^{p,p'}_{a,b,c}(L) \begin{Bmatrix} \boldsymbol{\mu}\ ; \boldsymbol{\nu} \\ \boldsymbol{\mu}^*; \boldsymbol{\nu}^* \end{Bmatrix} = F(\boldsymbol{u}^L, \boldsymbol{u}^R, L). \tag{7.13}$$

Proof: First consider $p' > 2p$ and $c = b + \Delta^R_{dR}$.

If $\sigma^R_{dR} = 0$ and $\lfloor bp/p' \rfloor = \lfloor cp/p' \rfloor$ with $0 < c < p'$ then $\gamma' = \gamma_0 - 2(L + \Delta^R_{dR}(a-b))$ by the definition in Section 1.10. This is also the case if $c = 0$ (when $b = 1$ and $\Delta^R_{dR} = -1$) or $c = p'$ (when $b = p'-1$ and $\Delta^R_{dR} = 1$). Otherwise $\gamma' = \gamma_0$. Expression (7.13) then follows from Theorem 7.1 and Lemma 7.15.

If b is interfacial then Table 2.1 shows that a path $h \in \mathcal{P}^{p,p'}_{a,b,c}(L)$ has the same generating function whether $c = b+1$ or $c = b-1$. The theorem then follows in the $p' > 2p$ case.

Now consider $p' < 2p$. If $h \in \mathcal{P}^{p,p'}_{a,b,c}(L)$, and $\hat{h} \in \mathcal{P}^{p'-p,p'}_{a,b,c}(L)$ is defined by $\hat{h}_i = h_i$ for $0 \leq i \leq L$, then $wt(\hat{h}) = \frac{1}{4}(L^2 - \alpha^2) - wt(h)$, where $\alpha = b - a$, by a direct analogue of Lemma 4.1 (here, as there, each scoring vertex of $\hat{h}$ corresponds to a non-scoring vertex of h and vice-versa). Since $p' > 2(p'-p)$, we thus obtain:

$$\chi^{p,p'}_{a,b,c}(L) \begin{Bmatrix} \boldsymbol{\mu}\ ; \boldsymbol{\nu} \\ \boldsymbol{\mu}^*; \boldsymbol{\nu}^* \end{Bmatrix} = q^{\frac{1}{4}(L^2-\alpha^2)} F^{p'-p,p'}(\boldsymbol{u}^L, \boldsymbol{u}^R, L; q^{-1}), \tag{7.14}$$

where $F^{p'-p,p'}(\boldsymbol{u}^L, \boldsymbol{u}^R, L; q^{-1})$ is $F(\boldsymbol{u}^L, \boldsymbol{u}^R, L; q^{-1})$ with all the quantities employed in its definition by (1.40) pertaining to the continued fraction of $p'/(p'-p)$. Below, we make use of $\boldsymbol{C}^{p'-p,p'}$ and $(\boldsymbol{u}^L_\flat + \boldsymbol{u}^R_\sharp)^{p'-p,p'}$ and $\gamma(\mathcal{X}^L, \mathcal{X}^R)^{p'-p,p'}$ defined similarly.

Now note that $\hat{\boldsymbol{m}}^T \boldsymbol{C}^{p'-p,p'} \hat{\boldsymbol{m}} = \hat{\boldsymbol{m}} \boldsymbol{C}^{p,p'} \hat{\boldsymbol{m}} + L^2 - 2Lm_1$,

$$(\boldsymbol{u}^L_\flat + \boldsymbol{u}^R_\sharp)^{p'-p,p'} \cdot \boldsymbol{m} = (\boldsymbol{u}^L + \boldsymbol{u}^R) \cdot \boldsymbol{m} - (\boldsymbol{u}^L_\flat + \boldsymbol{u}^R_\sharp)^{p,p'} \cdot \boldsymbol{m},$$

and $\gamma(\mathcal{X}^L, \mathcal{X}^R)^{p,p'} = -\gamma(\mathcal{X}^L, \mathcal{X}^R)^{p'-p,p'} - \alpha_0^2$ by (1.27). From these expressions together with Lemma 2.8, we obtain:

$$F^{p'-p,p'}(\boldsymbol{u}^L, \boldsymbol{u}^R, L; q^{-1}) = q^{-\frac{1}{4}(L^2-\alpha_0^2)} F^{p,p'}(\boldsymbol{u}^L, \boldsymbol{u}^R, L; q). \tag{7.15}$$

Together with (7.14), this proves (7.13) in the $p' < 2p$ case, having noted, via Lemma 7.11, that $\alpha = \alpha^{p,p'}_{a,b} = \alpha_0$. □

7.5. The other direction

Theorem 7.16 deals with either b being interfacial or c taking the specific value $b + \Delta^R_{d^R}$. In this section, we deal with the other cases in which b is not interfacial and $c = b - \Delta^R_{d^R}$. The generating functions then involve the fermion-like expressions defined in Section 1.15.

As in the previous section, we fix naive runs $\mathcal{X}^L = \{\tau^L_j, \sigma^L_j, \Delta^L_j\}^{d^L}_{j=1}$ and $\mathcal{X}^R = \{\tau^R_j, \sigma^R_j, \Delta^R_j\}^{d^R}_{j=1}$, and make use of the definitions of Section 7.1. In addition, for convenience, we set $\sigma = \sigma^R_{d^R}$ and $\Delta = \Delta^R_{d^R}$. It will also be useful to set

$$\tau = \begin{cases} \tau^R_{d^R} & \text{if } d^R > 1; \\ t_n & \text{if } d^R = 1 \text{ and } \sigma < t_n; \\ t & \text{if } d^R = 1 \text{ and } \sigma \geq t_n. \end{cases} \tag{7.16}$$

Recall that $\sigma < t_n$ implies that $d^R_0 = 0$, and $\sigma \geq t_n$ implies that $d^R_0 = 1$.

Theorem 7.17. *Let all parameters be as in Section 7.1. Let b be non-interfacial and set $c = b - \Delta^R_{d^R}$. If $L \equiv b - a \ (mod 2)$ then:*

$$\chi^{p,p'}_{a,b,c}(L) \begin{Bmatrix} \boldsymbol{\mu} \ ; \boldsymbol{\nu} \\ \boldsymbol{\mu}^*; \boldsymbol{\nu}^* \end{Bmatrix} = \widetilde{F}(\boldsymbol{u}^L, \boldsymbol{u}^R, L). \tag{7.17}$$

Proof: We first consider $p' > 2p$. Since b is not interfacial, Corollary 7.13 implies that $\sigma \leq t_1$.

In the subcases for which $d^R > d^R_0$ note that $\boldsymbol{\nu} \neq () \neq \boldsymbol{\nu}^*$. Then, the definitions of Section 7.1 yield $\nu_{d^R-1} = b + \Delta\kappa_\sigma$ and $\nu^*_{d^R-1} = b - \Delta(\kappa_\tau - \kappa_\sigma)$. In the subcase for which $d^R = d^R_0$, note that $\boldsymbol{\nu} = \boldsymbol{\nu}^* = ()$.

For $\sigma = 0$, when $d^R > d^R_0$ we have $\nu_{d^R-1} = b + \Delta$, whereupon the first identity of Lemma 5.11(2) gives:

$$\begin{aligned} \chi^{p,p'}_{a,b,c}(L) \begin{Bmatrix} \boldsymbol{\mu} \ ; \boldsymbol{\nu} \\ \boldsymbol{\mu}^*; \boldsymbol{\nu}^* \end{Bmatrix} &= q^{\frac{1}{2}(L+\Delta(a-b))} \chi^{p,p'}_{a,b,b+\Delta}(L) \begin{Bmatrix} \boldsymbol{\mu} \ ; \boldsymbol{\nu} \\ \boldsymbol{\mu}^*; \boldsymbol{\nu}^* \end{Bmatrix} \\ &= q^{\frac{1}{2}(L+\Delta(a-b))} F(\boldsymbol{u}^L, \boldsymbol{u}^R, L), \end{aligned} \tag{7.18}$$

the second equality following from Theorem 7.16. When $d^R = d^R_0$ so that $\boldsymbol{\nu} = \boldsymbol{\nu}^* = ()$, (7.18) also follows from Lemma 5.11(2) and Theorem 7.16. The definition (1.45) then gives the required result in this $\sigma = 0$ case.

Next, we consider $0 \leq \sigma < \tau - 1$. Set $b' = b - \Delta$. When $d^R > d^R_0$, we find that $b' - \nu_{d^R-1} = -\Delta(\kappa_\sigma + 1)$ and $b' - \nu^*_{d^R-1} = \Delta(\kappa_\tau - \kappa_\sigma - 1)$. In particular, b' is strictly between ν_{d^R-1} and $\nu^*_{d^R-1}$. The same is true of $b' - \Delta$ unless $\sigma + 2 = \tau \leq t_1 + 1$, in which case $b' - \Delta = \nu^*_{d^R-1}$. Substituting $b \to b'$ and $L \to L + 1$ into the second identity of Lemma 5.11(1) and rearranging, yields:

$$\begin{aligned} & q^{\frac{1}{2}(L+1-\Delta(a-b'))} \chi^{p,p'}_{a,b'+\Delta,b'}(L) \begin{Bmatrix} \boldsymbol{\mu} \ ; \boldsymbol{\nu} \\ \boldsymbol{\mu}^*; \boldsymbol{\nu}^* \end{Bmatrix} \\ &\qquad = \chi^{p,p'}_{a,b',b'+\Delta}(L+1) \begin{Bmatrix} \boldsymbol{\mu} \ ; \boldsymbol{\nu} \\ \boldsymbol{\mu}^*; \boldsymbol{\nu}^* \end{Bmatrix} - \chi^{p,p'}_{a,b'-\Delta,b'}(L) \begin{Bmatrix} \boldsymbol{\mu} \ ; \boldsymbol{\nu}^+ \\ \boldsymbol{\mu}^*; \boldsymbol{\nu}^{+*} \end{Bmatrix}. \end{aligned} \tag{7.19}$$

where if $b'-\Delta \neq \nu^*_{d^R-1}$ then $\boldsymbol{\nu}^+ = \boldsymbol{\nu}$ and $\boldsymbol{\nu}^{+*} = \boldsymbol{\nu}^*$; and if $b'-\Delta = \nu^*_{d^R-1}$ then $d^+ \geq d^R_0$ is the smallest value such that $b'-\Delta = \nu^*_{d^+}$, and then $\boldsymbol{\nu}^+ = (\nu_{d^R_0}, \ldots, \nu_{d^+-1})$ and $\boldsymbol{\nu}^{+*} = (\nu^*_{d^R_0}, \ldots, \nu^*_{d^+-1})$. The term on the left of (7.19) is (apart from the prefactor) that on the left of (7.17). The first term on the right is equal to $F(\boldsymbol{u}^L, \boldsymbol{u}^{R+}, L+1)$ by Theorem 7.16, after noting that with $\mathcal{X}^{R+}$ in place of $\mathcal{X}^R$, the definitions of Section 7.1 lead to the same mazy-pair $\boldsymbol{\nu}, \boldsymbol{\nu}^*$. If, on the one hand, $b' - \Delta \neq \nu^*_{d^R-1}$ the second term on the right is likewise equal to $F(\boldsymbol{u}^L, \boldsymbol{u}^{R++}, L)$ by using $\mathcal{X}^{R++}$ in place of $\mathcal{X}^R$. Therefore:

$$(7.20)\quad q^{\frac{1}{2}(L-\Delta(a-b))}\chi^{p,p'}_{a,b'+\Delta,b'}(L)\left\{\begin{matrix}\boldsymbol{\mu}\ ;\boldsymbol{\nu}\\ \boldsymbol{\mu}^*;\boldsymbol{\nu}^*\end{matrix}\right\} = F(\boldsymbol{u}^L, \boldsymbol{u}^{R+}, L+1) - F(\boldsymbol{u}^L, \boldsymbol{u}^{R++}, L),$$

from which (7.17) follows in this case via the definition (1.45).

On the other hand, if $b' - \Delta = \nu^*_{d^R-1}$, note in the subcase for which $d^R > 1$, that the mazy pair $\boldsymbol{\nu}^+, \boldsymbol{\nu}^{+*}$ arises from the run $\mathcal{X}' = \{\tau^R_j, \sigma^R_j, \Delta^R_j\}_{j=1}^{d^*}$ where $d^* = \max\{d^+, 1\}$. Also note that $\nu^*_{d^*} = \nu^*_{d^*+1} = \cdots = \nu^*_{d^R-1} = b - 2\Delta$ implies that $\tau^R_j = \sigma^R_j$ for $d^* < j < d$ and $\tau^R_{d^R} = \sigma^R_{d^R} + 2$ whereupon $\boldsymbol{u}(\mathcal{X}') = \boldsymbol{u}^{R++}$ as defined in Section 1.15, and (1.24) gives $\boldsymbol{\Delta}(\mathcal{X}') = \boldsymbol{\Delta}(\mathcal{X}^{++})$. In the subcase for which $d^R = 1$ and $d^R_0 = 0$, the mazy pair $\boldsymbol{\nu}^+ = \boldsymbol{\nu}^{+*} = ()$ arises from $\mathcal{X}' = \{\tau^R_1, t_n, \Delta^R_1\}$ whereupon again $\boldsymbol{u}(\mathcal{X}') = \boldsymbol{u}^{R++}$ and $\boldsymbol{\Delta}(\mathcal{X}') = \boldsymbol{\Delta}(\mathcal{X}^{++})$. Thus, even when $b' - \Delta = \nu^*_{d^R-1}$, the second term on the right of (7.19) is equal to $F(\boldsymbol{u}^L, \boldsymbol{u}^{R++}, L)$ by Theorem 7.16, and (7.20) and then (7.17) follow.

For this $0 \leq \sigma < \tau - 1$ case, it remains to consider the subcase for which $d^R = d^R_0 = 1$. Here $\boldsymbol{\nu} = \boldsymbol{\nu}^* = \boldsymbol{\nu}^+ = \boldsymbol{\nu}^{+*} = ()$. The instance $\sigma = t_1$ may be excluded because then $n = 1$ and b may be shown to be interfacial. Otherwise Lemma 5.11(1) and Theorem 7.16 lead to (7.19), (7.20) and (7.17) in a similar fashion to the $b' - \Delta \neq \nu^*_{d^R-1}$ case considered above after noting that $\sigma < t - 1$ ensures that $1 \leq b', b' - \Delta < p'$.

We now consider $0 < \sigma = \tau - 1$. In the subcase where $d^R > d^R_0$ we have $b - \Delta = \nu^*_{d^R-1}$ and $b + \Delta\kappa_\sigma = \nu_{d^R-1}$, so that certainly both b and $b + \Delta$ are strictly between $\nu^*_{d^R-1}$ and ν_{d^R-1}. Eliminating the terms that appear second on the right sides of the two identities in Lemma 5.11(1), yields:

$$(7.21)\quad \begin{aligned} &q^{\frac{1}{2}(L-\Delta(a-b))}\chi^{p,p'}_{a,b,b-\Delta}(L)\left\{\begin{matrix}\boldsymbol{\mu}\ ;\boldsymbol{\nu}\\ \boldsymbol{\mu}^*;\boldsymbol{\nu}^*\end{matrix}\right\}\\ &\qquad = \chi^{p,p'}_{a,b,b+\Delta}(L)\left\{\begin{matrix}\boldsymbol{\mu}\ ;\boldsymbol{\nu}\\ \boldsymbol{\mu}^*;\boldsymbol{\nu}^*\end{matrix}\right\} + (q^L-1)\chi^{p,p'}_{a,b-\Delta,b}(L-1)\left\{\begin{matrix}\boldsymbol{\mu}\ ;\boldsymbol{\nu}^+\\ \boldsymbol{\mu}^*;\boldsymbol{\nu}^{+*}\end{matrix}\right\},\end{aligned}$$

where if $d^+ \geq d^R_0$ is the smallest value such that $b - \Delta = \nu^*_{d^+}$, then $\boldsymbol{\nu}^+ = (\nu_{d^R_0}, \ldots, \nu_{d^+-1})$ and $\boldsymbol{\nu}^{+*} = (\nu^*_{d^R_0}, \ldots, \nu^*_{d^+-1})$.

The first term on the right of (7.21) is equal to $F(\boldsymbol{u}^L, \boldsymbol{u}^R, L)$ by Theorem 7.16. Now note that in the subcase for which $d^R > 1$, the mazy pair $\boldsymbol{\nu}^+, \boldsymbol{\nu}^{+*}$ arises from the run $\mathcal{X}' = \{\tau^R_j, \sigma^R_j, \Delta^R_j\}_{j=1}^{d^*}$ where $d^* = \max\{d^+, 1\}$. Also note that $\nu^*_{d^*} = \nu^*_{d^*+1} = \cdots = \nu^*_{d^R-1} = b - \Delta$ implies that $\tau^R_j = \sigma^R_j$ for $d^* < j < d^R$ and $\tau^R_{d^R} = \sigma^R_{d^R} + 1$ whereupon $\boldsymbol{u}(\mathcal{X}') = \boldsymbol{u}^{R+}$ as defined in Section 1.15, and (1.24) gives $\boldsymbol{\Delta}(\mathcal{X}') = \boldsymbol{\Delta}(\mathcal{X}^+)$. In the subcase for which $d^R = 1$ and $d^R_0 = 0$ the mazy pair $\boldsymbol{\nu}^+ = \boldsymbol{\nu}^{+*} = ()$ arises from $\mathcal{X}' = \{\tau^R_1, t_n, \Delta^R_1\}$ whereupon again $\boldsymbol{u}(\mathcal{X}') = \boldsymbol{u}^{R+}$ and $\boldsymbol{\Delta}(\mathcal{X}') = \boldsymbol{\Delta}(\mathcal{X}^+)$. Thus whenever $d^R > d^R_0$, Theorem 7.16 implies that the second term on the right of (7.21) is equal to $(q^L - 1)F(\boldsymbol{u}^L, \boldsymbol{u}^{R+}, L-1)$. Thus, (7.21) gives:

$$(7.22) \qquad q^{\frac{1}{2}(L-\Delta(a-b))}\chi^{p,p'}_{a,b,b-\Delta}(L)\begin{Bmatrix}\boldsymbol{\mu}\ ;\boldsymbol{\nu}\\ \boldsymbol{\mu}^*;\boldsymbol{\nu}^*\end{Bmatrix} = F(\boldsymbol{u}^L, \boldsymbol{u}^R, L) + (q^L-1)F(\boldsymbol{u}^L, \boldsymbol{u}^{R+}, L-1).$$

The definition (1.45) then yields the required (7.17). For the subcase for which $d^R = d^R_0 = 1$, Lemma 5.11(1) and Theorem 7.16 lead to (7.21), (7.22) and (7.17) in a similar fashion after noting that $\boldsymbol{\nu} = \boldsymbol{\nu}^* = \boldsymbol{\nu}^+ = \boldsymbol{\nu}^{+*} = ()$, and $0 < \sigma < t$ implies that $1 \le b \pm 1 < p'$.

The proof for the $p' < 2p$ cases differs from that of the $p' < 2p$ cases above only in that parts 3) and 4) of Lemma 5.11 are used as appropriate, yielding different coefficients, but yielding (7.17) as required. □

Note 7.18. *As indicated in the above proof, if $0 = \sigma < \tau - 1$ then the polynomials defined by the first and second cases of (1.45) are actually equal. The same comment applies to the first and second cases of (1.46).*

We prove the following result, which is not needed until Section 9, along lines similar to the proof of Theorem 7.17 above.

Lemma 7.19. *Let all parameters be as in Section 7.1.*

1) Let $p' > 2p$, let $0 \le \sigma < \tau - 1$ and let $b - \Delta^R_{d^R}$ be non-interfacial. If b is interfacial then:

$$F(\boldsymbol{u}^L, \boldsymbol{u}^R, L) = \widetilde{F}(\boldsymbol{u}^L, \boldsymbol{u}^{R+}, L+1) - q^{\frac{1}{2}(L+2+\Delta(a-b))}F(\boldsymbol{u}^L, \boldsymbol{u}^{R++}, L).$$

If b is non-interfacial then:

$$\widetilde{F}(\boldsymbol{u}^L, \boldsymbol{u}^R, L) = \widetilde{F}(\boldsymbol{u}^L, \boldsymbol{u}^{R+}, L+1) - q^{\frac{1}{2}(L+2+\Delta(a-b))}F(\boldsymbol{u}^L, \boldsymbol{u}^{R++}, L).$$

2) If $p' < 2p$ and $\sigma_d = 0$ then

$$\widetilde{F}(\boldsymbol{u}^L, \boldsymbol{u}^R, L) = F(\boldsymbol{u}^L, \boldsymbol{u}^{R+}, L-1).$$

Proof: If $d^R > d^R_0$, note that $\boldsymbol{\nu} \ne () \ne \boldsymbol{\nu}^*$. Then, the definitions of Section 7.1 yield $\nu_{d^R-1} = b + \Delta\kappa_\sigma$ and $\nu^*_{d^R-1} = b - \Delta(\kappa_\tau - \kappa_\sigma)$. If $d^R = d^R_0$, note that $\boldsymbol{\nu} = \boldsymbol{\nu}^* = ()$.

1) Set $b' = b - \Delta$. When $d^R > d^R_0$ we have $b' - \nu_{d^R-1} = -\Delta(\kappa_\sigma + 1)$ and $b' - \nu^*_{d^R-1} = \Delta(\kappa_\tau - \kappa_\sigma - 1)$. In particular, b' is strictly between ν_{d^R-1} and $\nu^*_{d^R-1}$. The same is true of $b' - \Delta$ unless $\sigma + 2 = \tau \le t_1 + 1$, in which case $b' - \Delta = \nu^*_{d^R-1}$. Substituting $b \to b'$ and $L \to L+1$ into the first identity of Lemma 5.11(1) and rearranging, yields:

$$(7.23) \qquad \chi^{p,p'}_{a,b'+\Delta,b'}(L)\begin{Bmatrix}\boldsymbol{\mu}\ ;\boldsymbol{\nu}\\ \boldsymbol{\mu}^*;\boldsymbol{\nu}^*\end{Bmatrix} = \chi^{p,p'}_{a,b',b'-\Delta}(L+1)\begin{Bmatrix}\boldsymbol{\mu}\ ;\boldsymbol{\nu}\\ \boldsymbol{\mu}^*;\boldsymbol{\nu}^*\end{Bmatrix} - q^{\frac{1}{2}(L+1+\Delta(a-b'))}\chi^{p,p'}_{a,b'-\Delta,b'}(L)\begin{Bmatrix}\boldsymbol{\mu}\ ;\boldsymbol{\nu}^+\\ \boldsymbol{\mu}^*;\boldsymbol{\nu}^{+*}\end{Bmatrix}.$$

where if $b' - \Delta \ne \nu^*_{d^R-1}$ then $\boldsymbol{\nu}^+ = \boldsymbol{\nu}$ and $\boldsymbol{\nu}^{+*} = \boldsymbol{\nu}^*$; and if $b' - \Delta = \nu^*_{d^R-1}$ then $d^+ \ge d^R_0$ is the smallest value such that $b' - \Delta = \nu^*_{d^+}$, and then $\boldsymbol{\nu}^+ = (\nu_{d^R_0}, \ldots, \nu_{d^+-1})$ and $\boldsymbol{\nu}^{+*} = (\nu^*_{d^R_0}, \ldots, \nu^*_{d^+-1})$. If b is interfacial then the left side of (7.23) is equal to $F(\boldsymbol{u}^L, \boldsymbol{u}^R, L)$ by Theorem 7.16, whereas if b is non-interfacial, it is equal to $\widetilde{F}(\boldsymbol{u}^L, \boldsymbol{u}^R, L)$ by Theorem 7.17. The first term on the right is equal to $\widetilde{F}(\boldsymbol{u}^L, \boldsymbol{u}^{R+}, L+1)$ by Theorem 7.17, after noting that with $\mathcal{X}^{R+}$ in place of $\mathcal{X}^R$, the definitions of Section 7.1 lead to the same mazy-pair $\boldsymbol{\nu}, \boldsymbol{\nu}^*$. Regardless of whether $b' - \Delta \neq \nu^*_{d^R-1}$ or $b' - \Delta = \nu^*_{d^R-1}$ (see the corresponding argument which applies to the second term on the right of (7.19) in the proof of Theorem 7.17), the second term on the right is equal to $F(\boldsymbol{u}^L, \boldsymbol{u}^{R++}, L)$ via Theorem 7.16. The required result follows in the $d^R > d^R_0$ case.

In the subcase for which $d^R = d^R_0 = 1$, we have $\boldsymbol{\nu} = \boldsymbol{\nu}^* = \boldsymbol{\nu}^+ = \boldsymbol{\nu}^{+*} = ()$. Lemma 5.11(1) and Theorems 7.16 and 7.17 then lead to (7.23) and the required result in a similar fashion to the $b' - \Delta \neq \nu^*_{d^R-1}$ case considered above after noting that $\sigma < t-1$ ensures that $1 \leq b', b' - \Delta < p'$.

2) In the case for which $d^R > d^R_0$ we have $b + \Delta = \nu_{d^R-1}$ here, whereupon the second identity of Lemma 5.11(4) gives:

$$\chi^{p,p'}_{a,b,b-\Delta}(L) \left\{ \begin{matrix} \boldsymbol{\mu} \,; \boldsymbol{\nu} \\ \boldsymbol{\mu}^*; \boldsymbol{\nu}^* \end{matrix} \right\} = \chi^{p,p'}_{a,b-\Delta,b}(L-1) \left\{ \begin{matrix} \boldsymbol{\mu} \,; \boldsymbol{\nu}^+ \\ \boldsymbol{\mu}^*; \boldsymbol{\nu}^{+*} \end{matrix} \right\}. \tag{7.24}$$

where if $b - \Delta \neq \nu^*_{d^R-1}$ then $\boldsymbol{\nu}^+ = \boldsymbol{\nu}$ and $\boldsymbol{\nu}^{+*} = \boldsymbol{\nu}^*$; and if $b - \Delta = \nu^*_{d^R-1}$ then $d^+ \geq d^R_0$ is the smallest value such that $b - \Delta = \nu^*_{d^+}$, and then $\boldsymbol{\nu}^+ = (\nu_{d^R_0}, \ldots, \nu_{d^+-1})$ and $\boldsymbol{\nu}^{+*} = (\nu^*_{d^R_0}, \ldots, \nu^*_{d^+-1})$. The left side of (7.24) is equal to $\widetilde{F}(\boldsymbol{u}^L, \boldsymbol{u}^R, L)$ by Theorem 7.17. If $b - \Delta \neq \nu^*_{d^R-1}$, the right side of (7.24) is equal to $F(\boldsymbol{u}^L, \boldsymbol{u}^{R+}, L-1)$ by Theorem 7.16, after noting that with $\mathcal{X}^{R+}$ in place of $\mathcal{X}^R$, the definitions of Section 7.1 lead to the same mazy-pair $\boldsymbol{\nu}, \boldsymbol{\nu}^*$. If $b - \Delta = \nu^*_{d^R-1}$, the argument which applies to the second term on the right of (7.21) in the proof of Theorem 7.17, equally applies here to show that the term on the right of (7.24) is equal to $F(\boldsymbol{u}^L, \boldsymbol{u}^{R+}, L-1)$ via Theorem 7.16. The required result follows in this $d^R > d^R_0$ case.

In the case for which $d^R = d^R_0$, we have $b + \Delta \in \{0, p'\}$ whereupon Lemma 5.11(4) again implies (7.24). Again, Theorems 7.16 and 7.17 yield the required result. □

7.6. The mn-system

Equation (1.31) defines the vector $\boldsymbol{n}$ in terms of $\hat{\boldsymbol{m}}$ and $\boldsymbol{u} = \boldsymbol{u}^L + \boldsymbol{u}^R$. The summands of (1.40) have $m_i \equiv Q_i$ for $0 \leq i < t$ whereupon, via (1.33), it follows that $n_i \in \mathbb{Z}$ for $1 \leq i \leq t$. Moreover, on expressing (1.31) in the form $n_i = -\frac{1}{2}(\boldsymbol{C}^* \hat{\boldsymbol{m}} - \boldsymbol{u}^L - \boldsymbol{u}^R)_i$, we see that the non-zero terms in (1.40) have each $n_i \geq 0$. (On examining the proof of Lemma 7.11, we find that n_i is the number of particles inserted at the ith stage of the induction performed there.)

Equation (1.41) follows immediately from the following lemma.

Lemma 7.20. *For* $0 \leq j < t$,

$$\sum_{i=1}^{t} l_i C_{ij} = -\delta_{j0}.$$

Proof: In view of the definition (1.28), only $i = j-1$, $i = j$ and $i = j+1$ contribute in the summation. Since $C_{j+1,j} = -1$, the lemma holds for $j = 0$.

Now consider $j \geq 1$ and let k be such that $t_k < j \leq t_{k+1}$. The definition of $\{l_i\}_{i=1}^t$ shows that if $j < t_{k+1}$ then $l_{j+1} = l_j + y_k$ and if $j = t_{k+1}$ then $l_{j+1} = y_k$. Since $C_{jj} = 2$ in the former case and $C_{jj} = 1$ in the latter case, $l_j C_{jj} + l_{j+1} C_{j+1,j} =$

$l_j - y_k$ in both cases. For $j = 1$, we readily find $l_1 = y_k = 1$ thus giving the required result.

For $j > 1$, a similar argument shows that if $j - 1 > t_k$ then $l_j = l_{j-1} + y_k$ and if $j - 1 = t_k$ then $l_j = y_{k-1}$. In the former case, $C_{j-1,j} = -1$, whereupon $\sum_{i=j-1}^{j+1} l_i C_{i,j} = 0$. In the latter case, $C_{j-1,j} = 1$, whereupon $\sum_{i=j-1}^{j+1} l_i C_{i,j} = l_{j-1} + y_{k-1} - y_k = 0$, with the final equality following because $l_{t_k} = y_{k-2} + (t_k - t_{k-1} - 1)y_{k-1} = y_k - y_{k-1}$. □

The summation over $\boldsymbol{m} \equiv \boldsymbol{Q}(\boldsymbol{u}^L + \boldsymbol{u}^R)$ in (1.40) may therefore be carried out by finding all solutions to (1.42) with $n_i \in \mathbb{Z}_{\geq 0}$ for $1 \leq i \leq t$ (the set of solutions is clearly finite), and then using (1.32) to obtain the corresponding values $(m_1, m_2, \ldots, m_{t-1})$.

8. Collating the runs

In this section, we show that Takahashi trees for a and b enable us to obtain a set of $F(\boldsymbol{u}^L, \boldsymbol{u}^R, L)$ whose sum is $\chi^{p,p'}_{a,b,c}(L)$.

8.1. Pulling the drawstrings

This section provides a few preliminary results that are required to make use of the Takahashi trees.

Lemma 8.1. *Let $\boldsymbol{\mu}, \boldsymbol{\mu}^*, \boldsymbol{\nu}, \boldsymbol{\nu}^*$ be a mazy-four in the (p,p')-model sandwiching (a,b). If $d^L \geq 2$ then:*

$$\chi^{p,p'}_{a,b,c}(L) \begin{Bmatrix} \mu_1, \ldots, \mu_{d^L-1}; \boldsymbol{\nu} \\ \mu^*_1, \ldots, \mu^*_{d^L-1}; \boldsymbol{\nu}^* \end{Bmatrix}$$
$$= \chi^{p,p'}_{a,b,c}(L) \begin{Bmatrix} \mu_1, \ldots, \mu_{d^L-1}, \mu_{d^L}; \boldsymbol{\nu} \\ \mu^*_1, \ldots, \mu^*_{d^L-1}, \mu^*_{d^L}; \boldsymbol{\nu}^* \end{Bmatrix} + \chi^{p,p'}_{a,b,c}(L) \begin{Bmatrix} \mu_1, \ldots, \mu_{d^L-1}, \mu^*_{d^L}; \boldsymbol{\nu} \\ \mu^*_1, \ldots, \mu^*_{d^L-1}, \mu_{d^L}; \boldsymbol{\nu}^* \end{Bmatrix}.$$

If $d^R \geq 2$ then:

$$\chi^{p,p'}_{a,b,c}(L) \begin{Bmatrix} \boldsymbol{\mu}\ ; \nu_1, \ldots, \nu_{d^R-1} \\ \boldsymbol{\mu}^*; \nu^*_1, \ldots, \nu^*_{d^R-1} \end{Bmatrix}$$
$$= \chi^{p,p'}_{a,b,c}(L) \begin{Bmatrix} \boldsymbol{\mu}\ ; \nu_1, \ldots, \nu_{d^R-1}, \nu_{d^R} \\ \boldsymbol{\mu}^*; \nu^*_1, \ldots, \nu^*_{d^R-1}, \nu^*_{d^R} \end{Bmatrix} + \chi^{p,p'}_{a,b,c}(L) \begin{Bmatrix} \boldsymbol{\mu}\ ; \nu_1, \ldots, \nu_{d^R-1}, \nu^*_{d^R} \\ \boldsymbol{\mu}^*; \nu^*_1, \ldots, \nu^*_{d^R-1}, \nu_{d^R} \end{Bmatrix}.$$

Proof: In the first case, since a is between μ_{d^L} and $\mu^*_{d^L}$, and μ_{d^L} and $\mu^*_{d^L}$ are both between μ_{d^L-1} and $\mu^*_{d^L-1}$, it follows that each path h with $h_0 = a$ that attains $\mu^*_{d^L-1}$, necessarily attains μ_{d^L} or $\mu^*_{d^L}$. Thence:

$$\mathcal{P}^{p,p'}_{a,b,c}(L) \begin{Bmatrix} \mu_1, \ldots, \mu_{d^L-1}; \boldsymbol{\nu} \\ \mu^*_1, \ldots, \mu^*_{d^L-1}; \boldsymbol{\nu}^* \end{Bmatrix}$$
$$= \mathcal{P}^{p,p'}_{a,b,c}(L) \begin{Bmatrix} \mu_1, \ldots, \mu_{d^L-1}, \mu_{d^L}; \boldsymbol{\nu} \\ \mu^*_1, \ldots, \mu^*_{d^L-1}, \mu^*_{d^L}; \boldsymbol{\nu}^* \end{Bmatrix} \cup \mathcal{P}^{p,p'}_{a,b,c}(L) \begin{Bmatrix} \mu_1, \ldots, \mu_{d^L-1}, \mu^*_{d^L}; \boldsymbol{\nu} \\ \mu^*_1, \ldots, \mu^*_{d^L-1}, \mu_{d^L}; \boldsymbol{\nu}^* \end{Bmatrix}.$$

Then, since the latter two sets are disjoint, the first part of the lemma follows. The second part follows in an analogous way. □

In the above lemma, it is essential that $d^L \geq 2$ for the first expression, and $d^R \geq 2$ for the second expression, since the two expressions don't necessarily hold if $d^L = 1$ and $d^R = 1$ respectively. The following two lemmas deal with that extra case.

Lemma 8.2. *Let $0 \le \mu^* < a < \mu \le p'$ and $0 \le \nu^* < b < \nu \le p'$. If $\mu \le \nu^*$ or $\nu \le \mu^*$ then:*

$$\chi^{p,p'}_{a,b,c}(L) = \chi^{p,p'}_{a,b,c}(L)\begin{Bmatrix}\mu\ ;\nu\\ \mu^*;\nu^*\end{Bmatrix} + \chi^{p,p'}_{a,b,c}(L)\begin{Bmatrix}\mu\ ;\nu^*\\ \mu^*;\nu\end{Bmatrix} + \chi^{p,p'}_{a,b,c}(L)\begin{Bmatrix}\mu^*;\nu\\ \mu\ ;\nu^*\end{Bmatrix} + \chi^{p,p'}_{a,b,c}(L)\begin{Bmatrix}\mu^*;\nu^*\\ \mu\ ;\nu\end{Bmatrix}.$$

Proof: Let $h \in \mathcal{P}^{p,p'}_{a,b,c}(L)$ whence $h_0 = a$ and $h_L = b$. In the first case $a < \mu \le \nu^* < b$, so that $h_i = \mu$ for some i and $h_{i'} = \nu^*$ for some i'. Thus h is certainly an element of one of $\mathcal{P}^{p,p'}_{a,b,c}(L)\left\{\begin{smallmatrix}\mu\ ;\nu\\ \mu^*;\nu^*\end{smallmatrix}\right\}$, $\mathcal{P}^{p,p'}_{a,b,c}(L)\left\{\begin{smallmatrix}\mu\ ;\nu^*\\ \mu^*;\nu\end{smallmatrix}\right\}$, $\mathcal{P}^{p,p'}_{a,b,c}(L)\left\{\begin{smallmatrix}\mu^*;\nu\\ \mu\ ;\nu^*\end{smallmatrix}\right\}$, and $\mathcal{P}^{p,p'}_{a,b,c}(L)\left\{\begin{smallmatrix}\mu^*;\nu^*\\ \mu\ ;\nu\end{smallmatrix}\right\}$. Since these four sets are disjoint, the lemma follows when $a < \mu \le \nu^* < b$.

When $\nu \le \mu*$, the lemma follows in an analogous way. □

Lemma 8.3. *Let $0 \le \mu^* < a, b < \mu \le p'$ and $\hat{p}' = \mu - \mu^*$. If $\hat{p}$ is such that the sth band of the $(\hat{p}, \hat{p}')$-model is of the same parity as the $(s + \mu^*)$th band of the (p, p')-model for $1 \le s \le \hat{p}' - 2$,*[1] *then:*

$$\chi^{p,p'}_{a,b,c}(L) = \chi^{p,p'}_{a,b,c}(L)\begin{Bmatrix}\mu\ ;\mu\\ \mu^*;\mu^*\end{Bmatrix} + \chi^{p,p'}_{a,b,c}(L)\begin{Bmatrix}\mu\ ;\mu^*\\ \mu^*;\mu\end{Bmatrix} + \chi^{p,p'}_{a,b,c}(L)\begin{Bmatrix}\mu^*;\mu\\ \mu\ ;\mu^*\end{Bmatrix} + \chi^{p,p'}_{a,b,c}(L)\begin{Bmatrix}\mu^*;\mu^*\\ \mu\ ;\mu\end{Bmatrix} + \chi^{\hat{p},\hat{p}'}_{\hat{a},\hat{b},\hat{c}}(L),$$

where $\hat{a} = a - u^*$, $\hat{b} = b - u^*$ and

$$\hat{c} = \begin{cases} 2 & \text{if } c = u^* > 0 \text{ and } \left\lfloor \frac{(b+1)p}{p'} \right\rfloor = \left\lfloor \frac{(b-1)p}{p'} \right\rfloor + 1; \\ \hat{p}' - 2 & \text{if } c = u < p' \text{ and } \left\lfloor \frac{(b+1)p}{p'} \right\rfloor = \left\lfloor \frac{(b-1)p}{p'} \right\rfloor + 1; \\ c - u^* & \text{otherwise.} \end{cases}$$

Proof: Let $h \in \mathcal{P}^{p,p'}_{a,b,c}(L)$ whence $h_0 = a$ and $h_L = b$. If $h_i = \mu$ for some i with $0 \le i \le L$, or $h_{i'} = \mu^*$ for some i' with $0 \le i' \le L$, then h is an element of one of the disjoint sets $\mathcal{P}^{p,p'}_{a,b,c}(L)\left\{\begin{smallmatrix}\mu\ ;\mu\\ \mu^*;\mu^*\end{smallmatrix}\right\}$, $\mathcal{P}^{p,p'}_{a,b,c}(L)\left\{\begin{smallmatrix}\mu\ ;\mu^*\\ \mu^*;\mu\end{smallmatrix}\right\}$, $\mathcal{P}^{p,p'}_{a,b,c}(L)\left\{\begin{smallmatrix}\mu^*;\mu\\ \mu\ ;\mu^*\end{smallmatrix}\right\}$, and $\mathcal{P}^{p,p'}_{a,b,c}(L)\left\{\begin{smallmatrix}\mu^*;\mu^*\\ \mu\ ;\mu\end{smallmatrix}\right\}$.

Otherwise $\mu^* < h_i < \mu$ for $0 \le i \le L$, whereupon setting $h'_i = h_i - \mu^*$ for $0 \le i \le L$, defines an element of $\mathcal{P}^{\hat{p},\hat{p}'}_{\hat{a},\hat{b},c-\mu^*}(L)$. Clearly, every element of $\mathcal{P}^{\hat{p},\hat{p}'}_{\hat{a},\hat{b},c-\mu^*}(L)$ arises uniquely in this way. The sequence of band parities between heights $\mu^* + 1$ and $\mu - 1$ of the (p, p')-model being identical to that between heights 1 and $\hat{p}' - 1$ of the $(\hat{p}, \hat{p}')$-model, guarantees that the contribution of the ith vertex to the weight of the path is identical in the two cases h and h', except possibly when $i = L$ and either $c = u^*$ or $c = u$. After noting the assignment of parities to the 0th and $(\hat{p}' - 1)$th bands as specified in Note 2.1, we see that this discrepancy is rectified by switching the direction of the final vertex whenever b is interfacial and either $c = u^*$ or $c = u$. Then $wt(h') = wt(h)$. The lemma follows. □

[1] In general, there is no guarantee that such a $\hat{p}$ exists.

Note 8.4. *Lemmas 8.2 and 8.3 will be employed with* $\mu^* = 0$, $\mu = p'$, $\nu^* = 0$ *or* $\nu = p'$. *Then, at least two of the terms on the right side of the expressions in Lemmas 8.2 and 8.3 are automatically zero.*

8.2. Constraints of the Takahashi tree

Let $1 \le a < p'$ and consider the Takahashi tree for a that is described in Section 1.6. Let $a_{i_1 i_2 \cdots i_d}$ be a particular leaf-node. From this leaf-node, obtain $\{\tau_j, \sigma_j, \Delta_j\}_{j=1}^d$ as in Section 1.7. The next result shows that $\{\tau_j, \sigma_j, \Delta_j\}_{j=1}^d$ satisfies the definition of a naive run that was given in Section 7.1. We will then refer to $\{\tau_j, \sigma_j, \Delta_j\}_{j=1}^d$ as the naive run corresponding to (the leaf-node) $a_{i_1 i_2 \cdots i_d}$.

Lemma 8.5. *For* $2 \le j < d$, *if* $\Delta_j = \Delta_{j+1}$ *then* $\sigma_j \le \tau_j$, *and if* $\Delta_j \ne \Delta_{j+1}$ *then* $\sigma_j < \tau_j$. *Also* $\sigma_d < \tau_d$.

Let $d > 1$. *If* $2 \le j < d$, *or both* $j = 1$ *and* $\sigma_1 \le t_n$, *then*

$$\begin{aligned} \tau_{j+1} &= t_{\zeta(\sigma_j - 1)} && \textit{if } \Delta_{j+1} = \Delta_j; \\ \tau_{j+1} &= t_{\zeta(\sigma_j)} && \textit{if } \Delta_{j+1} \ne \Delta_j, \end{aligned}$$

and if $t_n < \sigma_1 < t$ *then*

$$\begin{aligned} \tau_2 &= t_{n-1} && \textit{if } \Delta_2 = \Delta_1; \\ \tau_2 &= t_n - 1 && \textit{if } \Delta_2 \ne \Delta_1 \textit{ and } c_n > 1; \\ \tau_2 &= t_{n-2} && \textit{if } \Delta_2 \ne \Delta_1 \textit{ and } c_n = 1. \end{aligned}$$

Proof: For $2 \le j < d$, the description of Section 1.6 implies that there exists x such that:

$$\begin{aligned} a_{i_1 i_2 \cdots i_{j-1}} - a_{i_1 i_2 \cdots i_{j-1} i_{j-1}} &= (-1)^{i_{j-1}} \kappa_x; \\ a_{i_1 i_2 \cdots i_{j-1}} - a_{i_1 i_2 \cdots i_{j-1} \bar{i}_{j-1}} &= (-1)^{i_{j-1}} \kappa_{x+1}. \end{aligned} \tag{8.1}$$

The prescription of Section 1.7 then implies that $\sigma_j = x$ if $i_j \ne i_{j-1}$, and $\sigma_j = x+1$ if $i_j = i_{j-1}$. Since $\kappa_{\tau_j} = a_{i_1 i_2 \cdots i_{j-2} 1} - a_{i_1 i_2 \cdots i_{j-2} 0}$, $a_{i_1 i_2 \cdots i_{j-2} 0} < a < a_{i_1 i_2 \cdots i_{j-2} 1}$, and $a_{i_1 i_2 \cdots i_{j-1} 0} < a < a_{i_1 i_2 \cdots i_{j-1} 1}$, it follows that $\kappa_{x+1} \le \kappa_{\tau_j}$ and thus that $x + 1 \le \tau_j$. It now follows that if $\Delta_{j+1} \ne \Delta_j$ then $\sigma_j = x < \tau_j$, and if $\Delta_{j+1} = \Delta_j$ then $\sigma_j = x + 1 \le \tau_j$. That $\sigma_d \le \tau_d$ follows similarly: the case $\sigma_d = \tau_d$ is then excluded since otherwise $a_{i_1 i_2 \cdots i_{d-2} \bar{i}_{d-1}} = a_{i_1 i_2 \cdots i_d} = a$, and the former of these would label a leaf-node.

The prescription of Section 1.7 also implies, via (8.1), that $\kappa_{\tau_{j+1}} = \kappa_{x+1} - \kappa_x = y_{\zeta(x)} = \kappa_{t_{\zeta(x)}}$, so that $\tau_{j+1} = t_{\zeta(x)}$. This immediately yields:

$$\tau_{j+1} = \begin{cases} t_{\zeta(\sigma_j - 1)} & \text{if } \Delta_{j+1} = \Delta_j; \\ t_{\zeta(\sigma_j)} & \text{if } \Delta_{j+1} \ne \Delta_j, \end{cases}$$

as required in this $2 \le j < d$ case.

If $a < y_n + y_{n-1} = \kappa_{t_n + 1}$ then $a_0 = \kappa_x$ and $a_1 = \kappa_{x+1}$ for some x, whereupon $\tau_2 = t_{\zeta(x)}$. So with $\Delta_2 = -(-1)^{i_1}$ and $a_{\bar{i}_1} = \kappa_{\sigma_1}$, we obtain:

$$\tau_2 = \begin{cases} t_{\zeta(\sigma_1 - 1)} & \text{if } \Delta_2 = -1; \\ t_{\zeta(\sigma_1)} & \text{if } \Delta_2 = +1. \end{cases}$$

Since $\Delta_1 = -1$ in this case, this is as required. Note that there is exactly one case here where $\sigma_1 \not\le t_n$, and that is when $i_1 = 0$ and $\sigma_1 = t_n + 1$, whereupon $\Delta_2 = -1$ and $\zeta(\sigma_1 - 1) = n - 1$ as required.

If $a > (c_n - 1)y_n = p' - \kappa_{t_n+1}$ then $a_0 = p' - \kappa_{x+1}$ and $a_1 = p' - \kappa_x$ for some x, whereupon $\tau_2 = t_{\zeta(x)}$. So with $\Delta_2 = -(-1)^{i_1}$ and $a_{\bar{i}_1} = p' - \kappa_{\sigma_1}$, we obtain:

$$\tau_2 = \begin{cases} t_{\zeta(\sigma_1 - 1)} & \text{if } \Delta_2 = +1; \\ t_{\zeta(\sigma_1)} & \text{if } \Delta_2 = -1. \end{cases}$$

Since $\Delta_1 = +1$ in this case, this is as required. Note that there is exactly one case here where $\sigma_1 \not\leq t_n$, and that is when $i_1 = 1$ and $\sigma_0 = t_n + 1$, whereupon $\Delta_2 = +1$ and $\zeta(\sigma_1 - 1) = n - 1$ as required.

If $ky_n + y_{n-1} < a < (k+1)y_n$ for $1 \leq k \leq c_n - 2$ then $a_0 = \kappa_{t_n+k}$ and $a_1 = p' - \kappa_{t_{n+1}-k-1}$ so that $a_0 \in \mathcal{T}$ and $a_1 \in \mathcal{T}'$. Then:

$$\kappa_{\tau_2} = y_n - y_{n-1} = \begin{cases} \kappa_{t_n - 1} & \text{if } c_n > 1; \\ \kappa_{t_{n-2}} & \text{if } c_n = 1. \end{cases}$$

We now claim that $\Delta_1 \neq \Delta_2$ in this case as required. This follows because if $i_1 = 0$ then $\Delta_1 = +1$ and $\Delta_2 = -(-1)^0 = -1$ and if $i_1 = 1$ then $\Delta_1 = -1$ and $\Delta_2 = -(-1)^1 = +1$.

If $ky_n < a < ky_n + y_{n-1}$ for $2 \leq k \leq c_n - 2$ then $a_0 = p' - \kappa_{t_{n+1}-k}$ and $a_1 = \kappa_{t_n+k}$ so that $a_0 \in \mathcal{T}'$ and $a_1 \in \mathcal{T}$. Then $\kappa_{\tau_2} = y_{n-1} = \kappa_{t_{n-1}}$. We now claim that $\Delta_1 = \Delta_2$ in this case as required. This follows because if $i_1 = 0$ then $\Delta_1 = -1$ and $\Delta_2 = -(-1)^0 = -1$ and if $i_1 = 1$ then $\Delta_1 = +1$ and $\Delta_2 = -(-1)^1 = +1$. □

Lemma 8.6. *Let $\{\tau_j, \sigma_j, \Delta_j\}_{j=1}^d$ be the naive run corresponding to the leaf-node $a_{i_1 i_2 \cdots i_{d-1} 0}$ of the Takahashi tree for a. Then*

$$a_{i_1 i_2 \cdots i_{k-1} \bar{i}_k} = \sum_{m=2}^{k} \Delta_m (\kappa_{\tau_m} - \kappa_{\sigma_m}) \quad + \quad \begin{cases} \kappa_{\sigma_1} & \text{if } \Delta_1 = -1; \\ p' - \kappa_{\sigma_1} & \text{if } \Delta_1 = +1, \end{cases}$$

for $1 \leq k \leq d$, and

$$a_{i_1 i_2 \cdots i_{k-1} i_k} = a_{i_1 i_2 \cdots i_{k-1} \bar{i}_k} + \Delta_{k+1} \kappa_{\tau_{k+1}},$$

for $1 \leq k < d$.

Proof: The prescription of Section 1.7 implies that $\kappa_{\tau_{k+1}} = a_{i_1 i_2 \cdots i_{k-1} 1} - a_{i_1 i_2 \cdots i_{k-1} 0}$, whence $\Delta_{k+1} = -(-1)^{i_k}$ implies that $a_{i_1 i_2 \cdots i_k} = a_{i_1 i_2 \cdots \bar{i}_k} + \Delta_{k+1} \kappa_{\tau_{k+1}}$ for $1 \leq k < d$, thus giving the second expression.

By definition, $a_{\bar{i}_1} = \kappa_{\sigma_1}$ if $a_{\bar{i}_1} \in \mathcal{T}$ and thus $\Delta_1 = -1$; and $a_{\bar{i}_1} = p' - \kappa_{\sigma_1}$ if $a_{\bar{i}_1} \in \mathcal{T}'$ and thus $\Delta_1 = +1$. This gives the $k = 1$ case of the first expression.

The description of Section 1.6 implies that $-(-1)^{i_{k-1}}(a_{i_1 i_2 \cdots i_{k-1}} - a_{i_1 i_2 \cdots i_{k-1} i}) > 0$ for $i \in \{0, 1\}$ and $2 \leq k \leq d$. When $i = \bar{i}_k$, Section 1.7 specifies that this expression equals κ_{σ_k}. Thereupon, we obtain $a_{i_1 i_2 \cdots i_{k-1} \bar{i}_k} = a_{i_1 i_2 \cdots i_{k-1}} - \Delta_k \kappa_{\sigma_k}$. The lemma then follows by induction. □

8.3. Gathering in the Takahashi tree

As described in Section 1.7, each leaf-node $a_{i_1 i_2 \cdots i_{d-1} 0}$ of the Takahashi tree for a gives rise firstly to the naive run $\{\tau_j, \sigma_j, \Delta_j\}_{j=1}^d$ and then to a vector $\boldsymbol{u} \in \mathcal{U}(a)$ by (1.21). For the particular leaf-node $a_{i_1 i_2 \cdots i_{d-1} 0}$, we will denote this corresponding vector by $\boldsymbol{u}(a; i_1, i_2, \ldots, i_{d-1})$. Note that $\Delta(\boldsymbol{u}(a; i_1, i_2, \ldots, i_{d-1})) = \Delta_d$.

For vectors $\boldsymbol{u}(a; i_1, i_2, \ldots, i_{d^L-1})$ and $\boldsymbol{u}(b; j_1, j_2, \ldots, j_{d^R-1})$ arising from leaf-nodes of the Takahashi trees for a and b respectively, the following lemma identifies a set of paths for which $F(\boldsymbol{u}(a; i_1, i_2, \ldots, i_{d^L-1}), \boldsymbol{u}(b; j_1, j_2, \ldots, j_{d^R-1}), L)$ is the generating function. The lemma makes use of the following definition:

$$\Xi_0(s) = \begin{cases} 0 & \text{if } s < \kappa_{t_n}; \\ p' & \text{if } s > p' - \kappa_{t_n}; \end{cases}$$

$$\Xi_1(s) = \begin{cases} \kappa_{t_n} & \text{if } s < \kappa_{t_n}; \\ p' - \kappa_{t_n} & \text{if } s > p' - \kappa_{t_n}. \end{cases}$$

(Outside of the ranges given here, $\Xi_0(s)$ and $\Xi_1(s)$ are not defined and not needed.)

Lemma 8.7. *Let $1 \le a, b < p'$, let $a_{i_1 i_2 \cdots i_{d^L-1} 0}$ be a leaf-node of the Takahashi tree of a, and let $b_{j_1 j_2 \cdots j_{d^R-1} 0}$ be a leaf-node of the Takahashi tree of b. If b is interfacial in the (p,p')-model, set $c \in \{b \pm 1\}$. If b is not interfacial in the (p,p')-model, set $c = b + \Delta(\boldsymbol{u}(b; j_1, j_2, \ldots, j_{d^R-1}))$.*

(1) *If $\kappa_{t_n} \le a_{\bar{\imath}_1} \le p' - \kappa_{t_n}$ and $\kappa_{t_n} \le b_{\bar{\jmath}_1} \le p' - \kappa_{t_n}$ then:*

$$\chi^{p,p'}_{a,b,c}(L) \left\{ \begin{matrix} a_{i_1}, a_{i_1 i_2}, \ldots, a_{i_1 i_2 \cdots i_{d^L-1}}; b_{j_1}, b_{j_1 j_2}, \ldots, b_{j_1 j_2 \cdots j_{d^R-1}} \\ a_{\bar{\imath}_1}, a_{i_1 \bar{\imath}_2}, \ldots, a_{i_1 i_2 \cdots \bar{\imath}_{d^L-1}}; b_{\bar{\jmath}_1}, b_{j_1 \bar{\jmath}_2}, \ldots, b_{j_1 j_2 \cdots \bar{\jmath}_{d^R-1}} \end{matrix} \right\}$$

$$= F(\boldsymbol{u}(a; i_1, i_2, \ldots, i_{d^L-1}), \boldsymbol{u}(b; j_1, j_2, \ldots, j_{d^R-1}), L).$$

(2) *If $a_{\bar{\imath}_1} < \kappa_{t_n}$ or $a_{\bar{\imath}_1} > p' - \kappa_{t_n}$, and $\kappa_{t_n} \le b_{\bar{\jmath}_1} \le p' - \kappa_{t_n}$ then:*

$$\chi^{p,p'}_{a,b,c}(L) \left\{ \begin{matrix} \Xi_0(a_{\bar{\imath}_1}), a_{i_1}, a_{i_1 i_2}, \ldots, a_{i_1 i_2 \cdots i_{d^L-1}}; b_{j_1}, b_{j_1 j_2}, \ldots, b_{j_1 j_2 \cdots j_{d^R-1}} \\ \Xi_1(a_{\bar{\imath}_1}), a_{\bar{\imath}_1}, a_{i_1 \bar{\imath}_2}, \ldots, a_{i_1 i_2 \cdots \bar{\imath}_{d^L-1}}; b_{\bar{\jmath}_1}, b_{j_1 \bar{\jmath}_2}, \ldots, b_{j_1 j_2 \cdots \bar{\jmath}_{d^R-1}} \end{matrix} \right\}$$

$$= F(\boldsymbol{u}(a; i_1, i_2, \ldots, i_{d^L-1}), \boldsymbol{u}(b; j_1, j_2, \ldots, j_{d^R-1}), L).$$

(3) *If $\kappa_{t_n} \le a_{\bar{\imath}_1} \le p' - \kappa_{t_n}$ and, $b_{\bar{\jmath}_1} < \kappa_{t_n}$ or $b_{\bar{\jmath}_1} > p' - \kappa_{t_n}$ then:*

$$\chi^{p,p'}_{a,b,c}(L) \left\{ \begin{matrix} a_{i_1}, a_{i_1 i_2}, \ldots, a_{i_1 i_2 \cdots i_{d^L-1}}; \Xi_0(b_{\bar{\jmath}_1}), b_{j_1}, b_{j_1 j_2}, \ldots, b_{j_1 j_2 \cdots j_{d^R-1}} \\ a_{\bar{\imath}_1}, a_{i_1 \bar{\imath}_2}, \ldots, a_{i_1 i_2 \cdots \bar{\imath}_{d^L-1}}; \Xi_1(b_{\bar{\jmath}_1}), b_{\bar{\jmath}_1}, b_{j_1 \bar{\jmath}_2}, \ldots, b_{j_1 j_2 \cdots \bar{\jmath}_{d^R-1}} \end{matrix} \right\}$$

$$= F(\boldsymbol{u}(a; i_1, i_2, \ldots, i_{d^L-1}), \boldsymbol{u}(b; j_1, j_2, \ldots, j_{d^R-1}), L).$$

(4) *If $a_{\bar{\imath}_1} < \kappa_{t_n}$ or $a_{\bar{\imath}_1} > p' - \kappa_{t_n}$, and, $b_{\bar{\jmath}_1} < \kappa_{t_n}$ or $b_{\bar{\jmath}_1} > p' - \kappa_{t_n}$ then:*

$$\chi^{p,p'}_{a,b,c}(L) \left\{ \begin{matrix} \Xi_0(a_{\bar{\imath}_1}), a_{i_1}, a_{i_1 i_2}, \ldots, a_{i_1 i_2 \cdots i_{d^L-1}}; \Xi_0(b_{\bar{\jmath}_1}), b_{j_1}, b_{j_1 j_2}, \ldots, b_{j_1 j_2 \cdots j_{d^R-1}} \\ \Xi_1(a_{\bar{\imath}_1}), a_{\bar{\imath}_1}, a_{i_1 \bar{\imath}_2}, \ldots, a_{i_1 i_2 \cdots \bar{\imath}_{d^L-1}}; \Xi_1(b_{\bar{\jmath}_1}), b_{\bar{\jmath}_1}, b_{j_1 \bar{\jmath}_2}, \ldots, b_{j_1 j_2 \cdots \bar{\jmath}_{d^R-1}} \end{matrix} \right\}$$

$$= F(\boldsymbol{u}(a; i_1, i_2, \ldots, i_{d^L-1}), \boldsymbol{u}(b; j_1, j_2, \ldots, j_{d^R-1}), L).$$

Proof: Let the leaf-node $a_{i_1 i_2 \cdots i_{d^L-1} 0}$ of the Takahashi tree for a give rise to the run $\{\tau^L_j, \sigma^L_j, \Delta^L_j\}_{j=1}^{d^L}$ as in Section 1.7. Likewise, let the leaf-node $b_{j_1 j_2 \cdots j_{d^R-1} 0}$ of the Takahashi tree for b give rise to the run $\{\tau^R_j, \sigma^R_j, \Delta^R_j\}_{j=1}^{d^R}$. Lemma 8.5 states that $\{\tau^L_j, \sigma^L_j, \Delta^L_j\}_{j=1}^{d^L}$ and $\{\tau^R_j, \sigma^R_j, \Delta^R_j\}_{j=1}^{d^R}$ are both naive runs.

Using these naive runs, define μ^*_k and μ_k for $0 \le k < d^L$, ν^*_k and ν_k for $0 \le k < d^R$, and d^L_0 and d^R_0 as in Section 7.1. Then, Lemma 8.6 states that

$a_{i_1 i_2 \cdots i_{k-1} \bar{\imath}_k} = \mu_k^*$ and $a_{i_1 i_2 \cdots i_{k-1} i_k} = \mu_k$ for $1 \leq k < d^L$, and $b_{j_1 j_2 \cdots j_{k-1} \bar{\jmath}_k} = \nu_k^*$ and $b_{j_1 j_2 \cdots j_{k-1} j_k} = \nu_k$ for $1 \leq k < d^R$.

Now consider $\kappa_{t_n} \leq a_{\bar{\imath}_1} \leq p' - \kappa_{t_n}$. If $a_{\bar{\imath}_1} \in \mathcal{T}$ then $\kappa_{\sigma_1^L} \geq \kappa_{t_n}$ which implies that $\sigma_1^L \geq t_n$. If $a_{\bar{\imath}_1} \in \mathcal{T}'$ then $p' - \kappa_{\sigma_1^L} \leq p' - \kappa_{t_n}$ which also implies that $\sigma_1^L \geq t_n$. Similarly $\kappa_{t_n} \leq b_{\bar{\jmath}_1} \leq p' - \kappa_{t_n}$ implies that $\sigma_1^R \geq t_n$. Then $d_0^L = d_0^R = 1$, whence the first part of the lemma follows directly from Theorem 7.16 after noting that $\Delta_{d^R}^R = \Delta(\boldsymbol{u}(b; j_1, j_2, \ldots, j_{d^R-1}))$.

If $a_{\bar{\imath}_1} < \kappa_{t_n}$ then $a_{\bar{\imath}_1} \in \mathcal{T}$ whereupon $\sigma_1^L < t_n$, $\Delta_1^L = -1$ and $d_0^L = 0$. If $a_{\bar{\imath}_1} > p' - \kappa_{t_n}$ then $a_{\bar{\imath}_1} \in \mathcal{T}'$ whereupon $\sigma_1^L < t_n$, $\Delta_1^L = +1$ and $d_0^L = 0$. The second part of the lemma now follows directly from Theorem 7.16 after again noting that $\Delta_{d^R}^R = \Delta(\boldsymbol{u}(b; j_1, j_2, \ldots, j_{d^R-1}))$.

The remaining two parts follow analogously. □

Lemma 8.8. *Let $1 \leq a < p'$, and $2 \leq b < p' - 1$ with b interfacial in the (p, p')-model, $a \notin \mathcal{T} \cup \mathcal{T}'$ and $b \notin \mathcal{T} \cup \mathcal{T}'$. Set $c \in \{b \pm 1\}$.*

(1) *If $\kappa_{t_n} \leq a_{\bar{\imath}_1} \leq p' - \kappa_{t_n}$ and $\kappa_{t_n} \leq b_{\bar{\jmath}_1} \leq p' - \kappa_{t_n}$ then:*

$$\chi_{a,b,c}^{p,p'}(L) \left\{ \begin{matrix} a_{i_1}; b_{j_1} \\ a_{\bar{\imath}_1}; b_{\bar{\jmath}_1} \end{matrix} \right\} = \sum_{\substack{\boldsymbol{u}(a; i_1', i_2', \ldots, i_{d^L-1}') \in \mathcal{U}(a): i_1' = i_1 \\ \boldsymbol{u}(b; j_1', j_2', \ldots, j_{d^R-1}') \in \mathcal{U}(b): j_1' = j_1}} F(\boldsymbol{u}(a; i_1', i_2', \ldots, i_{d^L-1}'), \boldsymbol{u}(b; j_1', j_2', \ldots, j_{d^R-1}'), L).$$

(2) *If $a_{\bar{\imath}_1} < \kappa_{t_n}$ or $a_{\bar{\imath}_1} > p' - \kappa_{t_n}$, and $\kappa_{t_n} \leq b_{\bar{\jmath}_1} \leq p' - \kappa_{t_n}$ then:*

$$\chi_{a,b,c}^{p,p'}(L) \left\{ \begin{matrix} \Xi_0(a_{\bar{\imath}_1}), a_{i_1}; b_{j_1} \\ \Xi_1(a_{\bar{\imath}_1}), a_{\bar{\imath}_1}; b_{\bar{\jmath}_1} \end{matrix} \right\} = \sum_{\substack{\boldsymbol{u}(a; i_1', i_2', \ldots, i_{d^L-1}') \in \mathcal{U}(a): i_1' = i_1 \\ \boldsymbol{u}(b; j_1', j_2', \ldots, j_{d^R-1}') \in \mathcal{U}(b): j_1' = j_1}} F(\boldsymbol{u}(a; i_1', i_2', \ldots, i_{d^L-1}'), \boldsymbol{u}(b; j_1', j_2', \ldots, j_{d^R-1}'), L).$$

(3) *If $\kappa_{t_n} \leq a_{\bar{\imath}_1} \leq p' - \kappa_{t_n}$ and, $b_{\bar{\jmath}_1} < \kappa_{t_n}$ or $b_{\bar{\jmath}_1} > p' - \kappa_{t_n}$ then:*

$$\chi_{a,b,c}^{p,p'}(L) \left\{ \begin{matrix} a_{i_1}; \Xi_0(b_{\bar{\jmath}_1}), b_{j_1} \\ a_{\bar{\imath}_1}; \Xi_1(b_{\bar{\jmath}_1}), b_{\bar{\jmath}_1} \end{matrix} \right\} = \sum_{\substack{\boldsymbol{u}(a; i_1', i_2', \ldots, i_{d^L-1}') \in \mathcal{U}(a): i_1' = i_1 \\ \boldsymbol{u}(b; j_1', j_2', \ldots, j_{d^R-1}') \in \mathcal{U}(b): j_1' = j_1}} F(\boldsymbol{u}(a; i_1', i_2', \ldots, i_{d^L-1}'), \boldsymbol{u}(b; j_1', j_2', \ldots, j_{d^R-1}'), L).$$

(4) *If $a_{\bar{\imath}_1} < \kappa_{t_n}$ or $a_{\bar{\imath}_1} > p' - \kappa_{t_n}$, and, $b_{\bar{\jmath}_1} < \kappa_{t_n}$ or $b_{\bar{\jmath}_1} > p' - \kappa_{t_n}$ then:*

$$\chi_{a,b,c}^{p,p'}(L) \left\{ \begin{matrix} \Xi_0(a_{\bar{\imath}_1}), a_{i_1}; \Xi_0(b_{\bar{\jmath}_1}), b_{j_1} \\ \Xi_1(a_{\bar{\imath}_1}), a_{\bar{\imath}_1}; \Xi_1(b_{\bar{\jmath}_1}), b_{\bar{\jmath}_1} \end{matrix} \right\} = \sum_{\substack{\boldsymbol{u}(a; i_1', i_2', \ldots, i_{d^L-1}') \in \mathcal{U}(a): i_1' = i_1 \\ \boldsymbol{u}(b; j_1', j_2', \ldots, j_{d^R-1}') \in \mathcal{U}(b): j_1' = j_1}} F(\boldsymbol{u}(a; i_1', i_2', \ldots, i_{d^L-1}'), \boldsymbol{u}(b; j_1', j_2', \ldots, j_{d^R-1}'), L).$$

(Note that in each of the above sums, the values of d^L and d^R vary over the elements of $\mathcal{U}(a)$ and $\mathcal{U}(b)$.)

Proof: Since $a \notin \mathcal{T} \cup \mathcal{T}'$ and $b \notin \mathcal{T} \cup \mathcal{T}'$, the Takahashi trees of a and b each have more than one leaf-node. The results then follow from Lemma 8.7, and repeated use of Lemma 8.1. □

Note 8.9. *If either $a \in \mathcal{T} \cup \mathcal{T}'$ or $b \in \mathcal{T} \cup \mathcal{T}'$, then we obtain analogues of the four results of Lemma 8.8 (obtaining sixteen cases in all), which we won't write out in full. Each such case is obtained from the corresponding case of Lemma 8.8 by, if $a \in \mathcal{T} \cup \mathcal{T}'$ (so that $a_0 = a_1 = a$), omitting the column containing a_{i_1} and $a_{\bar{\imath}_1}$; and if $b \in \mathcal{T} \cup \mathcal{T}'$ (so that $b_0 = b_1 = b$), omitting the column containing b_{j_1} and $b_{\bar{\jmath}_1}$. For example, if $a \in \mathcal{T} \cup \mathcal{T}'$ and $b \notin \mathcal{T} \cup \mathcal{T}'$, then the appropriate analogue of Lemma 8.8(2) states that:*

If $a < \kappa_{t_n}$ or $a > p' - \kappa_{t_n}$, and $\kappa_{t_n} \leq b_{\bar{\jmath}_1} \leq p' - \kappa_{t_n}$ then:

$$\chi^{p,p'}_{a,b,c}(L) \left\{ \begin{matrix} \Xi_0(a); b_{j_1} \\ \Xi_1(a); b_{\bar{\jmath}_1} \end{matrix} \right\}$$

$$= \sum_{\substack{\boldsymbol{u}(a;i'_1,i'_2,\ldots,i'_{d^L-1}) \in \mathcal{U}(a): i'_1 = i_1 \\ \boldsymbol{u}(b;j'_1,j'_2,\ldots,j'_{d^R-1}) \in \mathcal{U}(b): j'_1 = j_1}} F(\boldsymbol{u}(a;i'_1,i'_2,\ldots,i'_{d^L-1}), \boldsymbol{u}(b;j'_1,j'_2,\ldots,j'_{d^R-1}), L).$$

Note further that if $a \in \mathcal{T} \cup \mathcal{T}'$ then $\mathcal{U}(a)$ contains just one element, and likewise, if $b \in \mathcal{T} \cup \mathcal{T}'$ then $\mathcal{U}(b)$ contains just one element.

The following two theorems make use of the following definition of $\hat{c}$ in terms of parameters that will be specified in those theorems:

$$\hat{c} = \begin{cases} 2 & \text{if } c = \xi_\eta > 0 \text{ and } \left\lfloor \frac{(b+1)p}{p'} \right\rfloor = \left\lfloor \frac{(b-1)p}{p'} \right\rfloor + 1; \\ \hat{p}' - 2 & \text{if } c = \xi_{\eta+1} < p' \text{ and } \left\lfloor \frac{(b+1)p}{p'} \right\rfloor = \left\lfloor \frac{(b-1)p}{p'} \right\rfloor + 1; \\ c - \xi_\eta & \text{otherwise.} \end{cases} \tag{8.2}$$

Note that if b is interfacial, then certainly $\left\lfloor \frac{(b+1)p}{p'} \right\rfloor = \left\lfloor \frac{(b-1)p}{p'} \right\rfloor + 1$.

Theorem 8.10. *Let $1 \leq a < p'$, and $2 \leq b < p' - 1$ with b interfacial in the (p,p')-model. Let η be such that $\xi_\eta \leq a < \xi_{\eta+1}$ and η' be such that $\xi_{\eta'} \leq b < \xi_{\eta'+1}$. Set $c \in \{b \pm 1\}$.*
1. If $\eta = \eta'$ and $\xi_\eta < a$ and $\xi_\eta < b$ then:

$$\chi^{p,p'}_{a,b,c}(L) = \sum_{\substack{\boldsymbol{u}^L \in \mathcal{U}(a) \\ \boldsymbol{u}^R \in \mathcal{U}(b)}} F(\boldsymbol{u}^L, \boldsymbol{u}^R, L) \quad + \quad \chi^{\hat{p},\hat{p}'}_{\hat{a},\hat{b},\hat{c}}(L),$$

where $\hat{p}' = \xi_{\eta+1} - \xi_\eta$, $\hat{p} = \tilde{\xi}_{\eta+1} - \tilde{\xi}_\eta$, $\hat{a} = a - \xi_\eta$, $\hat{b} = b - \xi_\eta$ and $\hat{c}$ is given by (8.2).
2. Otherwise:

$$\chi^{p,p'}_{a,b,c}(L) = \sum_{\substack{\boldsymbol{u}^L \in \mathcal{U}(a) \\ \boldsymbol{u}^R \in \mathcal{U}(b)}} F(\boldsymbol{u}^L, \boldsymbol{u}^R, L).$$

Proof: There are many cases to consider here.

For the moment, assume that $a \notin \mathcal{T} \cup \mathcal{T}'$ and $b \notin \mathcal{T} \cup \mathcal{T}'$ so that the Takahashi trees for a and b each have at least two leaf-nodes. In particular, we have $a_0 < a < a_1$ and $b_0 < b < b_1$, and certainly $\xi_\eta < a$ and $\xi_{\eta'} < b$. If $\kappa_{t_n} < a < p' - \kappa_{t_n}$ and $\kappa_{t_n} < b < p' - \kappa_{t_n}$ then $\kappa_{t_n} \le a_0 < a_1 \le p' - \kappa_{t_n}$ and $\kappa_{t_n} \le b_0 < b_1 \le p' - \kappa_{t_n}$. Thereupon, if $\eta \ne \eta'$, combining Lemma 8.2 and Lemma 8.8(1) gives the required result. If $\eta = \eta'$, combining Lemma 8.3 and Lemma 8.8(1) gives the required result after noting that Lemma E.11 implies that the band structure of the (p,p')-model between heights $\xi_\eta + 1$ and $\xi_{\eta+1} - 1$ is identical to the band structure of the $(\hat{p},\hat{p}')$-model.

If $a_0 < a_1 < \kappa_{t_n}$ then $\Xi_0(a_0) = \Xi_0(a_1) = 0$ and $\Xi_1(a_0) = \Xi_1(a_1) = \kappa_{t_n}$. When $\kappa_{t_n} \le b_0 < b_1 \le p' - \kappa_{t_n}$, use of Lemma 8.8(2) and Lemma 8.1 yields:

$$\chi^{p,p'}_{a,b,c}(L) \left\{ \begin{matrix} 0; & b_{j_1} \\ \kappa_{t_n}; & b_{\bar{j}_1} \end{matrix} \right\}$$

$$= \sum_{\substack{\boldsymbol{u}(a;i'_1,i'_2,\ldots,i'_{d^L-1}) \in \mathcal{U}(a) \\ \boldsymbol{u}(b;j'_1,j'_2,\ldots,j'_{d^R-1}) \in \mathcal{U}(b) : j'_1 = j_1}} F(\boldsymbol{u}(a;i'_1,i'_2,\ldots,i'_{d^L-1}), \boldsymbol{u}(b;j'_1,j'_2,\ldots,j'_{d^R-1}), L).$$

Then, since $0 < a < \kappa_{t_n} < b$, Lemma 8.2 gives the required result.

If $a_0 < a_1 = \kappa_{t_n}$ then we can use the same argument as in the above case, after noting that:

$$\chi^{p,p'}_{a,b,c}(L) \left\{ \begin{matrix} 0, & a_0; & b_{j_1} \\ \kappa_{t_n}, & a_1; & b_{\bar{j}_1} \end{matrix} \right\} = \chi^{p,p'}_{a,b,c}(L) \left\{ \begin{matrix} a_0; & b_{j_1} \\ a_1; & b_{\bar{j}_1} \end{matrix} \right\}$$

$$= \sum_{\substack{\boldsymbol{u}(a;i'_1,i'_2,\ldots,i'_{d^L-1}) \in \mathcal{U}(a) : i'_1 = 0 \\ \boldsymbol{u}(b;j'_1,j'_2,\ldots,j'_{d^R-1}) \in \mathcal{U}(b) : j'_1 = j_1}} F(\boldsymbol{u}(a;i'_1,i'_2,\ldots,i'_{d^L-1}), \boldsymbol{u}(b;j'_1,j'_2,\ldots,j'_{d^R-1}), L),$$

where the first equality is a consequence of (the analogue of) Lemma 5.6, and the second follows from Lemma 8.8(1).

The case $p' - \kappa_{t_n} \le a_0 < a_1$ is dealt with in a similar fashion, again using Lemmas 8.8(2), 8.1, and 8.2, and additionally if $a_0 = p' - \kappa_{t_n}$, Lemmas 5.6 and 8.8(1).

Each of the remaining cases for which $a \notin \mathcal{T} \cup \mathcal{T}'$ and $b \notin \mathcal{T} \cup \mathcal{T}'$ is dealt with in a similar fashion, using the appropriate part(s) of Lemma 8.8.

Now if either $a \in \mathcal{T} \cup \mathcal{T}'$ or $b \in \mathcal{T} \cup \mathcal{T}'$ (so that $d^L = 1$ or $d^R = 1$ respectively), then the result also follows in a similar way, making use of the appropriate analogue of Lemma 8.8 as described in Note 8.9. □

Note 8.11. *1. An analogue of Lemma 8.7 that deals with the case of b being non-interfacial in the (p,p')-model and $c = b - \Delta(\boldsymbol{u}(b;j_1,j_2,\ldots,j_{d^R-1}))$ is obtained upon replacing (in each case) $F(\boldsymbol{u}(a;i_1,i_2,\ldots,i_{d^L-1}), \boldsymbol{u}(b;j_1,j_2,\ldots,j_{d^R-1}), L)$ by $\tilde{F}(\boldsymbol{u}(a;i_1,i_2,\ldots,i_{d^L-1}), \boldsymbol{u}(b;j_1,j_2,\ldots,j_{d^R-1}), L)$. The proof of this result is identical to that of Lemma 8.7 except that Theorem 7.17 is used instead of Theorem 7.16.*

2. An analogue of Lemma 8.8 that deals with the case of b being non-interfacial in the (p,p')-model or $b \in \{1, p'-1\}$, is obtained upon, whenever the second summation index is such that $\Delta(\boldsymbol{u}(b; j'_1, j'_2, \ldots, j'_{d^R-1}), L) = b - c$, replacing the summand $F(\,\boldsymbol{u}(a\,; i'_1, i'_2, \ldots, i'_{d^L-1}\,),\, \boldsymbol{u}(b\,; j'_1, j'_2, \ldots, j'_{d^R-1}\,),\, L\,)$ by $\tilde{F}(\,\boldsymbol{u}(a\,; i'_1, i'_2, \ldots, i'_{d^L-1}\,),$ $\boldsymbol{u}(b\,; j'_1, j'_2, \ldots, j'_{d^R-1}\,),\, L\,)$. This follows from the use of Lemma 8.7, the modified form of Lemma 8.7 described above, and repeated use of Lemma 8.1 as in the proof of Lemma 8.8. The analogues of those results described in Note 8.9 also follow.

Theorem 8.12. *Let $1 \le a, b < p'$ with b non-interfacial in the (p,p')-model. Let η be such that $\xi_\eta \le a < \xi_{\eta+1}$ and η' be such that $\xi_{\eta'} \le b < \xi_{\eta'+1}$. Set $c \in \{b \pm 1\}$.*
1. If $\eta = \eta'$ and $\xi_\eta < a$ and $\xi_\eta < b$ then:

$$\chi^{p,p'}_{a,b,c}(L) = \sum_{\substack{\boldsymbol{u}^L \in \mathcal{U}(a) \\ \boldsymbol{u}^R \in \mathcal{U}^{c-b}(b)}} F(\boldsymbol{u}^L, \boldsymbol{u}^R, L) \;+\; \sum_{\substack{\boldsymbol{u}^L \in \mathcal{U}(a) \\ \boldsymbol{u}^R \in \mathcal{U}^{b-c}(b)}} \widetilde{F}(\boldsymbol{u}^L, \boldsymbol{u}^R, L) \;+\; \chi^{\hat{p},\hat{p}'}_{\hat{a},\hat{b},\hat{c}}(L),$$

where $\hat{p}' = \xi_{\eta+1} - \xi_\eta$, $\hat{p} = \tilde{\xi}_{\eta+1} - \tilde{\xi}_\eta$, $\hat{a} = a - \xi_\eta$, $\hat{b} = b - \xi_\eta$ and $\hat{c}$ is given by (8.2).
2. Otherwise:

$$\chi^{p,p'}_{a,b,c}(L) = \sum_{\substack{\boldsymbol{u}^L \in \mathcal{U}(a) \\ \boldsymbol{u}^R \in \mathcal{U}^{c-b}(b)}} F(\boldsymbol{u}^L, \boldsymbol{u}^R, L) \;+\; \sum_{\substack{\boldsymbol{u}^L \in \mathcal{U}(a) \\ \boldsymbol{u}^R \in \mathcal{U}^{b-c}(b)}} \widetilde{F}(\boldsymbol{u}^L, \boldsymbol{u}^R, L).$$

Proof: With Note 8.11 in mind, this theorem is proved in the same way as Theorem 8.10. □

Theorems 8.10 and 8.12 prove the expressions given in (1.37) and (1.43) respectively.

9. Fermionic character expressions

In this section we obtain fermionic expressions for the Virasoro characters $\chi^{p,p'}_{r,s}$. In the first instance, in Section 9.1, these character expressions are in terms of the Takahashi trees of s and b, where b is related to r by (1.10) for some $c \in \{b\pm1\}$. From these expressions, in Section 9.2, simpler fermionic expressions for $\chi^{p,p'}_{r,s}$ in terms of the Takahashi tree for s and the truncated Takahashi tree for r are derived.

9.1. Taking the limit $L \to \infty$.

In order to obtain fermionic expressions for the characters $\chi^{p,p'}_{r,s}$, we take the $L \to \infty$ limit in (1.37) and (1.43). However, this cannot be done immediately: it is first necessary to change variables in (1.40). To this end, given $\mathcal{X} = \{\tau_j, \sigma_j, \Delta_j\}_{j=1}^{d}$, we define $\boldsymbol{N}(\mathcal{X}) = (N_1, N_2, \ldots, N_{t_1})$, where for $1 \le i \le t_1$:

$$N_i = \begin{cases} 0 & \text{if } 1 \le i \le \sigma_d; \\ i - \sigma_d & \text{if } \sigma_d \le i \le \tau_d; \\ \tau_d - \sigma_d & \text{if } \tau_d \le i \le t_1. \end{cases}$$

Then define:

$$F^*(\boldsymbol{u}^L, \boldsymbol{u}^R, L) = \sum q^{\tilde{\boldsymbol{n}}^T \boldsymbol{B} \tilde{\boldsymbol{n}} - \boldsymbol{N}(\mathcal{X}^R)\cdot\tilde{\boldsymbol{n}} + \frac{1}{4}\hat{\boldsymbol{m}}^T \boldsymbol{C} \hat{\boldsymbol{m}} - \frac{1}{2}(\boldsymbol{u}^L_\flat + \boldsymbol{u}^R_\sharp)\cdot \boldsymbol{m} + \frac{1}{4}\gamma(\mathcal{X}^L, \mathcal{X}^R)}$$
$$\times \prod_{j=1}^{t_1} \begin{bmatrix} n_j + m_{t_1} + 2\sum_{i=j+1}^{t_1}(i-j)\tilde{n}_i \\ n_j \end{bmatrix}_q$$
$$\times \prod_{j=t_1+1}^{t-1} \begin{bmatrix} m_j - \frac{1}{2}\left(\sum_{i=t_1}^{t-1} C_{ji} m_i - (\boldsymbol{u}^L)_j - (\boldsymbol{u}^R)_j\right) \\ m_j \end{bmatrix}_q,$$

where the sum is over all $(m_{t_1}, m_{t_1+1}, \ldots, m_{t-1}) \equiv (Q_{t_1}, Q_{t_1+1}, \ldots, Q_{t-1})$, and over all $\boldsymbol{n} = (n_1, n_2, \ldots, n_{t_1})$ such that

$$\sum_{i=1}^{t_1} i(2n_i - u_i) = L - m_{t_1} - t_1 m_{t_1+1}, \tag{9.1}$$

where $(u_1, u_2, \ldots, u_t) = \boldsymbol{u}^L + \boldsymbol{u}^R$. Here, $\tilde{\boldsymbol{n}}$ is defined by (1.18), and the t-dimensional vector $\hat{\boldsymbol{m}}$ and the $(t-1)$-dimensional vector $\boldsymbol{m}$ are defined by:

$$\begin{aligned} \hat{\boldsymbol{m}} &= (0, 0, 0, \ldots, 0, m_{t_1+1}, m_{t_1+2}, \ldots, m_{t-1}); \\ \boldsymbol{m} &= (0, 0, \ldots, 0, m_{t_1+1}, m_{t_1+2}, \ldots, m_{t-1}), \end{aligned} \tag{9.2}$$

where the appropriate number of zeros have been included. In addition, $\mathcal{X}^L = \mathcal{X}(a, \boldsymbol{u}^L)$ and $\mathcal{X}^R = \mathcal{X}(b, \boldsymbol{u}^R)$, as defined in Section 1.7.

For $\boldsymbol{u} \in \mathcal{U}(b)$ and $\boldsymbol{u} = \boldsymbol{u}(\mathcal{X})$ with $\mathcal{X} = \{\tau_j, \sigma_j, \Delta_j\}_{j=1}^{d}$, we also define $\sigma(\boldsymbol{u}) = \sigma_d$ and $\Delta(\boldsymbol{u}) = \Delta_d$.

The results of this section involve the subset $\overline{\mathcal{U}}(b)$ of $\mathcal{U}(b)$ defined by:

$$\overline{\mathcal{U}}(b) = \{\boldsymbol{u} \in \mathcal{U}(b) : \begin{array}{l} \sigma(\boldsymbol{u}) = 0, \\ 0 < b + \Delta(\boldsymbol{u}) < p' \implies \lfloor bp/p' \rfloor \neq \lfloor (b + \Delta(\boldsymbol{u}))p/p' \rfloor, \\ b + \Delta(\boldsymbol{u}) \in \{0, p'\} \implies p' < 2p\}. \end{array} \tag{9.3}$$

Correspondingly, we define $\overline{\mathcal{U}}^{\Delta}(b) = \mathcal{U}^{\Delta}(b) \cap \overline{\mathcal{U}}(b)$.

The results that follow are valid in both cases $p' > 2p$ and $p' < 2p$.

Lemma 9.1. *Let $1 \leq a, b < p'$, and let $\boldsymbol{u}^L \in \mathcal{U}(a)$ and $\boldsymbol{u}^R \in \mathcal{U}(b)$. Then set $(Q_1, Q_2, \ldots, Q_{t-1}) = \boldsymbol{Q}(\boldsymbol{u}^L + \boldsymbol{u}^R)$.*

1. If $\boldsymbol{u}^R \in \overline{\mathcal{U}}(b)$ then:

$$F(\boldsymbol{u}^L, \boldsymbol{u}^R, L) = q^{\frac{1}{2}L} F^*(\boldsymbol{u}^L, \boldsymbol{u}^R, L).$$

2. If $\boldsymbol{u}^R \notin \overline{\mathcal{U}}(b)$ then:

$$F(\boldsymbol{u}^L, \boldsymbol{u}^R, L) = F^*(\boldsymbol{u}^L, \boldsymbol{u}^R, L).$$

Proof: In the expression (1.40) for $F(\boldsymbol{u}^L, \boldsymbol{u}^R, L)$, the summation is over variables $\{m_i\}_{i=1}^{t-1}$ with $m_0 = L$. Here we exchange $\{m_i\}_{i=0}^{t_1-1}$ for $\{n_i\}_{i=1}^{t_1}$ using the $\boldsymbol{mn}$-system. Since $m_0 = L$ is fixed, a constraint amongst the $\{n_i\}_{i=1}^{t_1}$ is required. In accordance with (1.29) and (1.30), set $n_i = \frac{1}{2}(m_{i-1} - 2m_i + m_{i+1} + u_i)$ for $1 \leq i < t_1$, and set $n_{t_1} = \frac{1}{2}(m_{t_1-1} - m_{t_1} - m_{t_1+1} + u_{t_1})$. For $1 \leq j \leq t_1$, we immediately obtain:

$$m_{j-1} - m_j = m_{t_1+1} + \sum_{i=j}^{t_1} (2n_i - u_i) = 2 \sum_{i=j}^{t_1} \tilde{n}_i.$$

Thereupon, for $0 \leq j \leq t_1$,

$$m_j = m_{t_1} + 2 \sum_{i=j+1}^{t_1} (i - j)\tilde{n}_i. \tag{9.4}$$

The $j = 0$ case here yields the constraint (9.1). Additionally, if $\hat{\boldsymbol{m}}_* = (m_0, m_1, m_2, \ldots, m_{t-1})$, we obtain:

$$\hat{\boldsymbol{m}}_*^T \boldsymbol{C} \hat{\boldsymbol{m}}_* - L^2 = \sum_{k=1}^{t_1} (m_{k-1} - m_k)^2 + \hat{\boldsymbol{m}}^T \boldsymbol{C} \hat{\boldsymbol{m}} = 4\tilde{\boldsymbol{n}}^T \boldsymbol{B} \tilde{\boldsymbol{n}} + \hat{\boldsymbol{m}}^T \boldsymbol{C} \hat{\boldsymbol{m}}.$$

Now restrict to the case $p' > 2p$ so that $t_1 > 0$ (all the above is valid in the $t_1 = 0$ case, although it's mainly degenerate). Let $\mathcal{X}^R = \{\tau_j, \sigma_j, \Delta_j\}_{j=1}^d$ and $\boldsymbol{N} = \boldsymbol{N}(\mathcal{X}^R)$. If $\sigma_d \leq t_1 \leq \tau_d$ then $(\boldsymbol{u}_\flat^L + \boldsymbol{u}_\sharp^R)_i = \delta_{\sigma_d,i} - \delta_{t_1,i}$ for $1 \leq i \leq t_1$. If $\sigma_d < \tau_d \leq t_1$ then $(\boldsymbol{u}_\flat^L + \boldsymbol{u}_\sharp^R)_i = \delta_{\sigma_d,i} - \delta_{\tau_d,i}$ for $1 \leq i \leq t_1$. Thereupon, for $0 < \sigma_d \leq t_1$,

$$\begin{aligned} (\boldsymbol{u}_\flat^L + \boldsymbol{u}_\sharp^R) \cdot (m_1, m_2, \ldots, m_{t-1}) &= (\boldsymbol{u}_\flat^L + \boldsymbol{u}_\sharp^R) \cdot \boldsymbol{m} + m_{\sigma_d} - \begin{cases} m_{\tau_d} & \text{if } \tau_d \leq t_1 \\ m_{t_1} & \text{if } \tau_d \geq t_1 \end{cases} \\ &= (\boldsymbol{u}_\flat^L + \boldsymbol{u}_\sharp^R) \cdot \boldsymbol{m} + 2\boldsymbol{N} \cdot \tilde{\boldsymbol{n}}, \end{aligned}$$

and for $\sigma_d = 0$,

$$(\boldsymbol{u}_\flat^L + \boldsymbol{u}_\sharp^R)\cdot(m_1, m_2, \ldots, m_{t-1}) = (\boldsymbol{u}_\flat^L + \boldsymbol{u}_\sharp^R)\cdot \boldsymbol{m} - \begin{cases} m_{\tau_d} & \text{if } \tau_d \le t_1 \\ m_{t_1} & \text{if } \tau_d \ge t_1 \end{cases}$$
$$= (\boldsymbol{u}_\flat^L + \boldsymbol{u}_\sharp^R)\cdot \boldsymbol{m} + 2\boldsymbol{N}\cdot\tilde{\boldsymbol{n}} - L,$$

having used (9.4) in each case. In the case $p' > 2p$ and $\sigma_d \le t_1$, the lemma then follows from the definition of $F(\boldsymbol{u}^L, \boldsymbol{u}^R, L)$, after noting that the definition of Section 1.10 implies that $\gamma'(\mathcal{X}^L, \mathcal{X}^R) = \gamma(\mathcal{X}^L, \mathcal{X}^R)$ unless both $\sigma_d = 0$ and $\boldsymbol{u}^R \notin \mathcal{U}(b)$, in which case $\gamma'(\mathcal{X}^L, \mathcal{X}^R) = \gamma(\mathcal{X}^L, \mathcal{X}^R) - 2L$.

In the case $p' > 2p$ and $\sigma_d > t_1$, the lemma follows after noting that

$$(\boldsymbol{u}_\flat^L + \boldsymbol{u}_\sharp^R)\cdot(m_1, m_2, \ldots, m_{t-1}) = (\boldsymbol{u}_\flat^L + \boldsymbol{u}_\sharp^R)\cdot \boldsymbol{m}, \tag{9.5}$$

$\boldsymbol{N} = 0$, and $\gamma'(\mathcal{X}^L, \mathcal{X}^R) = \gamma(\mathcal{X}^L, \mathcal{X}^R)$ in this case.

In the case $p' < 2p$ so that $t_1 = 0$, eq. (9.5) also holds. The lemma then follows in this case after noting from Section 1.10 that $\gamma'(\mathcal{X}^L, \mathcal{X}^R) = \gamma(\mathcal{X}^L, \mathcal{X}^R) + 2L$ if $\sigma_d = 0$ and either $\lfloor (b+\Delta_d)p/p' \rfloor \ne \lfloor bp/p' \rfloor$ or $b + \Delta_d \in \{0, p'\}$, whereas otherwise, $\gamma'(\mathcal{X}^L, \mathcal{X}^R) = \gamma(\mathcal{X}^L, \mathcal{X}^R)$. □

Note that although the set of vectors $\mathcal{U}(b)$ is identical in the cases of the (p,p')-model and the $(p'-p,p')$-model, those $\boldsymbol{u}^R$ which fall into the first category in Lemma 9.1 differ in the two cases.

We can now obtain $\lim_{L\to\infty} F(\boldsymbol{u}^L, \boldsymbol{u}^R, L)$. The result involves $F^*(\boldsymbol{u}^L, \boldsymbol{u}^R)$ defined by:

(9.6)
$$F^*(\boldsymbol{u}^L, \boldsymbol{u}^R) = \sum q^{\tilde{\boldsymbol{n}}^T \boldsymbol{B} \tilde{\boldsymbol{n}} - \boldsymbol{N}(\mathcal{X}^R)\cdot\tilde{\boldsymbol{n}} + \frac{1}{4}\hat{\boldsymbol{m}}^T \boldsymbol{C} \hat{\boldsymbol{m}} - \frac{1}{2}(\boldsymbol{u}_\flat^L + \boldsymbol{u}_\sharp^R)\cdot \boldsymbol{m} + \frac{1}{4}\gamma(\mathcal{X}^L, \mathcal{X}^R)}$$
$$\times \frac{1}{(q)_{m_{t_1+1}}} \prod_{j=1}^{t_1} \frac{1}{(q)_{n_j}} \prod_{j=t_1+2}^{t-1} \begin{bmatrix} m_j - \frac{1}{2}(\overline{\boldsymbol{C}}^* \overline{\boldsymbol{m}} - \overline{\boldsymbol{u}}^L - \overline{\boldsymbol{u}}^R)_j \\ m_j \end{bmatrix}_q,$$

where the sum is over all vectors $\overline{\boldsymbol{m}} = (m_{t_1+1}, m_{t_1+2}, \ldots, m_{t-1})$ such that $\overline{\boldsymbol{m}} \equiv (Q_{t_1+1}, Q_{t_1+2}, \ldots, Q_{t-1})$, and over all $\boldsymbol{n} = (n_1, n_2, \ldots, n_{t_1})$, and as above, we use (1.18) and (9.2). We set $\mathcal{X}^L = \mathcal{X}(a, \boldsymbol{u}^L)$. When $\boldsymbol{u}^R \in \mathcal{U}(b)$, we set $\mathcal{X}^R = \mathcal{X}(b, \boldsymbol{u}^R)$, but if $\boldsymbol{u}^R = \boldsymbol{u}^+$ with $\boldsymbol{u}^+ \in \mathcal{U}(b)$ then we set $\mathcal{X}^R = \mathcal{X}^+$ where $\mathcal{X} = \mathcal{X}(b, \boldsymbol{u})$. We also set $m_t = 0$.

Lemma 9.2. *Let $1 \le a, b < p'$, $\boldsymbol{u}^L \in \mathcal{U}(a)$ and $\boldsymbol{u}^R \in \mathcal{U}(b)$.*

1. *If $\boldsymbol{u}^R \in \overline{\mathcal{U}}(b)$ then:*
$$\lim_{L\to\infty} F(\boldsymbol{u}^L, \boldsymbol{u}^R, L) = 0.$$
2. *If $\boldsymbol{u}^R \notin \overline{\mathcal{U}}(b)$ then:*
$$\lim_{L\to\infty} F(\boldsymbol{u}^L, \boldsymbol{u}^R, L) = F^*(\boldsymbol{u}^L, \boldsymbol{u}^R).$$

Proof: We use the notation in the proof of Lemma 9.1. If $t_1 + 1 \le j < t$ note that $\sum_{i=t_1}^{t-1} C_{ji} m_i$ depends on m_{t_1} only if $j = t_1 + 1$, and then $C_{jt_1} = -1$. Otherwise, for $t_1 + 1 < j < t$, we have $\sum_{i=t_1}^{t-1} C_{ji} m_i = (\overline{\boldsymbol{C}}^* \overline{\boldsymbol{m}})_j$.

In view of (9.4) and since $m_0 = L$, taking the limit $L \to \infty$ effects $m_{t_1} \to \infty$. The current theorem then follows from Lemma 9.1 after noting that

$$\lim_{m\to\infty} \begin{bmatrix} m+n \\ n \end{bmatrix}_q = \lim_{m\to\infty} \begin{bmatrix} m+n \\ m \end{bmatrix}_q = \frac{1}{(q)_n}.$$

□

Theorem 9.3. *Let $1 \le a < p'$, and $2 \le b < p'-1$ with b interfacial in the (p,p')-model. Set $r = \lfloor (b+1)p/p' \rfloor$. Let η be such that $\xi_\eta \le a < \xi_{\eta+1}$ and η' be such that $\tilde{\xi}_{\eta'} \le r < \tilde{\xi}_{\eta'+1}$.*
1. If $\eta = \eta'$ and $\xi_\eta < a$ and $\tilde{\xi}_\eta < r$ then:

$$\chi^{p,p'}_{r,a} = \sum_{\substack{\boldsymbol{u}^L \in \mathcal{U}(a) \\ \boldsymbol{u}^R \in \mathcal{U}(b)\setminus\overline{\mathcal{U}}(b)}} F^*(\boldsymbol{u}^L, \boldsymbol{u}^R) \quad + \quad \chi^{\hat{p},\hat{p}'}_{\hat{r},\hat{a}}, \tag{9.7}$$

where $\hat{p}' = \xi_{\eta+1} - \xi_\eta$, $\hat{p} = \tilde{\xi}_{\eta+1} - \tilde{\xi}_\eta$, $\hat{a} = a - \xi_\eta$, $\hat{r} = r - \tilde{\xi}_\eta$.
2. Otherwise:

$$\chi^{p,p'}_{r,a} = \sum_{\substack{\boldsymbol{u}^L \in \mathcal{U}(a) \\ \boldsymbol{u}^R \in \mathcal{U}(b)\setminus\overline{\mathcal{U}}(b)}} F^*(\boldsymbol{u}^L, \boldsymbol{u}^R). \tag{9.8}$$

Proof: To prove these results, we take the $L \to \infty$ limit of the results given in Theorem 8.10. Since b is interfacial, we may choose either $c \in \{b \pm 1\}$. After checking that $0 < r < p$, (1.13) gives $\lim_{L\to\infty} \chi^{p,p'}_{a,b,c}(L) = \chi^{p,p'}_{r,a}$.

Now note that if $\boldsymbol{u}^R \in \overline{\mathcal{U}}(b)$, Lemma 9.2(1) implies that $\lim_{L\to\infty} F(\boldsymbol{u}^L, \boldsymbol{u}^R, L) = 0$. For the remaining terms where $\boldsymbol{u}^R \in \mathcal{U}(b)\setminus\overline{\mathcal{U}}(b)$, Lemma 9.2(2) implies that $\lim_{L\to\infty} F(\boldsymbol{u}^L, \boldsymbol{u}^R, L) = F^*(\boldsymbol{u}^L, \boldsymbol{u}^R)$. Then in the case $a = \xi_\eta$, taking the $L \to \infty$ limit of Theorem 8.10(2) immediately yields the required (9.8).

Hereafter, we assume that $a > \xi_\eta$. Define η'' such that $\xi_{\eta''} \le b < \xi_{\eta''+1}$ and set $\hat{b} = b - \xi_{\eta''}$. Since, by Lemma E.10, ξ_i is interfacial and neighbours the $\tilde{\xi}_i$th odd band for $1 \le i \le 2c_n - 2$, it follows that $\tilde{\xi}_{\eta''} \le r \le \tilde{\xi}_{\eta''+1}$.

Consider the case $\eta'' \ne \eta$. Now if $\eta' = \eta$ then $r = \tilde{\xi}_{\eta''+1}$ and thus $\eta' = \eta'' + 1$ so that $r = \tilde{\xi}_\eta$. So whether $\eta' \ne \eta$ or $\eta' = \eta$, we are required to prove (9.8). This follows immediately from taking the $L \to \infty$ limit of Theorem 8.10(2) which holds because $\eta'' \ne \eta$.

The case $\eta'' = \eta$ spawns three subcases: $r = \tilde{\xi}_\eta$; $r = \tilde{\xi}_{\eta+1}$; $\tilde{\xi}_\eta < r < \tilde{\xi}_{\eta+1}$. We consider each in turn.

If $r = \tilde{\xi}_\eta$ then it is required to prove (9.8). Since b is interfacial and neighbours the rth odd band, then $\eta > 0$ and either $b = \xi_\eta$ or $b = \xi_\eta + 1$. In the former case, we take the $L \to \infty$ limit of Theorem 8.10(2) to obtain the desired (9.8). In the latter case, $\hat{b} = 1$ and from (8.2), $\hat{c} = 2$. It then follows from (1.12) that $\lim_{L\to\infty} \chi^{\hat{p},\hat{p}'}_{\hat{a},\hat{b},\hat{c}}(L) = 0$, and thus Theorem 8.10(1) yields the desired (9.8).

If $r = \tilde{\xi}_{\eta+1}$ then $\eta' = \eta + 1$, so that again it is required to prove (9.8). Here, we necessarily have $b = \xi_{\eta+1} - 1$, and so $\hat{b} = \xi_{\eta+1} - 1 - \xi_\eta = \hat{p}' - 1$. Then $\hat{c} = \hat{p}' - 2$ from (8.2). Then (1.12) again implies that $\lim_{L\to\infty} \chi^{\hat{p},\hat{p}'}_{\hat{a},\hat{b},\hat{c}}(L) = 0$, and thus Theorem 8.10(1) yields the desired (9.8).

If $\tilde{\xi}_\eta < r < \tilde{\xi}_{\eta+1}$ then $\eta' = \eta$, and $\xi_\eta < b < \xi_{\eta+1}$. Thus Theorem 8.10(1) applies in this case, and we are required to prove (9.7). It suffices to show that $\lim_{L\to\infty} \chi^{\hat{p},\hat{p}'}_{\hat{a},\hat{b},\hat{c}}(L) = \chi^{\hat{p},\hat{p}'}_{\hat{r},\hat{a}}$. That this is so follows from (1.13) because there are $\tilde{\xi}_\eta$ odd bands in the (p,p')-model below height $\xi_\eta + 1$, and therefore $\hat{b}$ neighbours the $(r - \tilde{\xi}_\eta)$th odd band in the $(\hat{p},\hat{p}')$-model. □

Theorem 9.3 certainly provides a fermionic expression for $\chi^{p,p'}_{r,s}$ when b with the required property can be found. However a more succinct expression for $\chi^{p,p'}_{r,s}$, and which works for all r, is obtained in Section 9.2 in terms of the truncated Takahashi tree of r.

Lemma 9.4. *Let $p' > 2p$ and $1 \le a, b < p'$ with b non-interfacial in the (p,p')-model. If $\boldsymbol{u}^L \in \mathcal{U}(a)$ and $\boldsymbol{u}^R \in \mathcal{U}(b)$ then:*

$$\lim_{L\to\infty} \widetilde{F}(\boldsymbol{u}^L, \boldsymbol{u}^R, L) = 0.$$

Proof: Let $\{\tau_j, \sigma_j, \Delta_j\}_{j=1}^d = \mathcal{X}(b, \boldsymbol{u}^R)$ so that $\sigma_d = \sigma(\boldsymbol{u}^R)$ and $\Delta_d = \Delta(\boldsymbol{u}^R)$. In the case $\sigma_d = 0$, the required result follows from the definition (1.45) after noting via Theorem 7.16 that $F(\boldsymbol{u}^L, \boldsymbol{u}^R, L)$ is the generating function for certain paths.

Now consider $\sigma_d > 0$. Let $\mathcal{X}^- = \{\tau_j, \sigma_j^-, \Delta_j\}_{j=1}^d$, with $\sigma_j^- = \sigma_j$ for $1 \le j < d$ and $\sigma_d^- = \sigma_d - 1$. Then let $\boldsymbol{u}^{R-} = \boldsymbol{u}(\mathcal{X}^-)$, using (1.21). Note that $\sigma(\boldsymbol{u}^{R-}) = \sigma(\boldsymbol{u}^R) - 1$. We then have $0 \le \sigma(\boldsymbol{u}^{R-}) < \tau - 1$, where τ is specified by (1.44). Substituting $\boldsymbol{u}^R \to \boldsymbol{u}^{R-}$ (so that $\boldsymbol{u}^{R+} \to \boldsymbol{u}^R$ and $b \to b + \Delta$) into Lemma 7.19(1) and rearranging, yields:

$$(9.9)\quad \lim_{L\to\infty} \widetilde{F}(\boldsymbol{u}^L, \boldsymbol{u}^R, L) = \begin{cases} \lim_{L\to\infty} \widetilde{F}(\boldsymbol{u}^L, \boldsymbol{u}^{R-}, L) & \text{if } b+\Delta \text{ is non-interfacial;} \\ \lim_{L\to\infty} F(\boldsymbol{u}^L, \boldsymbol{u}^{R-}, L) & \text{if } b+\Delta \text{ is interfacial.} \end{cases}$$

Since b is not interfacial, Corollary 7.13 implies that $\sigma_d \le t_1$. Since p'/p has continued fraction $[c_0, c_1, \ldots, c_n]$, and $t_1 = c_0 - 1$, neighbouring odd bands in the (p,p')-model are separated by t_1 or $t_1 + 1$ even bands. If $b_{j_1 j_2 \cdots j_{d-1} 0}$ is the leaf-node of the Takahashi tree for b that corresponds to the run $\{\tau_j, \sigma_j, \Delta_j\}_{j=1}^d$ then, from Lemma 8.6, $b = b_{j_1 j_2 \cdots j_{d-1}} - \Delta \kappa_{\sigma_d} = b_{j_1 j_2 \cdots j_{d-1}} - \Delta(\sigma_d + 1)$, and $b_{j_1 j_2 \cdots j_{d-1}}$ is interfacial. It follows that if b is non-interfacial, then $b + \Delta$ is interfacial only if $\sigma_d = 1$. In this case that $b + \Delta$ is interfacial then $\sigma(\boldsymbol{u}^{R-}) = 0$ whereupon it follows that $\boldsymbol{u}^{R-} \in \overline{\mathcal{U}}(b + \Delta)$ and Lemma 9.2(1) gives $\lim_{L\to\infty} F(\boldsymbol{u}^L, \boldsymbol{u}^{R-}, L) = 0$.

In the case that $\sigma_d = 1$ and b is non-interfacial, the use of (1.45) again yields $\lim_{L\to\infty} \widetilde{F}(\boldsymbol{u}^L, \boldsymbol{u}^{R-}, L) = 0$. Thus the required result holds whenever $\sigma(\boldsymbol{u}^R) = 1$. For $\sigma(\boldsymbol{u}^R) > 1$, the required result then follows by repeated use of (9.9). □

Lemma 9.5. *Let $p' < 2p$ and $1 \le a, b < p'$ with b non-interfacial in the (p,p')-model. Let $\boldsymbol{u}^L \in \mathcal{U}(a)$ and $\boldsymbol{u}^R \in \mathcal{U}(b)$. Then:*

$$\lim_{L\to\infty} \widetilde{F}(\boldsymbol{u}^L, \boldsymbol{u}^R, L) = F^*(\boldsymbol{u}^L, \boldsymbol{u}^{R+}).$$

Proof: In the case $\sigma(\boldsymbol{u}^R) > 0$, this follows directly from the definition (1.46) and Lemma 9.2(2). In the case $\sigma(\boldsymbol{u}^R) = 0$, it follows from Lemma 7.19(2) combined with Lemma 9.2(2), on noting that $\sigma(\boldsymbol{u}^{R+}) > 0$ □

Theorem 9.6. *Let $p' < 2p$ and $1 \le a, b < p'$ with b non-interfacial in the (p,p')-model. Let $c \in \{b \pm 1\}$, let r be given by (2.3) and assume that $0 < r < p$. Let η be such that $\xi_\eta \le a < \xi_{\eta+1}$ and η' be such that $\tilde{\xi}_{\eta'} \le r < \tilde{\xi}_{\eta'+1}$.*
1. If $\eta = \eta'$ and $\xi_\eta < a$ and $\tilde{\xi}_\eta < r$ then:

$$(9.10)\quad \chi^{p,p'}_{r,a} = \sum_{\substack{\boldsymbol{u}^L \in \mathcal{U}(a) \\ \boldsymbol{u}^R \in \mathcal{U}^{c-b}(b) \backslash \overline{\mathcal{U}}^{c-b}(b)}} F^*(\boldsymbol{u}^L, \boldsymbol{u}^R) + \sum_{\substack{\boldsymbol{u}^L \in \mathcal{U}(a) \\ \boldsymbol{u}^R \in \mathcal{U}^{b-c}(b)}} F^*(\boldsymbol{u}^L, \boldsymbol{u}^{R+}) + \chi^{\hat{p},\hat{p}'}_{\hat{r},\hat{a}}.$$

where $\hat{p}' = \xi_{\eta+1} - \xi_\eta$, $\hat{p} = \tilde{\xi}_{\eta+1} - \tilde{\xi}_\eta$, $\hat{a} = a - \xi_\eta$, $\hat{r} = r - \tilde{\xi}_\eta$.
2. Otherwise:

$$(9.11)\quad \chi^{p,p'}_{r,a} = \sum_{\substack{\boldsymbol{u}^L \in \mathcal{U}(a) \\ \boldsymbol{u}^R \in \mathcal{U}^{c-b}(b) \backslash \overline{\mathcal{U}}^{c-b}(b)}} F^*(\boldsymbol{u}^L, \boldsymbol{u}^R) + \sum_{\substack{\boldsymbol{u}^L \in \mathcal{U}(a) \\ \boldsymbol{u}^R \in \mathcal{U}^{b-c}(b)}} F^*(\boldsymbol{u}^L, \boldsymbol{u}^{R+}).$$

Proof: To prove these results, we take the $L \to \infty$ limit of the results given in Theorem 8.12. With r as specified, (1.13) gives $\lim_{L\to\infty} \chi^{p,p'}_{a,b,c}(L) = \chi^{p,p'}_{r,a}$.

If $\boldsymbol{u}^R \in \overline{\mathcal{U}}^{c-b}(b)$ (so that $c = b + \Delta(\boldsymbol{u}^R)$) then Lemma 9.2(1) implies that $\lim_{L\to\infty} F(\boldsymbol{u}^L, \boldsymbol{u}^R, L) = 0$. If $\boldsymbol{u}^R \in \mathcal{U}^{c-b}(b)\backslash\overline{\mathcal{U}}^{c-b}(b)$, then Lemma 9.2(2) implies that $\lim_{L\to\infty} F(\boldsymbol{u}^L, \boldsymbol{u}^R, L) = F^*(\boldsymbol{u}^L, \boldsymbol{u}^R)$. For the remaining cases, where $\boldsymbol{u}^R \in \mathcal{U}^{c-b}(b)$, Lemma 9.5 implies that $\lim_{L\to\infty} \widetilde{F}(\boldsymbol{u}^L, \boldsymbol{u}^R, L) = F^*(\boldsymbol{u}^L, \boldsymbol{u}^{R+})$. Then in the case $a = \xi_\eta$, taking the $L \to \infty$ limit of Theorem 8.12(2) immediately yields the required (9.11).

Hereafter, we assume that $a > \xi_\eta$. The proof is now very similar to that of Theorem 9.3. Note that since $p' < 2p$ and b is non-interfacial, b lies between two odd bands. Define η'' such that $\xi_{\eta''} \le b < \xi_{\eta''+1}$ and set $\hat{b} = b - \xi_{\eta''}$. Since, by Lemma E.10, ξ_i is interfacial and neighbours the $\tilde{\xi}_i$th odd band for $1 \le i \le 2c_n - 2$, it follows that $\tilde{\xi}_{\eta''} \le r \le \tilde{\xi}_{\eta''+1}$ and $\xi_{\eta''} < b < \xi_{\eta''+1}$.

Consider the case $\eta'' \ne \eta$. Now if $\eta' = \eta$ then $r = \tilde{\xi}_{\eta''+1}$ and thus $\eta' = \eta'' + 1$ so that $r = \tilde{\xi}_\eta$. So whether $\eta' \ne \eta$ or $\eta' = \eta$, we are required to prove (9.11). This follows immediately from taking the $L \to \infty$ limit of Theorem 8.12(2) which holds because $\eta'' \ne \eta$.

For the case $\eta'' = \eta$, Theorem 8.12(1) holds. We consider the three subcases: $r = \tilde{\xi}_\eta$; $r = \tilde{\xi}_{\eta+1}$; $\tilde{\xi}_\eta < r < \tilde{\xi}_{\eta+1}$, in turn.

If $r = \tilde{\xi}_\eta$ then it is required to prove (9.11). Since b neighbours the rth odd band, then $b = \xi_\eta + 1$ and $c = \xi_\eta$ so that $\hat{b} = 1$ and from (8.2), $\hat{c} = 0$. It then follows from (2.3) and (1.12) that $\lim_{L\to\infty} \chi^{\hat{p},\hat{p}'}_{\hat{a},\hat{b},\hat{c}}(L) = 0$, and thus Theorem 8.12(1) yields the desired (9.11).

If $r = \tilde{\xi}_{\eta+1}$ then $\eta' = \eta + 1$, so that again it is required to prove (9.11). Here, we necessarily have $b = \xi_{\eta+1} - 1$ and $c = \xi_{\eta+1}$, and so $\hat{b} = \xi_{\eta+1} - 1 - \xi_\eta = \hat{p}' - 1$ and $\hat{c} = \hat{p}'$ from (8.2). Again, $\lim_{L\to\infty} \chi^{\hat{p},\hat{p}'}_{\hat{a},\hat{b},\hat{c}}(L) = 0$ follows from (2.3) and (1.12), and thus Theorem 8.12(1) yields the desired (9.11).

If $\tilde{\xi}_\eta < r < \tilde{\xi}_{\eta+1}$ then $\eta' = \eta$. Thus we are required to prove (9.10). Note that (8.2) implies that $\hat{c} = c - \xi_\eta$. To prove (9.10), it suffices to show that $\lim_{L\to\infty} \chi^{\hat{p},\hat{p}'}_{\hat{a},\hat{b},\hat{c}}(L) = \chi^{\hat{p},\hat{p}'}_{\hat{r},\hat{a}}$. That this is so follows from (1.13) after noting that because b and c straddle the rth odd band in the (p,p')-model and there are $\tilde{\xi}_\eta$ odd bands in the (p,p')-model below height $\xi_\eta+1$, then $\hat{b}$ and $\hat{c}$ straddle the $(r-\tilde{\xi}_\eta)$th odd band in the $(\hat{p},\hat{p}')$-model. □

Theorem 9.6 provides fermionic expressions for those $\chi^{p,p'}_{r,s}$ that are not covered by Theorem 9.3. In Section 9.2, these two Theorems are subsumed into a more succinct expression for $\chi^{p,p'}_{r,s}$ which makes use of the truncated Takahashi tree of r.

Lemma 9.7. *Let $1\le a,b<p'$. If $\boldsymbol{u}^L\in\mathcal{U}(a)$ and $\boldsymbol{u}\in\mathcal{U}(b)\backslash\overline{\mathcal{U}}(b)$ with $\sigma(\boldsymbol{u})\le t_1$ then:*

$$F^*(\boldsymbol{u}^L,\boldsymbol{u})=F^*(\boldsymbol{u}^L,\boldsymbol{u}^+).$$

Proof: Let $\mathcal{X}^L=\mathcal{X}(a,\boldsymbol{u}^L)$, $\boldsymbol{\Delta}^L=\boldsymbol{\Delta}(\mathcal{X}^L)$, $\mathcal{X}=\mathcal{X}(b,\boldsymbol{u})$, $\sigma=\sigma(\boldsymbol{u})$ and $\Delta=\Delta(\boldsymbol{u})$, and let $\mathcal{X}^+$ and $\boldsymbol{u}^+$ be defined as in Section 1.15. Using (1.24), set $\boldsymbol{\Delta}=\boldsymbol{\Delta}(\mathcal{X})$ and $\boldsymbol{\Delta}^+=\boldsymbol{\Delta}(\mathcal{X}^+)$. On exchanging $\boldsymbol{u}^R=\boldsymbol{u}$ and $\mathcal{X}^R=\mathcal{X}$ for $\boldsymbol{u}^R=\boldsymbol{u}^+$ and $\mathcal{X}^R=\mathcal{X}^+$ in the definition (9.6), the only possible changes are in the terms $\tilde{\boldsymbol{n}}^T\boldsymbol{B}\tilde{\boldsymbol{n}}$, $\boldsymbol{N}(\boldsymbol{u}^R)\cdot\tilde{\boldsymbol{n}}$ and $\gamma(\mathcal{X}^L,\mathcal{X}^R)$ (by virtue of $\boldsymbol{\Delta}^R$ changing from $\boldsymbol{\Delta}(\mathcal{X})$ to $\boldsymbol{\Delta}(\mathcal{X}^+)$).

In the case $\sigma=t_1$, the first two of these are clearly unchanged. From (1.25), (1.26) and (1.27), it is easily shown that $\gamma_{t_1}=-\gamma_{t_1+1}-\alpha^2_{t_1+1}-2\alpha_{t_1+1}(\beta_{t_1+1}+(\boldsymbol{\Delta}^L)_{t_1+1})$. Then since $(\boldsymbol{\Delta}^R)_j=0$ for $0\le j\le t_1$ and β'_j is unchanged for $0\le j\le t_1$, it follows that γ_0 is unchanged. The lemma follows in the $\sigma=t_1>0$ case. When $\sigma=t_1=0$ (here $p'<2p$), the lemma follows after noting that $\boldsymbol{u}\in\mathcal{U}(b)\backslash\overline{\mathcal{U}}(b)$ implies that $\gamma(\mathcal{X}^L,\mathcal{X})=\gamma_0$ and thus $\gamma(\mathcal{X}^L,\mathcal{X})=\gamma(\mathcal{X}^L,\mathcal{X}^+)$.

For $0\le\sigma<t_1$, the value of $\tilde{\boldsymbol{n}}^T\boldsymbol{B}\tilde{\boldsymbol{n}}$ is seen to change by $\frac{1}{4}-(\tilde{n}_{\sigma+1}+\cdots+\tilde{n}_{t_1})$, and the value of $\boldsymbol{N}(\boldsymbol{u}^R)\cdot\tilde{\boldsymbol{n}}$ is seen to change by $-(\tilde{n}_{\sigma+1}+\cdots+\tilde{n}_{t_1})$. Thus to prove the lemma it remains to show that $\gamma(\boldsymbol{u}^L,\boldsymbol{u}^R)$ changes by -1.

For $0\le j<t_1$, we have $\gamma_j=\gamma_{j+1}+2\alpha_{j+1}(\boldsymbol{\Delta}^R)_{j+1}-\beta_j^2$. Combining the $j=\sigma$ and $j=\sigma-1$ instances of this and noting that $\alpha_j=\alpha_{j+1}+\beta_j$, gives:

$$\gamma_{\sigma-1}=\gamma_{\sigma+1}+2\alpha_{\sigma+1}((\boldsymbol{\Delta}^R)_{\sigma+1}+(\boldsymbol{\Delta}^R)_\sigma)+2\beta_\sigma(\boldsymbol{\Delta}^R)_\sigma-\beta_\sigma^2-\beta_{\sigma-1}^2$$

for $0<\sigma<t_1$. In passing from $\boldsymbol{u}$ to $\boldsymbol{u}^+$, we have that $(\boldsymbol{\Delta}^R)_{\sigma+1}+(\boldsymbol{\Delta}^R)_\sigma$ is unchanged, $(\boldsymbol{\Delta}^R)_\sigma$ changes from Δ to 0 and β_σ decreases by Δ, with β_j unchanged for $0\le j<\sigma$. Thereupon $\gamma(\mathcal{X}^L,\mathcal{X}^R)=\gamma_0$ changes by -1 for $0<\sigma<t_1$.

By the definition of Section 1.10, for $\sigma=0$ and $p'>2p$, we have on the one hand $\gamma(\mathcal{X}^L,\mathcal{X})=\gamma_0+2\Delta\alpha_0$ whereas $\gamma(\mathcal{X}^L,\mathcal{X}^+)=\gamma_0$ on the other. As above, $\gamma_0=\gamma_1+2\alpha_1(\boldsymbol{\Delta}^R)_1-\beta_0^2$. We then see that γ_0 changes by $2\alpha_1\Delta+2\beta_0\Delta-1=2\alpha_0\Delta-1$, whereupon the required change of -1 in $\gamma(\mathcal{X}^L,\mathcal{X}^R)$ follows. □

9.2. Assimilating the trees

In Theorems 9.3 and 9.6 of the previous section, we obtained fermionic expressions for the characters $\chi^{p,p'}_{r,s}$ in terms of the Takahashi trees for s and b, where b is related to r by (1.10) for some $c\in\{b\pm1\}$. In this section, we obviate the need to find such b and c, by obtaining more succinct expressions for $\chi^{p,p'}_{r,s}$ in terms of the truncated Takahashi tree for r and the Takahashi tree for s. We do this via a detailed examination of the relationship between the truncated Takahashi tree for r and the Takahashi tree for b, where r and b are related as above.

As an example, consider Fig. 9.1, where we give the Takahashi trees for both $b=119$ and $b=120$ in the case where $(p,p')=(69,223)$. Each of these values of b is interfacial and, in fact, $\rho^{69,223}(b)=37$ in both cases. As will be shown in this section, for any s, the fermionic expressions for $\chi_{37,s}^{69,223}$ that result from these two trees are identical (up to trivial differences), and furthermore are identical to the fermionic expression which results from the use of the truncated Takahashi tree for $r = 37$ that was given in Fig. 1.3. (The leaf-nodes in Fig. 9.1 that are affixed with an asterisk are those for which the corresponding vector $\boldsymbol{u}^R \in \overline{\mathcal{U}}(b)$. By virtue of Theorem 9.3, these nodes do not contribute terms to the fermionic expression (9.7) or (9.8) for $\chi_{37,s}^{69,223}$.)

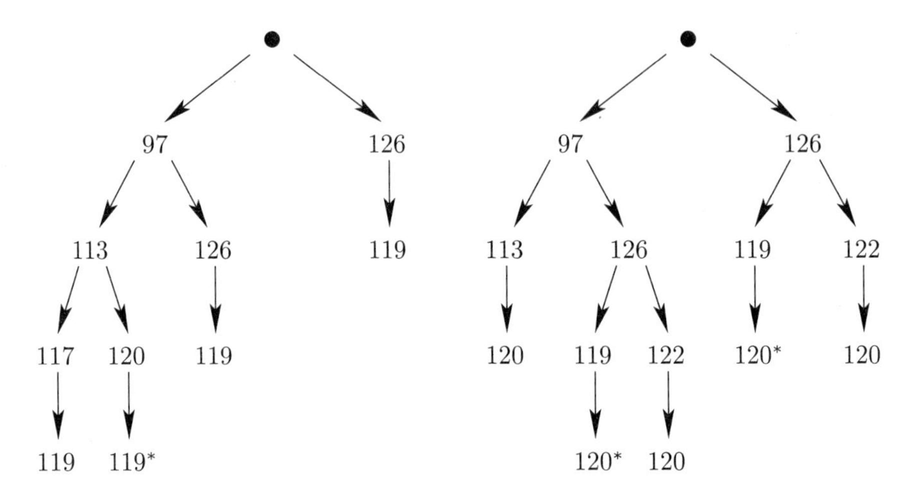

FIGURE 9.1. Takahashi trees for $b = 119$ and $b = 120$ when $(p, p') = (69, 223)$.

In this section, we will sometimes refer to the Takahashi tree for b as the b-tree and the truncated Takahashi tree for r as the r-tree. Since p and p' will be fixed throughout this section, we abbreviate $\rho^{p,p'}(\cdot)$ to $\rho(\cdot)$.

First, we need to extend the results on the b-tree contained in Lemma 8.6.

Lemma 9.8. *Let $\{\tau_j, \sigma_j, \Delta_j\}_{j=1}^d$ be the naive run corresponding to the leaf-node $b_{i_1 i_2 \cdots i_{d-1} 0}$ of the Takahashi tree for b. Then, for $1 \le k < d$, both $b_{i_1 i_2 \cdots i_{k-1} \bar{\imath}_k}$ and $b_{i_1 i_2 \cdots i_{k-1} i_k}$ are interfacial. Moreover,*

$$\rho(b_{i_1 i_2 \cdots i_{k-1} \bar{\imath}_k}) = \sum_{m=2}^{k} \Delta_m (\tilde{\kappa}_{\tau_m} - \tilde{\kappa}_{\sigma_m}) + \begin{cases} \tilde{\kappa}_{\sigma_1} & \text{if } \Delta_1 = -1; \\ p - \tilde{\kappa}_{\sigma_1} & \text{if } \Delta_1 = +1, \end{cases}$$

and

$$\rho(b_{i_1 i_2 \cdots i_{k-1} i_k}) = \rho(b_{i_1 i_2 \cdots i_{k-1} \bar{\imath}_k}) + \Delta_{k+1} \tilde{\kappa}_{\tau_{k+1}}.$$

If b is interfacial and $\sigma_d > 0$ then:

$$\rho(b) = \sum_{m=2}^{d} \Delta_m (\tilde{\kappa}_{\tau_m} - \tilde{\kappa}_{\sigma_m}) + \begin{cases} \tilde{\kappa}_{\sigma_1} & \text{if } \Delta_1 = -1; \\ p - \tilde{\kappa}_{\sigma_1} & \text{if } \Delta_1 = +1. \end{cases}$$

If b is not interfacial and $\sigma_d > 0$ then:

$$\left\lfloor \frac{bp}{p'} \right\rfloor = \sum_{m=2}^{d} \Delta_m (\tilde{\kappa}_{\tau_m} - \tilde{\kappa}_{\sigma_m}) - \frac{1}{2}(1 \pm \Delta_d) + \begin{cases} \tilde{\kappa}_{\sigma_1} & \text{if } \Delta_1 = -1; \\ p - \tilde{\kappa}_{\sigma_1} & \text{if } \Delta_1 = +1, \end{cases}$$

where the '$-$' sign applies when $p' > 2p$ and the '$+$' sign applies when $p' < 2p$.

Proof: First consider the case $p' > 2p$.

On comparing the definitions of Section 7.1 with the expressions stated in Lemma 8.6, we see that $b_{i_1 i_2 \cdots i_{k-1}\bar{\imath}_k} = \nu^*_k$ and $b_{i_1 i_2 \cdots i_{k-1} i_k} = \nu_k$ for $1 \le k < d$. Corollary 7.14 then shows that $b_{i_1 i_2 \cdots i_{k-1}\bar{\imath}_k}$ is interfacial and $\rho(b_{i_1 i_2 \cdots i_{k-1}\bar{\imath}_k}) = \tilde{\nu}^*_k$, the definition of which in Section 7.1 proves the required expression for $\rho(b_{i_1 i_2 \cdots i_{k-1}\bar{\imath}_k})$.

In the case of $b_{i_1 i_2 \cdots i_{k-1} i_k}$, first consider $\sigma_{k+1} \ne \tau_{k+1}$. Then Corollary 7.14 shows that $b_{i_1 i_2 \cdots i_{k-1}\bar{\imath}_k}$ is interfacial and $\rho(b_{i_1 i_2 \cdots i_{k-1} i_k}) - \rho(b_{i_1 i_2 \cdots i_{k-1}\bar{\imath}_k}) = \tilde{\nu}_k - \tilde{\nu}^*_k = \Delta_{k+1}\tilde{\kappa}_{\tau_{k+1}}$ as required.

If $\tau_{k+1} = \sigma_{k+1}$, consider any leaf-node of the form $b_{i_1 \cdots i_k \bar{\imath}_{k+1} i'_{k+2} \cdots i'_{d'}}$, where necessarily $d' > k+1$. Using Lemma 8.6, it is easily seen that the corresponding naive run $\{\tau'_j, \sigma'_j, \Delta'_j\}_{j=1}^{d'}$ is such that $\tau'_j = \tau_j$ for $1 \le j \le k+1$ and $\sigma'_j = \sigma_j$ for $1 \le j \le k$. Moreover, $\sigma'_{k+1} \ne \sigma_{k+1}$ and so $\sigma'_{k+1} \ne \tau'_{k+1}$. Then the above argument applied to this case shows that $b_{i_1 i_2 \cdots i_{k-1} i_k}$ is interfacial and yields the required expression for $\rho(b_{i_1 i_2 \cdots i_{k-1} i_k})$.

When $\sigma_d > 0$, Corollary 7.14 immediately gives the required expressions for $\rho(b)$ and $\lfloor bp/p' \rfloor$.

In the case where $p' < 2p$, note that $p' > 2(p'-p)$. We are then able to use the result established above for $(p,p') \to (p'-p,p')$. If $\{\kappa'_j\}_{j=1}^t$ and $\{\tilde{\kappa}'_j\}_{j=1}^t$ are respectively the Takahashi lengths and the truncated Takahashi lengths for the $(p'-p,p')$-model then Lemma E.2 yields $\kappa'_j = \kappa_j$ and $\tilde{\kappa}'_j = \kappa_j - \tilde{\kappa}_j$ for $1 \le j \le t$. Lemma C.4(2) implies that $b_{i_1 i_2 \cdots i_{k-1}\bar{\imath}_k}$ is interfacial in the (p,p')-model, and that

$$\begin{aligned}\rho^{p,p'}(b_{i_1 i_2 \cdots i_{k-1}\bar{\imath}_k}) &= b_{i_1 i_2 \cdots i_{k-1}\bar{\imath}_k} - \rho^{p'-p,p'}(b_{i_1 i_2 \cdots i_{k-1}\bar{\imath}_k}) \\ &= b_{i_1 i_2 \cdots i_{k-1}\bar{\imath}_k} - \sum_{m=2}^{k} \Delta_m(\tilde{\kappa}'_{\tau_m} - \tilde{\kappa}'_{\sigma_m}) - \begin{cases} \tilde{\kappa}'_{\sigma_1} & \text{if } \Delta_1 = -1; \\ p'-p-\tilde{\kappa}'_{\sigma_1} & \text{if } \Delta_1 = +1. \end{cases}\end{aligned}$$

The desired expression for $\rho^{p,p'}(b_{i_1 i_2 \cdots i_{k-1}\bar{\imath}_k})$ then results from the expression for $b_{i_1 i_2 \cdots i_{k-1}\bar{\imath}_k}$ given in Lemma 8.6 and $\kappa'_j = \kappa_j$ and $\tilde{\kappa}'_j = \kappa_j - \tilde{\kappa}_j$ for $1 \le j \le t$ (the $j = 0$ case is not required because $k < d$).

The other expressions follow in a similar way, noting in the final case that $\lfloor bp/p' \rfloor = b - 1 - \lfloor b(p'-p)/p' \rfloor$. □

In particular, this lemma implies that every node of the b-tree that is not a leaf-node is interfacial.

Now consider a leaf-node $r_{i_1 i_2 \cdots i_d}$ of the truncated Takahashi tree for r, and let $\{\tilde{\tau}_j, \tilde{\sigma}_j, \tilde{\Delta}_j\}_{j=1}^d$ be the corresponding run. That it is actually a naive run follows from the appropriate analogue of Lemma 8.5. We also have an analogue of Lemma 8.6, which for ease of reference, we state here in full:

Lemma 9.9. *Let $\{\tilde{\tau}_j, \tilde{\sigma}_j, \tilde{\Delta}_j\}_{j=1}^d$ be the naive run corresponding to the leaf-node $r_{i_1 i_2 \cdots i_{d-1} 0}$ of the truncated Takahashi tree for r. Then:*

$$r_{i_1 i_2 \cdots i_{k-1}\bar{\imath}_k} = \sum_{m=2}^{k} \tilde{\Delta}_m(\tilde{\kappa}_{\tilde{\tau}_m} - \tilde{\kappa}_{\tilde{\sigma}_m}) \quad + \quad \begin{cases} \tilde{\kappa}_{\tilde{\sigma}_1} & \text{if } \tilde{\Delta}_1 = -1; \\ p - \tilde{\kappa}_{\tilde{\sigma}_1} & \text{if } \tilde{\Delta}_1 = +1, \end{cases}$$

for $1 \le k \le d$, and

$$r_{i_1 i_2 \cdots i_{k-1} i_k} = r_{i_1 i_2 \cdots i_{k-1}\bar{\imath}_k} + \tilde{\Delta}_{k+1}\tilde{\kappa}_{\tilde{\tau}_{k+1}},$$

for $1 \leq k < d$.

Proof: This is proved in the same way as Lemma 8.6. □

For Lemmas 9.10 through 9.14, and Theorem 9.15 we let $1 \leq r < p$ and select b and c such that $b, c \in \{\lfloor rp'/p \rfloor, \lfloor rp'/p \rfloor + 1\}$ with $b \neq c$. This ensures that (1.10) is satisfied, and that b and c straddle the rth odd band. If b is interfacial then $\rho(b) = r$, and in this case we could actually choose either $c = b \pm 1$. If b is non-interfacial then necessarily $p' < 2p$. It will be useful to note that if $\{\tau_j, \sigma_j, \Delta_j\}_{j=1}^{d}$ is the naive run corresponding to a leaf-node of the b-tree then $\sigma_d \leq t_2$. This follows from Corollary 7.13 (replacing t_1 with t_2 because we are dealing with the $p' < 2p$ case). It will be also be useful to note that $\kappa_j = j + 1$ and $\tilde{\kappa}_j = j$ for $1 \leq j \leq t_2 + 1$ in this $p' < 2p$ case.

The next few lemmas will enable us to compare the b-tree with the r-tree.

Lemma 9.10. *For $1 \leq r < p$, let $b, c \in \{\lfloor rp'/p \rfloor, \lfloor rp'/p \rfloor + 1\}$ with $b \neq c$. For $k \geq 1$, let $b_{i_1 i_2 \cdots i_k}$ be a non-leaf-node of the b-tree with $\rho(b_{i_1 i_2 \cdots i_k}) = r$. Then $b_{i_1 i_2 \cdots i_k} = b - (-1)^{i_k}$ and $b_{i_1 i_2 \cdots i_k}$ is a through-node of the b-tree. If $\boldsymbol{u} \in \mathcal{U}(b)$ is the vector corresponding to the leaf-node $b_{i_1 i_2 \cdots i_k 0}$, then $\boldsymbol{u} \in \overline{\mathcal{U}}(b)$. Moreover, if b is not interfacial then $b - c = (-1)^{i_k}$ and $\boldsymbol{u} \in \overline{\mathcal{U}}^{c-b}(b)$*

Proof: $\rho(b_{i_1 i_2 \cdots i_k}) = r$ implies that $b_{i_1 i_2 \cdots i_k}$ borders the rth odd band. So does b and thus $|b_{i_1 i_2 \cdots i_k} - b| = 1$. The definition of the b-tree then stipulates that $b_{i_1 i_2 \cdots i_k}$ is a through-node, $b_{i_1 i_2 \cdots i_k 0}$ is a leaf-node and $b_{i_1 i_2 \cdots i_k} - b = -(-1)^{i_k} \kappa_{\sigma_{k+1}}$. Thereupon, $\kappa_{\sigma_{k+1}} = 1$ so that $\sigma_{k+1} = 0$.

For the vector $\boldsymbol{u}$, we then have $\sigma(\boldsymbol{u}) = 0$ and $\Delta(\boldsymbol{u}) = -(-1)^{i_k}$. Since b and $b_{i_1 i_2 \cdots i_k}$ are either side of the rth odd band, we have that $\lfloor bp/p' \rfloor \neq \lfloor (b+\Delta(\boldsymbol{u}))p/p' \rfloor$. Then $\boldsymbol{u} \in \overline{\mathcal{U}}(b)$ by definition. In the case that b is not interfacial, then necessarily $c = b_{i_1 i_2 \cdots i_k}$ so that $c - b = -(-1)^{i_k} = \Delta(\boldsymbol{u})$, giving $\boldsymbol{u} \in \overline{\mathcal{U}}^{c-b}(b)$ as required. □

Lemma 9.11. *For $1 \leq r < p$, let $b, c \in \{\lfloor rp'/p \rfloor, \lfloor rp'/p \rfloor + 1\}$ with $b \neq c$.*

1) For $d \geq 2$, let $b_{i_1 i_2 \cdots i_{d-1}}$ be a through-node of the b-tree with $\rho(b_{i_1 i_2 \cdots i_{d-1}}) \neq r$, and let $\boldsymbol{u} \in \mathcal{U}(b)$ be the vector corresponding to the leaf-node $b_{i_1 i_2 \cdots i_{d-1} 0}$. If b is interfacial then $\boldsymbol{u} \notin \overline{\mathcal{U}}(b)$, and if b is not interfacial then $\boldsymbol{u} \notin \overline{\mathcal{U}}^{c-b}(b)$. In addition, if $r_{i_1 i_2 \cdots i_{d-1}}$ is a node of the r-tree with $r_{i_1 i_2 \cdots i_{d-1}} = \rho(b_{i_1 i_2 \cdots i_{d-1}})$ then $r_{i_1 i_2 \cdots i_{d-1}}$ is a through-node.

2) Let b_0 be a leaf-node of the b-tree, and let $\boldsymbol{u} \in \mathcal{U}(b)$ be the corresponding vector. If b is interfacial then $\boldsymbol{u} \notin \overline{\mathcal{U}}(b)$, and if b is not interfacial then $\boldsymbol{u} \notin \overline{\mathcal{U}}^{c-b}(b)$. In addition, r_0 is a leaf-node of the r-tree.

Proof: 1) Lemma 8.6 implies that $b_{i_1 i_2 \cdots i_{d-1}} - b = \Delta_d \kappa_{\sigma_d}$, for $\Delta_d = -(-1)^{i_k}$ and some σ_d. Note that $\sigma(\boldsymbol{u}) = \sigma_d$ and $\Delta(\boldsymbol{u}) = \Delta_d$. If $\sigma_d > 0$ then $\boldsymbol{u} \notin \overline{\mathcal{U}}(b)$ by definition. If $\sigma_d = 0$ and b is interfacial, we claim that $\lfloor bp/p' \rfloor = \lfloor (b + \Delta_d)p/p' \rfloor$. Otherwise, $\sigma_d = 0$ and $\lfloor bp/p' \rfloor \neq \lfloor (b+\Delta_d)p/p' \rfloor$ imply that $b_{i_1 i_2 \cdots i_{d-1}}$ neighbours the rth odd band, as does b, whereupon noting that $b_{i_1 i_2 \cdots i_{d-1}}$ is interfacial (by Lemma 9.8), we would have $\rho(b_{i_1 i_2 \cdots i_{d-1}}) = r$, which is not the case. So $\boldsymbol{u} \notin \overline{\mathcal{U}}(b)$. If $\sigma_d = 0$ and b is not interfacial, we claim that $c = b - \Delta$. Otherwise $c = b + \Delta = b_{i_1 i_2 \cdots i_{d-1}}$, again implying that $\rho(b_{i_1 i_2 \cdots i_{d-1}}) = r$, which is not the case. Therefore $\boldsymbol{u} \in \mathcal{U}^{b-c}(b)$ so that certainly $\boldsymbol{u} \notin \overline{\mathcal{U}}^{c-b}(b)$ as required.

For $\sigma_d > 0$ and b interfacial, Lemma 9.8 implies that $r_{i_1 i_2 \cdots i_{d-1}} - r = \Delta_d \tilde{\kappa}_{\sigma_d}$. For $\sigma_d > 0$ and b non-interfacial (when necessarily $p' < 2p$), Lemma 9.8 implies that $r_{i_1 i_2 \cdots i_{d-1}} - r = \Delta_d \tilde{\kappa}_{\sigma_d}$ or $r_{i_1 i_2 \cdots i_{d-1}} - r = \Delta_d(\tilde{\kappa}_{\sigma_d} + 1) = \Delta_d(\tilde{\kappa}_{\sigma_d+1})$. In either case, the definition of the r-tree implies that $r_{i_1 i_2 \cdots i_{d-1}}$ is a through-node of the r-tree. For $\sigma_d = 0$, $b_{i_1 i_2 \cdots i_{d-1}} - b = \Delta_d$ implies that $r_{i_1 i_2 \cdots i_{d-1}} - r = \Delta_d$ whence $r_{i_1 i_2 \cdots i_{d-1}}$ is a through-node of the r-tree.

2) This case arises only if $b \in \mathcal{T} \cup \mathcal{T}'$ so that either $b = \kappa_{\sigma_1}$ or $b = p' - \kappa_{\sigma_1}$ for some σ_1. If $\sigma_1 > 0$ then immediately $\boldsymbol{u} \notin \overline{\mathcal{U}}(b)$. If $\sigma_1 > 0$ and b is interfacial then Lemma 9.8 also implies that either $r = \rho(b) = \tilde{\kappa}_\sigma$ or $r = \rho(b) = p - \tilde{\kappa}_\sigma$. It follows that r_0 is a leaf-node of the r-tree. If $\sigma_1 > 0$ and b is not interfacial then Lemma 9.8 implies that either $\lfloor bp/p' \rfloor = \tilde{\kappa}_{\sigma_1}$ or $\lfloor bp/p' \rfloor = p - 1 - \tilde{\kappa}_{\sigma_1}$. Since $r = \lfloor bp/p' \rfloor$ or $r = \lfloor bp/p' \rfloor + 1$, and noting that $p' < 2p$ and $\sigma_1 \le t_2$ here, it immediately follows that r_0 is a leaf-node of the r-tree.

If $\sigma_1 = 0$ then either $b = 1$ and $\Delta_1 = -1$, or $b = p' - 1$ and $\Delta_1 = 1$ (this case only arises when $p' < 2p$). That $1 \le r < p$ forces $c = 2$ or $c = p' - 2$ respectively: i.e. $c = b - \Delta_d$ so that $\boldsymbol{u} \in \mathcal{U}^{b-c}(b)$ and thus $\boldsymbol{u} \notin \overline{\mathcal{U}}^{c-b}(b)$ as required. We also have $r = 1$ or $r = p - 1$ respectively so that r_0 is a leaf-node of the r-tree. □

Lemma 9.12. *For $1 \le r < p$, let $b, c \in \{\lfloor rp'/p \rfloor, \lfloor rp'/p \rfloor + 1\}$ with $b \ne c$.*

1) For $k \ge 1$, let $b_{i_1 i_2 \cdots i_k}$ be a branch-node of the b-tree and $r_{i_1 i_2 \cdots i_k}$ a node of the r-tree such that $\rho(b_{i_1 i_2 \cdots i_k}) = r_{i_1 i_2 \cdots i_k} \ne r$. If $\rho(b_{i_1 i_2 \cdots i_k h}) = r$ for some $h \in \{0, 1\}$, then $r_{i_1 i_2 \cdots i_k}$ is a through-node of the r-tree.

2) Let the b-tree nodes b_0 and b_1 be such that $b_0 \ne b_1$ with $\rho(b_h) = r$ for either $h = 0$ or $h = 1$. Then $r \in \tilde{\mathcal{T}} \cup \tilde{\mathcal{T}}'$ and r_0 is a leaf-node. Set $k = 0$ for what follows.

3) Let $h, k, i_1, i_2, \ldots, i_k$ be as in (1) or (2) above. There exists a unique $d \ge k+2$ such that if we set $i_{k+1} = \cdots = i_{d-1} = \bar{h}$ then $b_{i_1 i_2 \cdots i_{d-1} 0}$ is a leaf-node. The vector $\boldsymbol{u} \in \mathcal{U}(b)$ corresponding to this leaf-node is such that $\boldsymbol{u} \notin \overline{\mathcal{U}}(b)$ when b is interfacial, and $\boldsymbol{u} \in \mathcal{U}^{b-c}(b)$ when b is not interfacial. In addition, if $k + 2 \le j \le d$ then $b_{i_1 i_2 \cdots i_{j-2} h} = b - (-1)^h$ and $b_{i_1 i_2 \cdots i_{j-2} h 0}$ is a leaf-node. In each of these instances, the corresponding vector $\boldsymbol{u} \in \mathcal{U}(b)$ is such that $\boldsymbol{u} \in \overline{\mathcal{U}}(b)$, and also such that $\boldsymbol{u} \in \overline{\mathcal{U}}^{c-b}(b)$ when b is not interfacial.

If $\{\tau_j, \sigma_j, \Delta_j\}_{j=1}^d$ is the naive run corresponding to the leaf-node $b_{i_1 i_2 \cdots i_{d-1} 0}$ then $\tau_j = \sigma_j$ for $k + 2 \le j < d$ and $\Delta_j = (-1)^h$ for $k + 2 \le j \le d$. In addition, $\sigma_d + 1 = \tau_d \le t_1 + 1$ in the $p' > 2p$ case, and $\sigma_d + 1 = \tau_d \le t_2 + 1$ in the $p' < 2p$ case. If b is interfacial then $t_1 \le \sigma_d + 1$.

Proof: 1) Consider a leaf-node of the form $b_{i_1 \cdots i_k \bar{h} \cdots}$. Lemma 8.6 shows that $b_{i_1 i_2 \cdots i_k h} = b_{i_1 i_2 \cdots i_k} + (-1)^{i_k} \kappa_\sigma$ for some σ. Lemma 9.8 then shows that $b_{i_1 i_2 \cdots i_k h}$ and $b_{i_1 i_2 \cdots i_k}$ are both interfacial with $\rho(b_{i_1 i_2 \cdots i_k h}) = \rho(b_{i_1 i_2 \cdots i_k}) + (-1)^{i_k} \tilde{\kappa}_\sigma$. Since $r_{i_1 i_2 \cdots i_k} = \rho(b_{i_1 i_2 \cdots i_k})$ and $\rho(b_{i_1 i_2 \cdots i_k h}) = r$, the definition of the r-tree stipulates that $r_{i_1 i_2 \cdots i_k}$ is a through-node.

2) Consider a leaf-node of the form $b_{\bar{h} \cdots}$. Lemma 8.6 shows that either $b_{i_1} = \kappa_\sigma$ or $b_{i_1} = p' - \kappa_\sigma$ for some σ. Lemma 9.8 shows that b_{i_1} is interfacial and $\rho(b_{i_1}) = \tilde{\kappa}_\sigma$ or $\rho(b_{i_1}) = p - \tilde{\kappa}_\sigma$ respectively. Then $r = \rho(b_{i_1}) \in \tilde{\mathcal{T}} \cup \tilde{\mathcal{T}}'$ and the definition of the r-tree stipulates that r_0 is a leaf-node.

3) $b_{i_1 i_2 \cdots i_k h}$ is a non-leaf-node of the b-tree. Use of Lemma 9.10 shows that it is a through-node and that the vector $\boldsymbol{u} \in \mathcal{U}(b)$ corresponding to the leaf-node $b_{i_1 i_2 \cdots i_k h 0}$ is such that $\boldsymbol{u} \in \overline{\mathcal{U}}(b)$, and also such that $\boldsymbol{u} \in \overline{\mathcal{U}}^{c-b}(b)$ when b is not

interfacial. With $\Delta = (-1)^h$, Lemma 9.10 also shows that $b_{i_1 i_2 \cdots i_k h} = b - \Delta$, and if b is non-interfacial then $\Delta = b - c$.

$b_{i_1 i_2 \cdots i_k \overline{h}}$ is a non-leaf-node of the b-tree and is thus interfacial. $\rho(b_{i_1 i_2 \cdots i_k \overline{h}}) \neq r$ since otherwise one of $b_{i_1 i_2 \cdots i_k h}$ and $b_{i_1 i_2 \cdots i_k \overline{h}}$ would necessarily be equal to b. So either $b_{i_1 i_2 \cdots i_k \overline{h}}$ is a through-node or a branch-node. In the former case, let $\boldsymbol{u} \in \mathcal{U}(b)$ be the vector corresponding to the leaf-node $b_{i_1 i_2 \cdots i_k \overline{h} 0}$. If b is interfacial, Lemma 9.11 implies that $\boldsymbol{u} \notin \overline{\mathcal{U}}(b)$, and if b is non-interfacial then $\Delta(\boldsymbol{u}) = -(-1)^{\overline{h}} = \Delta = b - c$ so that $\boldsymbol{u} \in \mathcal{U}^{b-c}(b)$, as required.

In the case that $b_{i_1 i_2 \cdots i_k \overline{h}}$ is a branch-node, there exists no σ such that $b_{i_1 i_2 \cdots i_k \overline{h}} - \Delta \kappa_\sigma = b$. However, if τ is such that $\kappa_\tau = b_{i_1 i_2 \cdots i_k 1} - b_{i_1 i_2 \cdots i_k 0}$, then $b_{i_1 i_2 \cdots i_k \overline{h}} - \Delta \kappa_\tau = b_{i_1 i_2 \cdots i_k h} = b - \Delta$, whereupon the definition of the b-tree stipulates that $b_{i_1 i_2 \cdots i_k \overline{h} h} = b_{i_1 i_2 \cdots i_k h}$ and $b_{i_1 i_2 \cdots i_k \overline{h}\,\overline{h}} = b_{i_1 i_2 \cdots i_k \overline{h}} - \Delta \kappa_{\tau - 1}$.

Since $\rho(b_{i_1 i_2 \cdots i_k \overline{h} h}) = r$, the node $b_{i_1 i_2 \cdots i_k \overline{h}}$ now satisfies the same criteria as did $b_{i_1 i_2 \cdots i_k}$. We then iterate the above argument until we encounter a through-node of the form $b_{i_1 i_2 \cdots i_k \overline{h}\,\overline{h} \cdots \overline{h}}$. This must eventually occur because the b-tree is finite. Then as above, the vector $\boldsymbol{u} \in \mathcal{U}(b)$ corresponding to the leaf-node $b_{i_1 i_2 \cdots i_k \overline{h}\,\overline{h} \cdots \overline{h} 0}$ is such that $\boldsymbol{u} \notin \overline{\mathcal{U}}(b)$ when b is interfacial, and $\boldsymbol{u} \in \mathcal{U}^{b-c}(b)$ when b is not interfacial. Set d to be the number of levels that this leaf-node occurs below the root node. Then also as above, for $k+2 \leq j \leq d$ we have $b_{i_1 i_2 \cdots i_k i_{k+1} \cdots i_{j-2} h} = b - \Delta$ and that the vector $\boldsymbol{u} \in \mathcal{U}(b)$ corresponding to the leaf-node $b_{i_1 i_2 \cdots i_k i_{k+1} \cdots i_{j-2} h 0}$ is such that $\boldsymbol{u} \in \overline{\mathcal{U}}(b)$, and also satisfies $\boldsymbol{u} \in \overline{\mathcal{U}}^{c-b}(b)$ when b is not interfacial. This accounts for all descendents of $b_{i_1 i_2 \cdots i_k}$.

Since $b_{i_1 \cdots i_{j-2} \overline{i}_{j-1}} = b_{i_1 \cdots i_{j-1} \overline{i}_j}$ for $k+2 \leq j < d$, it follows from Lemma 8.6 that $\tau_j = \sigma_j$. For $k+2 \leq j \leq d$, that $\Delta_j = (-1)^h$ follows immediately from $i_{j-1} = \overline{h}$

Finally, we have $b_{i_1 \cdots i_{d-2} \overline{i}_{d-1}} = b_{i_1 \cdots i_k h} = b - \Delta$, and from Lemma 8.6, we have $b_{i_1 \cdots i_{d-1}} - b_{i_1 \cdots i_{d-2} \overline{i}_{d-1}} = \Delta \kappa_{\tau_d}$ and $b = b_{i_1 \cdots i_{d-1}} - \Delta \kappa_{\sigma_d}$. Together, these imply that $\kappa_{\sigma_d} = \kappa_{\tau_d} - 1$. In the $p' > 2p$ case this can only occur if $\sigma_d + 1 = \tau_d \leq t_1 + 1$, whereas in the $p' < 2p$ case this can only occur if $\sigma_d + 1 = \tau_d \leq t_2 + 1$. In addition, since $b_{i_1 \cdots i_{d-1}}$ is interfacial with $\rho(b_{i_1 \cdots i_{d-1}}) \neq r$, and odd bands in the (p,p')-model are separated by at least t_1 even bands, $b = b_{i_1 \cdots i_{d-1}} - \Delta \kappa_{\sigma_d}$ implies that if b is interfacial then $t_1 \leq \kappa_{\sigma_d} = \sigma_d + 1$ which completes the proof. □

Lemma 9.13. *For $1 \leq r < p$, let $b, c \in \{\lfloor rp'/p \rfloor, \lfloor rp'/p \rfloor + 1\}$ with $b \neq c$.*

1) For $k \geq 1$, let $b_{i_1 i_2 \cdots i_k}$ be a branch-node of the b-tree and $r_{i_1 i_2 \cdots i_k}$ a node of the r-tree such that $\rho(b_{i_1 i_2 \cdots i_k}) = r_{i_1 i_2 \cdots i_k} \neq r$. If

$$\rho(b_{i_1 i_2 \cdots i_k 0}) < r < \rho(b_{i_1 i_2 \cdots i_k 1})$$

then $r_{i_1 i_2 \cdots i_k}$ is a branch-node of the r-tree and moreover $\rho(b_{i_1 i_2 \cdots i_k 0}) = r_{i_1 i_2 \cdots i_k 0}$ and $\rho(b_{i_1 i_2 \cdots i_k 1}) = r_{i_1 i_2 \cdots i_k 1}$.

2) Let b_0 and b_1 be nodes of the b-tree such that $\rho(b_0) < r < \rho(b_1)$. Then r_0 and r_1 are nodes of the r-tree with $\rho(b_0) = r_0$ and $\rho(b_1) = r_1$.

Proof: 1) The definition of the b-tree implies that there exists x such that:

$$b_{i_1 i_2 \cdots i_k 0} = b_{i_1 i_2 \cdots i_k} + (-1)^{i_k} \kappa_{x + i_k},$$
$$b_{i_1 i_2 \cdots i_k 1} = b_{i_1 i_2 \cdots i_k} + (-1)^{i_k} \kappa_{x + 1 - i_k}.$$

By choosing an appropriate leaf-node, Lemmas 8.6 and 9.8 then imply that

$$\rho(b_{i_1 i_2 \cdots i_k 0}) = \rho(b_{i_1 i_2 \cdots i_k}) + (-1)^{i_k} \tilde{\kappa}_{x+i_k},$$
$$\rho(b_{i_1 i_2 \cdots i_k 1}) = \rho(b_{i_1 i_2 \cdots i_k}) + (-1)^{i_k} \tilde{\kappa}_{x+1-i_k}.$$

By hypothesis, r is strictly between these values. Since $\rho(b_{i_1 i_2 \cdots i_k}) = r_{i_1 i_2 \cdots i_k} \neq r$, the definition of the r-tree stipulates that $r_{i_1 i_2 \cdots i_k}$ is a branch-node and that $r_{i_1 i_2 \cdots i_k 0} = \rho(b_{i_1 i_2 \cdots i_k 0})$ and $r_{i_1 i_2 \cdots i_k 1} = \rho(b_{i_1 i_2 \cdots i_k 1})$.

2) There are four cases to consider depending on which of the sets $\mathcal{T}$ and $\mathcal{T}'$ contains b_0 and b_1. We consider only the case where $b_0, b_1 \in \mathcal{T}$: the other cases are similar.

The definition of the b-tree shows that there exists σ_1 such that $b_0 = \kappa_{\sigma_1}$ and $b_1 = \kappa_{\sigma_1+1}$ Lemmas 8.6 and 9.8 imply that $\rho(b_0) = \tilde{\kappa}_{\sigma_1}$ and $\rho(b_1) = \tilde{\kappa}_{\sigma_1+1}$. We claim that $r_0 = \tilde{\kappa}_{\sigma_1}$ and $r_1 = \tilde{\kappa}_{\sigma_1+1}$. This is so because $\tilde{\kappa}_{\sigma_1} < r < \tilde{\kappa}_{\sigma_1+1}$ and if there were any elements of $\tilde{\mathcal{T}} \cup \tilde{\mathcal{T}}'$ strictly between $\tilde{\kappa}_{\sigma_1}$ and $\tilde{\kappa}_{\sigma_1+1}$ then, via Lemma E.5(1), there would be elements of $\mathcal{T} \cup \mathcal{T}'$ strictly between κ_{σ_1} and κ_{σ_1+1} whereupon the description of the b-tree would not then yield $b_0 = \kappa_{\sigma_1}$ and $b_1 = \kappa_{\sigma_1+1}$. $\square$

Lemma 9.14. *For $1 \leq r < p$, let $b, c \in \{\lfloor rp'/p \rfloor, \lfloor rp'/p \rfloor + 1\}$ with $b \neq c$. For $d \geq 1$, let $b_{i_1 i_2 \cdots i_{d-1} i_d}$ be a leaf-node of the b-tree and let $\{\tau_j, \sigma_j, \Delta_j\}_{j=1}^d$ be the corresponding naive run. Assume that $r_{i_1 i_2 \cdots i_{\tilde{d}-1} i_{\tilde{d}}}$ is a leaf-node of the r-tree for some $\tilde{d}$ with $1 \leq \tilde{d} \leq d$, and let $\{\tilde{\tau}_j, \tilde{\sigma}_j, \tilde{\Delta}_j\}_{j=1}^{\tilde{d}}$ be the corresponding naive run. Then $\tilde{\tau}_j = \tau_j$ and $\tilde{\Delta}_j = \Delta_j$ for $1 \leq j \leq \tilde{d}$, and $\tilde{\sigma}_j = \sigma_j$ for $1 \leq j < \tilde{d}$.*

If either both $\tilde{d} < d$ and $\rho(b_{i_1 i_2 \cdots i_{\tilde{d}-1} \bar{\imath}_{\tilde{d}}}) = r$, or both $\tilde{d} = d$ and b is interfacial, then $\tilde{\sigma}_{\tilde{d}} = \max\{\sigma_{\tilde{d}}, t_1 + 1\}$. If $\tilde{d} = d$ and b is not interfacial then

$$\tilde{\sigma}_d = \begin{cases} \sigma_d & \text{if } c - b = \Delta_d; \\ \sigma_d + 1 & \text{if } c - b \neq \Delta_d. \end{cases}$$

Proof: We first claim that $\rho(b_{i_1 \cdots i_k 0}) \neq r$ and $\rho(b_{i_1 \cdots i_k 1}) \neq r$ for $0 \leq k \leq \tilde{d} - 2$. Otherwise, let k be the smallest value for which the claim doesn't hold, so that $\rho(b_{i_1 \cdots i_k h}) = r$ for some $h \in \{0, 1\}$. If $k = 0$ (in which case $\tilde{d} \geq 2$), then Lemma 9.12(2) shows that r_0 is a leaf-node, whereas if $k \geq 1$ then Lemma 9.12(1) shows that $r_{i_1 \cdots i_k 0}$ is a leaf-node. This contradicts the hypothesis that $r_{i_1 i_2 \cdots i_{\tilde{d}-1} 0}$ is a leaf-node, thus establishing the claim.

Lemma 9.13 now shows that $\rho(b_{i_1 \cdots i_k 0}) = r_{i_1 \cdots i_k 0}$ and $\rho(b_{i_1 \cdots i_k 1}) = r_{i_1 \cdots i_k 1}$ for $0 \leq k \leq \tilde{d} - 2$. Comparison of Lemmas 9.8 and 9.9 then shows that $\tilde{\tau}_j = \tau_j$ for $1 \leq j \leq \tilde{d}$, and $\tilde{\sigma}_j = \sigma_j$ for $1 \leq j < \tilde{d}$. That $\tilde{\Delta}_j = \Delta_j$ for $1 \leq j \leq \tilde{d}$ is immediate from their definitions. It remains to prove the expressions for $\tilde{\sigma}_{\tilde{d}}$.

If $d > \tilde{d} > 1$ then Lemma 9.8 implies that $|\rho(b_{i_1 \cdots i_{\tilde{d}-1}}) - \rho(b_{i_1 \cdots i_{\tilde{d}-1} \bar{\imath}_{\tilde{d}}})| = \tilde{\kappa}_{\sigma_{\tilde{d}}}$. Thereupon, $|r_{i_1 \cdots i_{\tilde{d}-1}} - r| = \tilde{\kappa}_{\sigma_{\tilde{d}}}$ so that $\tilde{\kappa}_{\tilde{\sigma}_{\tilde{d}}} = \tilde{\kappa}_{\sigma_{\tilde{d}}}$. On noting that $\tilde{\kappa}_i = 1$ if and only if $0 \leq i \leq t_1 + 1$, the required result follows here.

If $d = \tilde{d} > 1$ with b interfacial (and thus $\rho(b) = r$), a similar argument applies when $\sigma_d > 0$. If $\sigma_d = 0$ then $|b_{i_1 \cdots i_{d-1}} - b| = 1$ so that, noting that $r_{i_1 \cdots i_{\tilde{d}-1}} \neq r$, then necessarily $|r_{i_1 \cdots i_{\tilde{d}-1}} - r| = 1$ and thus $\tilde{\sigma}_{\tilde{d}} = t_1 + 1$.

If $d > \tilde{d} = 1$ then Lemma 9.8 implies that $\rho(b_{\bar{\imath}_1}) = \tilde{\kappa}_{\sigma_1}$ or $\rho(b_{\bar{\imath}_1}) = p - \tilde{\kappa}_{\sigma_1}$. Since $r = \rho(b_{\bar{\imath}_1})$, we then immediately obtain $\tilde{\kappa}_{\tilde{\sigma}_{\tilde{d}}} = \tilde{\kappa}_{\sigma_{\tilde{d}}}$, from which again the desired result follows.

If $d = \tilde{d} = 1$ with b interfacial (and thus $\rho(b) = r$), a similar argument applies when $\sigma_1 > 0$. The case $\sigma_1 = 0$ is excluded here since then either $b_0 = 1$ or $b_0 = p' - 1$, neither of which is interfacial.

Now consider $\tilde{d} = d$ and b being non-interfacial (so that $p' < 2p$ and $\sigma_d \le t_2$). Here $r = \lfloor bp/p' \rfloor + \frac{1}{2}(c - b + 1)$. If $\sigma_d > 0$, Lemma 9.8 then yields:

$$r = \rho(b_{i_1 \cdots i_{d-1}}) - \Delta_d \tilde{\kappa}_{\sigma_d} + \frac{1}{2}(c - b - \Delta_d).$$

Then, since $r = r_{i_1 \cdots i_{d-1}} - \Delta_d \tilde{\kappa}_{\tilde{\sigma}_d} = \rho(b_{i_1 \cdots i_{d-1}}) - \Delta_d \tilde{\kappa}_{\tilde{\sigma}_d}$, we obtain $\Delta_d \tilde{\kappa}_{\tilde{\sigma}_d} = \Delta_d \tilde{\kappa}_{\sigma_d} - \frac{1}{2}(c - b - \Delta_d)$, from which the required expressions for $\tilde{\sigma}_d$ follow after noting that $\sigma_d \le t_2$.

If $\sigma_d = 0$ then $b_{i_1 \cdots i_{d-1}} = b + \Delta_d$, and we claim that $c = b - \Delta_d$. Otherwise $b_{i_1 \cdots i_{d-1}} = b - \Delta_d = c$ which is interfacial and thus $\rho(b_{i_1 \cdots i_{d-1}}) = r$, contradicting the fact that $r_{i_1 \cdots i_{d-1}} = \rho(b_{i_1 \cdots i_{d-1}})$ is not a leaf-node. It now follows that $|r_{i_1 \cdots i_{d-1}} - r| = 1$ and so $\tilde{\sigma}_d = t_1 + 1 = 1$ in this case, as required. □

Theorem 9.15. *For $1 \le r < p$, let $b, c \in \{\lfloor rp'/p \rfloor, \lfloor rp'/p \rfloor + 1\}$ with $b \ne c$.*

1) If b is interfacial then there is a bijection between $\mathcal{U}(b) \backslash \overline{\mathcal{U}}(b)$ and $\widetilde{\mathcal{U}}(r)$ such that if $\boldsymbol{u}$ maps to $\tilde{\boldsymbol{u}}$ under this bijection, then:

$$F^*(\boldsymbol{u}^L, \tilde{\boldsymbol{u}}) = F^*(\boldsymbol{u}^L, \boldsymbol{u}), \tag{9.12}$$

for all $\boldsymbol{u}^L \in \mathcal{U}(a)$ with $1 \le a < p'$.

2) If b is not interfacial then there is a bijection between $\mathcal{U}(b) \backslash \overline{\mathcal{U}}^{c-b}(b)$ and $\widetilde{\mathcal{U}}(r)$ such that if $\boldsymbol{u}$ maps to $\tilde{\boldsymbol{u}}$ under this bijection, then:

$$F^*(\boldsymbol{u}^L, \tilde{\boldsymbol{u}}) = \begin{cases} F^*(\boldsymbol{u}^L, \boldsymbol{u}) & \text{if } \boldsymbol{u} \in \mathcal{U}^{c-b}(b); \\ F^*(\boldsymbol{u}^L, \boldsymbol{u}^+) & \text{if } \boldsymbol{u} \in \mathcal{U}^{b-c}(b), \end{cases} \tag{9.13}$$

for all $\boldsymbol{u}^L \in \mathcal{U}(a)$ with $1 \le a < p'$.

Proof: We will describe a traversal of the b-tree level by level (breadth first) starting at level one. During the traversal, the r-tree is constructed level by level, and certain nodes of the b-tree are eliminated from further consideration.

At the kth level ($k \ge 1$) this process will ensure that we only examine non-leaf b-tree nodes $b_{i_1 \cdots i_k}$ for which $r_{i_1 \cdots i_k}$ is a node of the r-tree and for which $\rho(b_{i_1 \cdots i_k}) \ne r$. Consider such a node $b_{i_1 \cdots i_k}$. This node is either a through-node, a branch-node for which $\rho(b_{i_1 \cdots i_k h}) = r$ for some $h \in \{0, 1\}$, or a branch-node for which $\rho(b_{i_1 \cdots i_k 0}) < r < \rho(b_{i_1 \cdots i_k 1})$. We consider these three cases separately below. However, we will also deal simultaneously with the analogous three cases that arise when $k = 0$: the root node being a through-node, $\rho(b_h) = r$ for some $h \in \{0, 1\}$, or $\rho(b_0) < r < \rho(b_1)$.

In the case that $k = 0$ and the root node is a through-node, or $k > 0$ and $b_{i_1 \cdots i_k}$ is a through-node, set $d = k + 1$ so that $b_{i_1 \cdots i_{d-1} 0}$ is a leaf-node of the b-tree and, by Lemma 9.11, $r_{i_1 \cdots i_{d-1} 0}$ is a leaf-node of the r-tree. Let $\{\tau_j, \sigma_j, \Delta_j\}_{j=1}^d$ and $\{\tilde{\tau}_j, \tilde{\sigma}_j, \tilde{\Delta}_j\}_{j=1}^d$ be the respective corresponding naive runs, and let $\boldsymbol{u} \in \mathcal{U}(b)$ and $\tilde{\boldsymbol{u}} \in \widetilde{\mathcal{U}}(r)$ be the respective corresponding vectors. Lemma 9.11 shows that $\boldsymbol{u} \notin \overline{\mathcal{U}}(b)$ if b is interfacial, and $\boldsymbol{u} \notin \overline{\mathcal{U}}^{c-b}(b)$ if b is not interfacial. For the required bijection, we map $\boldsymbol{u} \mapsto \tilde{\boldsymbol{u}}$ with (9.12) and (9.13) yet to be demonstrated.

Lemma 9.14 shows that $\tau_j = \tilde{\tau}_j$ for $1 \leq j \leq d$ and $\sigma_j = \tilde{\sigma}_j$ for $1 \leq j < d$. In the case that b is interfacial, Lemma 9.14 gives $\tilde{\sigma}_d = \max\{\sigma_d, t_1+1\}$. If $\sigma_d \geq t_1+1$ then $\tilde{\boldsymbol{u}} = \boldsymbol{u}$ and (9.12) holds trivially. Otherwise, (9.12) follows by using Lemma 9.7 (possibly more than once). In the case that b is not interfacial, note first that $\boldsymbol{u} \in \mathcal{U}^{\Delta_d}(b)$. For $\Delta_d = c - b$, Lemma 9.14 shows that $\tilde{\sigma}_d = \sigma_d$ whence $\tilde{\boldsymbol{u}} = \boldsymbol{u}$ and (9.13) holds trivially. For $\Delta_d = b - c$, Lemma 9.14 shows that $\tilde{\sigma}_d = \sigma_d + 1$ whence $\tilde{\boldsymbol{u}} = \boldsymbol{u}^+$ and (9.13) holds trivially. We exclude examining the leaf-node $b_{i_1 \cdots i_{d-1}0}$ in our subsequent traversing of the b-tree.

We now deal with the cases in which either both $k = 0$ and the root node is a branch-node, or both $k > 0$ and $b_{i_1 \cdots i_k}$ is a branch-node. If, in this case, $\rho(b_{i_1 \cdots i_k h}) = r$ for some $h \in \{0,1\}$, Lemma 9.12(1,2) shows that $r_{i_1 \cdots i_{\tilde{d}-1}0}$ is a leaf-node of the r-tree where $\tilde{d} = k+1$. Let $\{\tilde{\tau}_j, \tilde{\sigma}_j, \tilde{\Delta}_j\}_{j=1}^{\tilde{d}}$ be the corresponding naive run, and let $\tilde{\boldsymbol{u}} \in \widetilde{\mathcal{U}}(r)$ be the corresponding vector. Lemma 9.12(3) shows that there is a unique $d \geq k+2$ for which $b_{i_1 \cdots i_{d-1}0}$ is a leaf-node with $i_{k+1} = \cdots = i_{d-1} = \bar{h}$. Let $\{\tau_j, \sigma_j, \Delta_j\}_{j=1}^{d}$ be the corresponding naive run, and let $\boldsymbol{u} \in \mathcal{U}(b)$ be the corresponding vector. Lemma 9.12(3) shows that $\boldsymbol{u} \notin \overline{\mathcal{U}}(b)$ if b is interfacial, and $\boldsymbol{u} \in \mathcal{U}^{b-c}(b)$ if b is not interfacial. For the required bijection, we map $\boldsymbol{u} \mapsto \tilde{\boldsymbol{u}}$ with (9.12) and (9.13) yet to be demonstrated.

Lemma 9.14 shows that $\tau_j = \tilde{\tau}_j$ for $1 \leq j \leq \tilde{d}$ and $\sigma_j = \tilde{\sigma}_j$ for $1 \leq j < \tilde{d}$ with $\tilde{\sigma}_{\tilde{d}} = \max\{\sigma_{\tilde{d}}, t_1+1\}$, on noting that $d > k+1 = \tilde{d}$. Then $\tilde{\sigma}_{\tilde{d}} = \sigma_{\tilde{d}}$ because otherwise $d > \tilde{d}$ implies that $\tau_d < \sigma_{\tilde{d}} \leq t_1$ which cannot occur in a naive run. Lemma 9.12(3) also shows that $\tau_j = \sigma_j$ for $\tilde{d}+1 \leq j < d$, and $\sigma_d + 1 = \tau_d \leq t_1 + 1$ for $p' > 2p$ and $\sigma_d + 1 = \tau_d \leq t_2 + 1$ for $p' < 2p$. Therefore, $\boldsymbol{u} = \tilde{\boldsymbol{u}} + \boldsymbol{u}_{\sigma_d, \sigma_d+1}$ so that $\boldsymbol{u}^+ = \tilde{\boldsymbol{u}}$. In the case of b being non-interfacial, (9.13) follows immediately. In the case of b being interfacial and $\sigma_d \leq t_1$, the use of Lemma 9.7 proves (9.12). We now claim that the case of b being interfacial and $\sigma_d > t_1$ does not arise here. First note that necessarily $p' < 2p$ so that $t_1 = 0$. Lemma 9.12(3) implies that $b_{i_1 \cdots i_{d-2}h} = b - \Delta_d$ so that both b and $b - \Delta_d$ are interfacial, and separated by an odd band. This implies that $t_2 = 1$ (by considering the model obtained by toggling the parity of each band). So $\sigma_d = t_2 = 1$ and $\tau_d = 2$. Now $b_{i_1 \cdots i_{d-2}\bar{h}} = b + \Delta_d \kappa_{\sigma_d} = b + 2\Delta_d$ is interfacial. The definition of a naive run implies that either $\tau_d = t_k$ for some k, or $\tau_d = t_n - 1$. Thus either $t_3 = 2$ or both $t_3 = 3$ and $n = 3$. Lemma E.12 now confirms the above claim.

Lemma 9.12(3) also shows that all leaf-nodes other than $b_{i_1 \cdots i_{d-1}0}$ that are descendents of $b_{i_1 \cdots i_k}$ yield vectors $\boldsymbol{u} \in \overline{\mathcal{U}}(b)$ when b is interfacial, and vectors $\boldsymbol{u} \in \overline{\mathcal{U}}^{c-b}(b)$ when b is non-interfacial. Thus we exclude examining all nodes that are descendents of $b_{i_1 \cdots i_k}$ in our subsequent traversing of the b-tree.

In the case in which either both $k = 0$ and the root node is a branch-node, or both $k > 0$ and $b_{i_1 \cdots i_k}$ is a branch-node, we consider $\rho(b_{i_1 \cdots i_k 0}) < r < \rho(b_{i_1 \cdots i_k 1})$, Lemma 9.13 shows that both $r_{i_1 \cdots i_k 0}$ and $r_{i_1 \cdots i_k 1}$ are non-leaf-nodes of the r-tree with the value of neither equal to r. Each will be examined in traversing the next level of the b-tree.

Once this recursive procedure has finished, the r-tree will have been completely constructed from the b-tree. In addition, in the case of b being interfacial, an explicit bijection will have been established between the leaf-nodes of the b-tree which yield vectors $\boldsymbol{u} \in \mathcal{U}(b) \backslash \overline{\mathcal{U}}(b)$ and the leaf-nodes of the r-tree. Since the corresponding bijection between $\boldsymbol{u} \in \mathcal{U}(b) \backslash \overline{\mathcal{U}}(b)$ and $\tilde{\boldsymbol{u}} \in \widetilde{\mathcal{U}}(r)$ satisfies (9.12) in each case, the theorem is proved for b interfacial. Similarly, in the case of b being non-interfacial, an explicit bijection will have been established between the leaf-nodes of the b-tree which yield vectors $\boldsymbol{u} \in \mathcal{U}(b) \backslash \overline{\mathcal{U}}^{c-b}(b)$ and the leaf-nodes of the r-tree. Since the corresponding bijection between $\boldsymbol{u} \in \mathcal{U}(b) \backslash \overline{\mathcal{U}}^{c-b}(b)$ and $\tilde{\boldsymbol{u}} \in \widetilde{\mathcal{U}}(r)$ satisfies (9.13) in each case, the theorem is proved for b non-interfacial. $\square$

We are now in a position to prove the fermionic expressions for all Virasoro characters $\chi^{p,p'}_{r,s}$ that were stated in (1.16). For convenience, the cases of the extra term $\chi^{\hat{p},\hat{p}'}_{\hat{r},\hat{s}}$ being present or not are stated separately.

Theorem 9.16. *Let $1 \le s < p'$ and $1 \le r < p$. Let η be such that $\xi_\eta \le s < \xi_{\eta+1}$ and η' be such that $\tilde{\xi}_{\eta'} \le r < \tilde{\xi}_{\eta'+1}$.*
1) If $\eta = \eta'$ and $\xi_\eta < s$ and $\tilde{\xi}_\eta < r$ then:

$$\chi^{p,p'}_{r,s} = \sum_{\substack{\boldsymbol{u}^L \in \mathcal{U}(s) \\ \boldsymbol{u}^R \in \widetilde{\mathcal{U}}(r)}} F(\boldsymbol{u}^L, \boldsymbol{u}^R) \quad + \quad \chi^{\hat{p},\hat{p}'}_{\hat{r},\hat{s}}, \tag{9.14}$$

where $\hat{p}' = \xi_{\eta+1} - \xi_\eta$, $\hat{p} = \tilde{\xi}_{\eta+1} - \tilde{\xi}_\eta$, $\hat{s} = s - \xi_\eta$, $\hat{r} = r - \tilde{\xi}_\eta$.
2) Otherwise:

$$\chi^{p,p'}_{r,s} = \sum_{\substack{\boldsymbol{u}^L \in \mathcal{U}(s) \\ \boldsymbol{u}^R \in \widetilde{\mathcal{U}}(r)}} F(\boldsymbol{u}^L, \boldsymbol{u}^R). \tag{9.15}$$

Proof: First note that if $\boldsymbol{u}^R \in \widetilde{\mathcal{U}}(r)$ then $\sigma(\boldsymbol{u}^R) \ge t_1 + 1$. In the notation used for (9.6), this guarantees that $\boldsymbol{N}(\boldsymbol{u}^R) = 0$ and $(\boldsymbol{u}^L_\flat + \boldsymbol{u}^R_\sharp) \cdot \boldsymbol{m} = (\overline{\boldsymbol{u}}^L_\flat + \overline{\boldsymbol{u}}^R_\sharp) \cdot \overline{\boldsymbol{m}}$. Thereupon $F^*(\boldsymbol{u}^L, \boldsymbol{u}^R) = F(\boldsymbol{u}^L, \boldsymbol{u}^R)$.

Let b be equal to one of $\lfloor rp'/p \rfloor$ and $\lfloor rp'/p \rfloor + 1$, and let c be equal to the other. If b is interfacial, the application of Theorem 9.15(1) to the two cases of Theorem 9.3 now yields (9.14) and (9.15). Similarly, if b is non-interfacial, the application of Theorem 9.15(2) to the two cases of Theorem 9.6 also yields (9.14) and (9.15). $\square$

10. Discussion

In this paper, we have shown how to write down constant-sign fermionic expressions for all Virasoro characters $\chi^{p,p'}_{r,s}$, where $1<p<p'$ with p and p' coprime, $1\le r<p$ and $1\le s<p'$. The first step in this process constructs a tree for s from the set $\mathcal{T}\cup\mathcal{T}'$ where $\mathcal{T}$ is the set of Takahashi lengths associated with the continued fraction of p'/p, and $\mathcal{T}'$ is a set of values complementary to $\mathcal{T}$. A tree for r is constructed in a similar way using the set $\tilde{\mathcal{T}}\cup\tilde{\mathcal{T}}'$, where $\tilde{\mathcal{T}}$ is the set of truncated Takahashi lengths associated with the continued fraction of p'/p, and $\tilde{\mathcal{T}}'$ is a set of values complementary to $\tilde{\mathcal{T}}$. A set $\mathcal{U}(s)$ of vectors is obtained from the leaf-nodes of the first of these trees, with each leaf-node giving rise to precisely one vector. Similarly, a set $\widetilde{\mathcal{U}}(r)$ of vectors is obtained from the leaf-nodes of the second of these trees. Each pair of vectors $\boldsymbol{u}^L\in\mathcal{U}(s)$ and $\boldsymbol{u}^R\in\widetilde{\mathcal{U}}(r)$ gives rise to a fundamental fermionic form $F(\boldsymbol{u}^L,\boldsymbol{u}^R)$. In most cases, taking the sum of such terms over all pairs $\boldsymbol{u}^L,\boldsymbol{u}^R$ yields the required fermionic expression for $\chi^{p,p'}_{r,s}$. This expression thus comprises $|\mathcal{U}(s)|\cdot|\widetilde{\mathcal{U}}(r)|$ fundamental fermionic forms. In the remaining cases, in addition to these terms, we require a further character $\chi^{\hat{p},\hat{p}'}_{\hat{r},\hat{s}}$. The process described above is now applied to this character. Thus in general, writing down the fermionic expression for $\chi^{p,p'}_{r,s}$ is a recursive procedure. It is interesting to note that the number of fundamental fermionic forms comprising the resulting fermionic expressions for $\chi^{p,p'}_{r,s}$, varies erratically as r and s run over their permitted ranges. A similar observation was made in [**12**] for the expressions given there.

Expressions of a similar nature are provided for the finitized characters $\chi^{p,p'}_{a,b,c}(L)$, where $1\le p<p'$ with p and p' coprime, $1\le a,b<p'$ and $c=b\pm1$. These finitized characters are generating functions for length L Forrester-Baxter paths. The fermionic expressions here involve a sum over all pairs $\boldsymbol{u}^L,\boldsymbol{u}^R$ with $\boldsymbol{u}^L\in\mathcal{U}(a)$ and $\boldsymbol{u}^R\in\mathcal{U}(b)$: the sets $\mathcal{U}(a)$ and $\mathcal{U}(b)$ having both been produced using the set $\mathcal{T}\cup\mathcal{T}'$. If b satisfies (1.36), each term in the sum is the fundamental fermionic form $F(\boldsymbol{u}^L,\boldsymbol{u}^R,L)$. For other values of b, each term in the sum is either $F(\boldsymbol{u}^L,\boldsymbol{u}^R,L)$ or $\widetilde{F}(\boldsymbol{u}^L,\boldsymbol{u}^R,L)$, where the latter is itself a linear combination of terms $F(\boldsymbol{u}^L,\boldsymbol{u},L')$ for various $\boldsymbol{u}$ and L'. The fermionic expression for $\chi^{p,p'}_{a,b,c}(L)$ also sometimes involves a term $\chi^{\hat{p},\hat{p}'}_{\hat{a},\hat{b},\hat{c}}(L)$ whence, like the fermionic expression for $\chi^{p,p'}_{r,s}$, it is recursive in nature. For b satisfying (1.36), the expressions obtained for $\chi^{p,p'}_{a,b,c}(L)$ are genuinely fermionic in that they are positive sums over manifestly positive definite $F(\boldsymbol{u}^L,\boldsymbol{u}^R,L)$. For b not satisfying (1.36), this is not the case. Nonetheless, whatever the value of b, after some work, the fermionic expressions for $\chi^{p,p'}_{r,s}$ described above emerge from these expressions on taking the $L\to\infty$ limit.

By using the Takahashi trees that give rise to the sets $\mathcal{U}(a)$ and $\mathcal{U}(b)$, it is possible to characterise the set of paths for which $F(\boldsymbol{u}^L, \boldsymbol{u}^R, L)$ or $\widetilde{F}(\boldsymbol{u}^L, \boldsymbol{u}^R, L)$ is the generating function. In the former case, this set of paths is specified in Lemma 8.7 as those which attain certain heights in a certain order. It is a similar set of paths in the latter case, as described in Note 8.11. The corresponding term $F(\boldsymbol{u}^L, \boldsymbol{u}^R)$ that forms a component of the fermionic expression for $\chi^{p,p'}_{r,s}$ may then be seen to be the generating function of a certain set of paths of infinite length. However, it is somewhat unfortunate that characterising these latter paths requires the construction of the Takahashi tree for a suitable b, whereas in forming the fermionic expression for $\chi^{p,p'}_{r,s}$, the construction of such a tree has been superseded by the construction of the truncated Takahashi tree for r.

In fact, this and many aspects of the proof of the fermionic expressions for $\chi^{p,p'}_{r,s}$ hint that perhaps $\chi^{p,p'}_{a,b,c}(L)$ is not the most natural finitization of $\chi^{p,p'}_{r,s}$. These aspects include the need to temporarily redefine the weight function to that of (2.5), the frequent need to treat the cases $p' > 2p$ and $p' < 2p$ separately, the appearance of $\delta^{p,p'}_{a,e}$ and $\delta^{p'-p,p'}_{a,e}$ in Corollaries 5.8 and 5.10, the appearance of terms $F(\boldsymbol{u}^L, \boldsymbol{u}^R, L)$ in the fermionic expression for $\chi^{p,p'}_{a,b,c}(L)$ that vanish as $L \to \infty$, and the difference between the definitions of $\gamma(\mathcal{X}^L, \mathcal{X}^R)$ and $\gamma'(\mathcal{X}^L, \mathcal{X}^R)$ in Section 1.10. A more natural finitization may lead to a better combinatorial description of the characters $\chi^{p,p'}_{r,s}$. This, in turn, would be of great benefit both in the analysis of the expressions given in this paper as well as in the construction of the corresponding irreducible representations of the Virasoro algebra.

It is also intriguing that the constant term $\frac{1}{4}\gamma(\mathcal{X}^L, \mathcal{X}^R)$ varies relatively little on running through the leaf-nodes of the Takahashi tree for s and the leaf-nodes of the truncated Takahashi tree for r. Moreover, these values are also close to the negative of the modular anomaly $\Delta^{p,p'}_{r,s} - \frac{1}{24}c^{p,p'}$. This hints at the fundamental fermionic forms $F(\boldsymbol{u}^L, \boldsymbol{u}^R)$ themselves having interesting properties under the action of the modular group.

Equating the fermionic expression for each $\chi^{p,p'}_{r,s}$ with the corresponding bosonic expression given by (1.2) yields bosonic-fermionic q-series identities. Furthermore, (1.4) or (1.5) provides a product expression for the character $\chi^{p,p'}_{r,s}$ in certain cases. Equating each of these with the corresponding fermionic expression given here yields identities of the Rogers-Ramanujan type. In addition, equating the fermionic expression for each finitized character $\chi^{p,p'}_{a,b,c}(L)$ with the corresponding bosonic expression (1.9) yields bosonic-fermionic polynomial identities.

Finally, we remark that further fermionic expressions and identities for $\chi^{p,p'}_{r,s}$ may be produced by instead of using the set $\mathcal{T} \cup \mathcal{T}'$ to obtain the Takahashi tree for s and the set $\tilde{\mathcal{T}} \cup \tilde{\mathcal{T}}'$ to obtain the truncated Takahashi tree for r, using either just the set $\mathcal{T} \cup \{\kappa_t\}$ or the set $\mathcal{T}' \cup \{p' - \kappa_t\}$ in the former case, or using just the set $\tilde{\mathcal{T}} \cup \{\tilde{\kappa}_t\}$ or the set $\tilde{\mathcal{T}}' \cup \{p - \tilde{\kappa}_t\}$ in the latter case. However, such expressions for $\chi^{p,p'}_{r,s}$ are not as efficient as those given in Section 1, in that they involve as least as many terms $F(\boldsymbol{u}^L, \boldsymbol{u}^R)$ and sometimes more. In addition, an extra term $\chi^{\hat{p},\hat{p}'}_{\hat{r},\hat{s}}$ might be present when the expression of Section 1 requires no such term. Similar comments pertain to the finitized characters $\chi^{p,p'}_{a,b,c}(L)$. The definitions of Section 1.13 must be suitably modified to accommodate these additional expressions.

Looking ahead, the question now arises as to whether the techniques developed in this paper can be extended to deal with the characters of the W_n algebras, the $n = 2$ case of which is the Virasoro algebra. The RSOS models that pertain to these algebras are described in [**29, 35, 32**]. As yet, fermionic expressions for the W_n algebra characters are available only in very special cases [**22, 39, 16, 46, 40, 7, 45**]. In particular, the expressions of [**7, 45**] have led to a number of elegant new identities of Rogers-Ramanujan type.

Acknowledgement

I wish to thank Omar Foda very much for his support and encouragement during this work. I also wish to thank Omar, Alexander Berkovich, Barry McCoy, Will Orrick, Anne Schilling and Ole Warnaar for useful discussions. I am grateful to Will and Ole for reading and commenting on a draft of part of the manuscript.

APPENDIX A

Examples

Here, we provide three examples of the constructions of the fermionic expressions that are described in Section 1.

A.1. Example 1

Let $p' = 109$, $p = 26$, $r = 9$ and $s = 51$. Then p'/p has continued fraction $[4,5,5]$, so that $n = 2$, $\{t_k\}_{k=0}^{3} = \{-1,3,8,13\}$, $t = 12$, $\{y_k\}_{k=-1}^{3} = \{0,1,4,21,109\}$ and $\{z_k\}_{k=-1}^{3} = \{1,0,1,5,26\}$. The sets $\mathcal{T}$, $\mathcal{T}'$, $\tilde{\mathcal{T}}$ and $\tilde{\mathcal{T}}'$ of Takahashi lengths, complementary Takahashi lengths, truncated Takahashi lengths and complementary Takahashi lengths are:

$$\mathcal{T} = \{\kappa_i\}_{i=0}^{11} = \{1,2,3,4,5,9,13,17,21,25,46,67\};$$
$$\mathcal{T}' = \{p' - \kappa_i\}_{i=0}^{11} = \{108,107,106,105,104,100,96,92,88,84,63,42\};$$
$$\tilde{\mathcal{T}} = \{\tilde{\kappa}_i\}_{i=4}^{11} = \{1,2,3,4,5,6,11,16\};$$
$$\tilde{\mathcal{T}}' = \{p - \kappa_i\}_{i=4}^{11} = \{25,24,23,22,21,20,15,10\}.$$

Using these, we obtain the Takahashi tree for $s = 51$ and the truncated Takahashi tree for $r = 9$ shown in Fig. A.1.

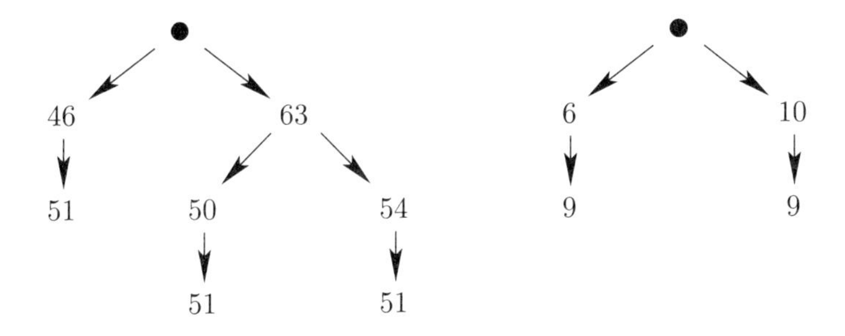

FIGURE A.1. Takahashi tree for $s = 51$ and truncated Takahashi tree for $r = 9$ when $p' = 109$ and $p = 26$.

Since these trees have three and two leaf-nodes respectively, $|\mathcal{U}(s)| = 3$ and $|\widetilde{\mathcal{U}}(r)| = 2$. Thus for the current example, the summation in (1.16) will be over six terms $F(\boldsymbol{u}^L, \boldsymbol{u}^R)$. To determine whether the additional term $\chi_{\hat{r},\hat{s}}^{\hat{p},\hat{p}'}$ is present, we obtain following the prescription of Section 1.13,

$$\{\xi_\ell\}_{\ell=0}^{9} = \{0,21,25,42,46,63,67,84,88,109\};$$
$$\{\tilde{\xi}_\ell\}_{\ell=0}^{9} = \{0,5,6,10,11,15,16,20,21,26\}.$$

Then, since $\eta(s) = 4 \neq \tilde{\eta}(r) = 2$, that additional term is not present.

As described in Section 1.7 each leaf-node of the Takahashi tree for s gives rise to a run $\mathcal{X}$. For the three leaf-nodes present when $s = 51$, we denote the corresponding runs by $\mathcal{X}_1^L, \mathcal{X}_2^L, \mathcal{X}_3^L$. Similarly, as described in Section 1.7, each leaf node of the truncated Takahashi tree for r gives rise to a run $\mathcal{X}$: for the two leaf-nodes present when $r = 9$, we denote the corresponding runs by $\mathcal{X}_1^R, \mathcal{X}_2^R$. These sets are given by:

$$\mathcal{X}_1^L = \{\{13,7\},\{10,4\},\{1,-1\}\} \qquad \mathcal{X}_1^R = \{\{13,7\},\{11,6\},\{1,-1\}\},$$
$$\mathcal{X}_2^L = \{\{13,7,3\},\{10,5,0\},\{-1,1,-1\}\} \qquad \mathcal{X}_2^R = \{\{13,7\},\{9,4\},\{-1,1\}\},$$
$$\mathcal{X}_3^L = \{\{13,7,3\},\{10,6,2\},\{-1,1,1\}\}.$$

From these, using (1.21), we obtain $\boldsymbol{u}_i^L = \boldsymbol{u}(\mathcal{X}_i^L)$ for $i = 1,2,3$, and $\boldsymbol{u}_j^R = \boldsymbol{u}(\mathcal{X}_j^R)$ for $j = 1,2$:

$$\boldsymbol{u}_1^L = (0,0,0,1,0,0,-1,0,0,1,0,1), \qquad \boldsymbol{u}_1^R = (0,0,0,0,0,1,-1,0,0,0,1,1),$$
$$\boldsymbol{u}_2^L = (0,0,-1,0,1,0,-1,0,0,1,0,0), \qquad \boldsymbol{u}_2^R = (0,0,0,1,0,0,-1,0,1,0,0,0),$$
$$\boldsymbol{u}_3^L = (0,1,-1,0,0,1,-1,0,0,1,0,0).$$

So $\mathcal{U}(s) = \{\boldsymbol{u}_1^L, \boldsymbol{u}_2^L, \boldsymbol{u}_3^L\}$ and $\widetilde{\mathcal{U}}(r) = \{\boldsymbol{u}_1^R, \boldsymbol{u}_2^R\}$.

Setting $\boldsymbol{\Delta}_i^L = \boldsymbol{\Delta}(\mathcal{X}_i^L)$ and $\boldsymbol{\Delta}_j^R = \boldsymbol{\Delta}(\mathcal{X}_j^R)$, we also obtain:

$$\boldsymbol{\Delta}_1^L = (0,0,0,-1,0,0,1,0,0,1,0,-1), \qquad \boldsymbol{\Delta}_1^R = (0,0,0,0,0,-1,1,0,0,0,1,-1),$$
$$\boldsymbol{\Delta}_2^L = (0,0,1,0,1,0,-1,0,0,-1,0,0), \qquad \boldsymbol{\Delta}_2^R = (0,0,0,1,0,0,-1,0,-1,0,0,0),$$
$$\boldsymbol{\Delta}_3^L = (0,1,-1,0,0,1,-1,0,0,-1,0,0).$$

For $i \in \{1,2,3\}$ and $j \in \{1,2\}$, let $\boldsymbol{m}_{ij}^{(1)} = (\overline{\boldsymbol{u}}_{i\flat}^L + \overline{\boldsymbol{u}}_{j\sharp}^R)\cdot\boldsymbol{m}$ for $\boldsymbol{m} = (m_4, m_5, \ldots, m_{11})$, and $\gamma_{ij} = \gamma(\mathcal{X}_i^L, \mathcal{X}_j^R)$. These are respectively minus twice the linear term and four times the constant term that appears in the exponent in (1.40). Via the prescriptions of Section 1.9 and 1.10, we obtain:

$$\boldsymbol{m}_{11}^{(1)} = m_4 - m_7 + m_{11}, \qquad \gamma_{11} = -42;$$
$$\boldsymbol{m}_{21}^{(1)} = m_5 - m_7 + m_{11}, \qquad \gamma_{21} = -42;$$
$$\boldsymbol{m}_{31}^{(1)} = m_6 - m_7 + m_{11}, \qquad \gamma_{31} = -43;$$
$$\boldsymbol{m}_{12}^{(1)} = m_4 - m_7 + m_9, \qquad \gamma_{12} = -42;$$
$$\boldsymbol{m}_{22}^{(1)} = m_5 - m_7 + m_9, \qquad \gamma_{22} = -42;$$
$$\boldsymbol{m}_{32}^{(1)} = m_6 - m_7 + m_9, \qquad \gamma_{32} = -43.$$

Let $\boldsymbol{n}^{(2)} = \tilde{\boldsymbol{n}}^T\boldsymbol{B}\tilde{\boldsymbol{n}} + \frac{1}{4}\boldsymbol{m}^T\boldsymbol{C}\boldsymbol{m}$ with $\tilde{\boldsymbol{n}} = (\tilde{n}_1, \tilde{n}_2, \tilde{n}_3)$ and $\boldsymbol{m} = (m_4, m_5, \ldots, m_{11})$. Then, via the descriptions of $\overline{\boldsymbol{C}}$ and $\boldsymbol{B}$ given in Section 1.11, we obtain:

$$\begin{aligned}\boldsymbol{n}^{(2)} &= (\tilde{n}_1 + \tilde{n}_2 + \tilde{n}_3)^2 + (\tilde{n}_2 + \tilde{n}_3)^2 + \tilde{n}_3^2 \\ &\quad + \tfrac{1}{4}\left(m_4^2 + (m_4 - m_5)^2 + (m_5 - m_6)^2 + (m_6 - m_7)^2 + (m_7 - m_8)^2 \right. \\ &\qquad \left. + m_9^2 + (m_9 - m_{10})^2 + (m_{10} - m_{11})^2 + m_{11}^2\right).\end{aligned}$$

Now let $\boldsymbol{u}_{ij} = \boldsymbol{u}_i^L + \boldsymbol{u}_j^R$ for $i \in \{1,2,3\}$ and $j \in \{1,2\}$. These together with the values of $\overline{\boldsymbol{Q}}_{ij} = \overline{\boldsymbol{Q}}(\boldsymbol{u}_{ij})$ obtained using $\boldsymbol{C}^*$ or $\overline{\boldsymbol{C}}^*$ (or even via (1.29) and (1.30)) as described in Section 1.12 are:

$$\begin{aligned}
\overline{\boldsymbol{u}}_{11} &= (0,1,-2,0,0,1,1,2), & \overline{\boldsymbol{Q}}_{11} &= (1,1,1,0,1,1,1,0);\\
\overline{\boldsymbol{u}}_{21} &= (1,1,-2,0,0,1,1,1), & \overline{\boldsymbol{Q}}_{21} &= (0,0,1,1,1,0,1,1);\\
\overline{\boldsymbol{u}}_{31} &= (0,2,-2,0,0,1,1,1), & \overline{\boldsymbol{Q}}_{31} &= (1,1,1,1,1,0,1,1);\\
\overline{\boldsymbol{u}}_{12} &= (0,0,-2,0,1,1,0,1), & \overline{\boldsymbol{Q}}_{12} &= (1,1,1,1,1,0,0,1);\\
\overline{\boldsymbol{u}}_{22} &= (1,0,-2,0,1,1,0,0), & \overline{\boldsymbol{Q}}_{22} &= (0,0,1,0,1,1,0,0);\\
\overline{\boldsymbol{u}}_{32} &= (0,1,-2,0,1,1,0,0), & \overline{\boldsymbol{Q}}_{32} &= (1,1,1,0,1,1,0,0).
\end{aligned}$$

In this case, (1.16) yields:

$$\chi_{9,51}^{26,109} = \sum_{i=1}^{3}\sum_{j=1}^{2} F_{ij},$$

where

$$\begin{aligned}
F_{ij} = \sum_{\substack{n_1,n_2,n_3\\ \boldsymbol{m}\equiv\overline{\boldsymbol{Q}}_{ij}}} & \frac{q^{\boldsymbol{n}^{(2)}-\frac{1}{2}\boldsymbol{m}_{ij}^{(1)}+\frac{1}{4}\gamma_{ij}}}{(q)_{n_1}(q)_{n_2}(q)_{n_3}(q)_{m_4}} \begin{bmatrix} \frac{1}{2}(m_4+m_6+(\boldsymbol{u}_{ij})_5) \\ m_5 \end{bmatrix}_q \\
&\times \begin{bmatrix} \frac{1}{2}(m_5+m_7+(\boldsymbol{u}_{ij})_6) \\ m_6 \end{bmatrix}_q \begin{bmatrix} \frac{1}{2}(m_6+m_8+(\boldsymbol{u}_{ij})_7) \\ m_7 \end{bmatrix}_q \\
&\times \begin{bmatrix} \frac{1}{2}(m_7+m_8-m_9+(\boldsymbol{u}_{ij})_8) \\ m_8 \end{bmatrix}_q \begin{bmatrix} \frac{1}{2}(m_8+m_{10}+(\boldsymbol{u}_{ij})_9) \\ m_9 \end{bmatrix}_q \\
&\times \begin{bmatrix} \frac{1}{2}(m_9+m_{11}+(\boldsymbol{u}_{ij})_{10}) \\ m_{10} \end{bmatrix}_q \begin{bmatrix} \frac{1}{2}(m_{10}+(\boldsymbol{u}_{ij})_{11}) \\ m_{11} \end{bmatrix}_q,
\end{aligned}$$

where $\boldsymbol{m} = (m_4, m_5, \ldots, m_{11})$ and $\tilde{n}_1 = n_1 - \frac{1}{2}(\boldsymbol{u}_{ij})_1$, $\tilde{n}_2 = n_2 - \frac{1}{2}(\boldsymbol{u}_{ij})_2$, $\tilde{n}_3 = n_3 - \frac{1}{2}(\boldsymbol{u}_{ij})_3 + \frac{1}{2}m_4$. The summation here extends over an infinite number of terms because the values that n_1, n_2, n_3, m_4 can take are unbounded.

We now seek a fermionic expression for a finitization $\chi_{a,b,c}^{26,109}(L)$ of $\chi_{9,51}^{26,109}$. Certainly, we require $a = 51$, but a number of values of b and c satisfy (1.10). The simplest fermionic expressions (1.37) are obtained when b is interfacial. In this case, either $b = 37$ or $b = 38$ is interfacial, whereupon either $c = b \pm 1$ satisfies (1.10). Here we select $b = 38$ and $c = 39$. Thus we use (1.37) to obtain a fermionic expression for $\chi_{51,38,39}^{26,109}(L)$.

The Takahashi trees for a and b are given in Fig. A.2.

The three leaf-nodes of the Takahashi tree for $a = 51$ give rise to the runs $\mathcal{X}_1^L, \mathcal{X}_2^L, \mathcal{X}_3^L$ which are, of course, exactly as above. The two leaf-nodes of the Takahashi tree for $b = 38$ give rise to runs $\mathcal{X}_1^R, \mathcal{X}_2^R$, which are *not* exactly as above.

$$\begin{aligned}
\mathcal{X}_1^L &= \{\{13,7\},\{10,4\},\{1,-1\}\} & \mathcal{X}_1^R &= \{\{13,7\},\{11,6\},\{1,-1\}\},\\
\mathcal{X}_2^L &= \{\{13,7,3\},\{10,5,0\},\{-1,1,-1\}\} & \mathcal{X}_2^R &= \{\{13,7\},\{9,3\},\{-1,1\}\},\\
\mathcal{X}_3^L &= \{\{13,7,3\},\{10,6,2\},\{-1,1,1\}\}.
\end{aligned}$$

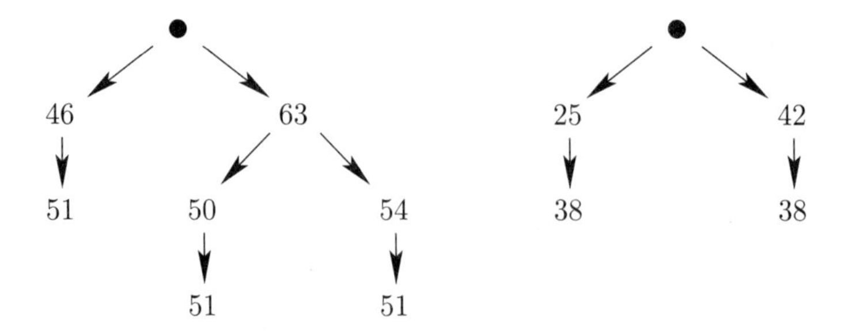

FIGURE A.2. Takahashi trees for $a = 51$ and $b = 38$ when $p' = 109$ and $p = 26$.

From these, using (1.21), we obtain $\boldsymbol{u}_i^L = \boldsymbol{u}(\mathcal{X}_i^L)$ for $i = 1, 2, 3$, and $\boldsymbol{u}_j^R = \boldsymbol{u}(\mathcal{X}_j^R)$ for $j = 1, 2$:

$$\begin{aligned}
\boldsymbol{u}_1^L &= (0,0,0,1,0,0,-1,0,0,1,0,1), & \boldsymbol{u}_1^R &= (0,0,0,0,0,1,-1,0,0,0,1,1),\\
\boldsymbol{u}_2^L &= (0,0,-1,0,1,0,-1,0,0,1,0,0), & \boldsymbol{u}_2^R &= (0,0,0,0,0,0,-1,0,1,0,0,0),\\
\boldsymbol{u}_3^L &= (0,1,-1,0,0,1,-1,0,0,1,0,0). & &
\end{aligned}$$

So $\mathcal{U}(a) = \{\boldsymbol{u}_1^L, \boldsymbol{u}_2^L, \boldsymbol{u}_3^L\}$ and $\mathcal{U}(b) = \{\boldsymbol{u}_1^R, \boldsymbol{u}_2^R\}$.

For $i \in \{1,2,3\}$ and $j \in \{1,2\}$, defining $\boldsymbol{m}_{ij}^{(1)} = (\boldsymbol{u}_{i\flat}^L + \boldsymbol{u}_{j\sharp}^R) \cdot \boldsymbol{m}$ for $\boldsymbol{m} = (L, m_1, m_2, \ldots, m_{11})$, and $\gamma'_{ij} = \gamma(\mathcal{X}_i^L, \mathcal{X}_j^R)$, we obtain exactly the values of $\boldsymbol{m}_{ij}^{(1)}$ obtained above, and $\gamma'_{ij} = \gamma_{ij}$ given above.

Let $\boldsymbol{m}^{(2)} = \hat{\boldsymbol{m}}^T \boldsymbol{C} \hat{\boldsymbol{m}} - L^2$. Then, having calculated $\boldsymbol{C}$ as in Section 1.11, we obtain:

$$\begin{aligned}
\boldsymbol{m}^{(2)} = (L - m_1)^2 &+ (m_1 - m_2)^2 + (m_2 - m_3)^2 + m_4^2\\
&+ (m_4 - m_5)^2 + (m_5 - m_6)^2 + (m_6 - m_7)^2 + (m_7 - m_8)^2 + m_9^2\\
&+ (m_9 - m_{10})^2 + (m_{10} - m_{11})^2 + m_{11}^2.
\end{aligned}$$

Then take $\boldsymbol{u}_{ij} = \boldsymbol{u}_i^L + \boldsymbol{u}_j^R$ for $i \in \{1,2,3\}$ and $j \in \{1,2\}$. These together with the values of $\boldsymbol{Q}_{ij} = \boldsymbol{Q}(\boldsymbol{u}_{ij})$ obtained using $\boldsymbol{C}^*$ from Section 1.11 (or alternatively, obtained via (1.29) and (1.30)) are:

$$\begin{aligned}
\boldsymbol{u}_{11} &= (0,0,0,1,0,1,-2,0,0,1,1,2), & \boldsymbol{Q}_{11} &= (0,1,0,1,1,1,0,1,1,1,0);\\
\boldsymbol{u}_{21} &= (0,0,-1,0,1,1,-2,0,0,1,1,1), & \boldsymbol{Q}_{21} &= (0,1,0,0,0,1,1,1,0,1,1);\\
\boldsymbol{u}_{31} &= (0,1,-1,0,0,2,-2,0,0,1,1,1), & \boldsymbol{Q}_{31} &= (0,1,1,1,1,1,1,1,0,1,1);\\
\boldsymbol{u}_{12} &= (0,0,0,1,0,0,-2,0,1,1,0,1), & \boldsymbol{Q}_{12} &= (0,1,0,1,1,1,1,1,0,0,1);\\
\boldsymbol{u}_{22} &= (0,0,-1,0,1,0,-2,0,1,1,0,0), & \boldsymbol{Q}_{22} &= (0,1,0,0,0,1,0,1,1,0,0);\\
\boldsymbol{u}_{32} &= (0,1,-1,0,0,1,-2,0,1,1,0,0), & \boldsymbol{Q}_{32} &= (0,1,1,1,1,1,0,1,1,0,0).
\end{aligned}$$

Putting all of this into (1.37) and (1.40) produces:

$$\chi_{51,38,37}^{26,109}(L) = \chi_{51,38,39}^{26,109}(L) = \sum_{i=1}^{3} \sum_{j=1}^{2} F_{ij}(L),$$

where

$$F_{ij}(L) = \sum_{\boldsymbol{m}\equiv \boldsymbol{Q}_{ij}} q^{\frac{1}{4}\boldsymbol{m}^{(2)}-\frac{1}{2}\boldsymbol{m}_{ij}^{(1)}+\frac{1}{4}\gamma'_{ij}} \begin{bmatrix} \frac{1}{2}(L+m_2+(\boldsymbol{u}_{ij})_1) \\ m_1 \end{bmatrix}_q \begin{bmatrix} \frac{1}{2}(m_1+m_3+(\boldsymbol{u}_{ij})_2) \\ m_2 \end{bmatrix}_q$$
$$\times \begin{bmatrix} \frac{1}{2}(m_2+m_3-m_4+(\boldsymbol{u}_{ij})_3) \\ m_3 \end{bmatrix}_q \begin{bmatrix} \frac{1}{2}(m_3+m_5+(\boldsymbol{u}_{ij})_4) \\ m_4 \end{bmatrix}_q \begin{bmatrix} \frac{1}{2}(m_4+m_6+(\boldsymbol{u}_{ij})_5) \\ m_5 \end{bmatrix}_q$$
$$\times \begin{bmatrix} \frac{1}{2}(m_5+m_7+(\boldsymbol{u}_{ij})_6) \\ m_6 \end{bmatrix}_q \begin{bmatrix} \frac{1}{2}(m_6+m_8+(\boldsymbol{u}_{ij})_7) \\ m_7 \end{bmatrix}_q \begin{bmatrix} \frac{1}{2}(m_7+m_8-m_9+(\boldsymbol{u}_{ij})_8) \\ m_8 \end{bmatrix}_q$$
$$\times \begin{bmatrix} \frac{1}{2}(m_8+m_{10}+(\boldsymbol{u}_{ij})_9) \\ m_9 \end{bmatrix}_q \begin{bmatrix} \frac{1}{2}(m_9+m_{11}+(\boldsymbol{u}_{ij})_{10}) \\ m_{10} \end{bmatrix}_q \begin{bmatrix} \frac{1}{2}(m_{10}+(\boldsymbol{u}_{ij})_{11}) \\ m_{11} \end{bmatrix}_q.$$

Note that the summation here comprises a finite number of non-zero terms. It might be better carried out by finding all solutions $\{n_i\}_{i=1}^{12}$ of (1.42), where in this case, $\{l_i\}_{i=1}^{12} = \{1,2,3,1,5,9,13,17,4,25,46,67\}$, and then obtaining $\{m_i\}_{i=1}^{12}$ via (1.32), or via (1.29) and (1.30). Note that the Gaussian polynomial terms are then best expressed in the form $\left[{m_j+n_j \atop n_j}\right]_q$.

The $L\to\infty$ limit of the above expression yields the Virasoro character $\chi_{9,51}^{26,109}$. In fact, we have $\lim_{L\to\infty} F_{ij}(L) = F_{ij}$ for $i=1,2,3$ and $j=1,2$.

A.2. Example 2

Let $p'=75$, $p=53$, $r=17$ and $s=72$. Then p'/p has continued fraction $[1,2,2,2,4]$, so that $n=4$, $\{t_k\}_{k=0}^{5} = \{-1,0,2,4,6,10\}$, $t=9$, $\{y_k\}_{k=-1}^{5} = \{0,1,1,3,7,17,75\}$ and $\{z_k\}_{k=-1}^{5} = \{1,0,1,2,5,12,53\}$. The set $\mathcal{T}$ of Takahashi lengths is then $\mathcal{T} = \{\kappa_i\}_{i=0}^{8} = \{1,2,3,4,7,10,17,24,41\}$ and the set $\tilde{\mathcal{T}}$ of truncated Takahashi lengths is $\tilde{\mathcal{T}} = \{\kappa_i\}_{i=1}^{8} = \{1,2,3,5,7,12,17,29\}$. Since $p'-s=3\in\mathcal{T}$, the Takahashi tree for $s=72$ is trivial, comprising just one node in addition to the root node. This leaf-node gives rise to the run $\mathcal{X}^L = \{\{10\},\{2\},\{1\}\}$. Similarly, since $17\in\tilde{\mathcal{T}}$, the truncated Takahashi tree for $r=17$ is trivial. Its single leaf-node gives rise to the run $\mathcal{X}^R = \{\{10\},\{7\},\{-1\}\}$. From these, we obtain $\boldsymbol{u}^L = \boldsymbol{u}(\mathcal{X}^L)$ and $\boldsymbol{u}^R = \boldsymbol{u}(\mathcal{X}^R)$ given by:

$$\boldsymbol{u}^L = (0,0,0,-1,0,-1,0,0,1), \qquad \boldsymbol{u}^R = (0,0,0,0,0,0,1,0,0);$$

and $\boldsymbol{\Delta}^L = \boldsymbol{\Delta}(\mathcal{X}^L)$ and $\boldsymbol{\Delta}^R = \boldsymbol{\Delta}(\mathcal{X}^R)$ given by:

$$\boldsymbol{\Delta}^L = (0,0,0,-1,0,-1,0,0,1), \qquad \boldsymbol{\Delta}^R = (0,0,0,0,0,0,-1,0,0).$$

The procedure of Section 1.10 then yields $\gamma(\mathcal{X}^L,\mathcal{X}^R) = -1624$. The linear term is specified by $\boldsymbol{m}^{(1)} = (\boldsymbol{u}_\flat^L + \boldsymbol{u}_\sharp^R)\cdot\boldsymbol{m}$ for $\boldsymbol{m}=(m_1,m_2,\ldots,m_8)$, whereupon $\boldsymbol{m}^{(1)} = -m_6+m_7$.

Let $\boldsymbol{n}^{(2)} = \frac{1}{4}\boldsymbol{m}^T\boldsymbol{C}\boldsymbol{m}$. Then, via the descriptions of $\overline{\boldsymbol{C}}$ and $\boldsymbol{B}$ given in Section 1.11, we obtain:

$$\boldsymbol{n}^{(2)} = \tfrac{1}{4}\left(m_1^2 + (m_1-m_2)^2 + m_3^2 + (m_3-m_4)^2 \right.$$
$$\left. + m_5^2 + (m_5-m_6)^2 + m_7^2 + (m_7-m_8)^2 + m_8^2\right).$$

Setting $\boldsymbol{u} = \boldsymbol{u}^L + \boldsymbol{u}^R$ and $\overline{\boldsymbol{Q}} = \overline{\boldsymbol{Q}}(\boldsymbol{u})$ obtained using $\boldsymbol{C}^*$ or $\overline{\boldsymbol{C}}^*$ (or even via (1.29) and (1.30)), gives:

$$\overline{\boldsymbol{u}} = (0,0,-1,0,-1,1,0,1), \qquad \overline{\boldsymbol{Q}} = (0,0,0,0,1,0,0,1).$$

We can now substitute all these values into (1.16). We note that the sum comprises just one term $F(\boldsymbol{u}^L, \boldsymbol{u}^R)$. To determine whether the extra term is present, we calculate:

$$\{\xi_\ell\}_{\ell=0}^{9} = \{0, 17, 24, 34, 41, 51, 58, 75\}; \qquad \{\tilde{\xi}_\ell\}_{\ell=0}^{9} = \{0, 12, 17, 24, 29, 36, 41, 53\}.$$

Then, since $\eta(s) = 6 \neq \tilde{\eta}(r) = 2$, that additional term is not present. Thereupon:

$$\chi^{53,75}_{17,72} = \sum_{\substack{n_1,n_2\\ \boldsymbol{m}\equiv\overline{\boldsymbol{Q}}}} \frac{q^{\boldsymbol{n}^{(2)}-\frac{1}{2}\boldsymbol{m}^{(1)}-\frac{1}{4}\cdot 1624}}{(q)_{m_1}} \begin{bmatrix} \frac{1}{2}(m_1+m_2-m_3) \\ m_2 \end{bmatrix}_q \begin{bmatrix} \frac{1}{2}(m_2+m_4) \\ m_3 \end{bmatrix}_q$$
$$\times \begin{bmatrix} \frac{1}{2}(m_3+m_4-m_5-1) \\ m_4 \end{bmatrix}_q \begin{bmatrix} \frac{1}{2}(m_4+m_6) \\ m_5 \end{bmatrix}_q$$
$$\times \begin{bmatrix} \frac{1}{2}(m_5+m_6-m_7-1) \\ m_6 \end{bmatrix}_q \begin{bmatrix} \frac{1}{2}(m_6+m_8+1) \\ m_7 \end{bmatrix}_q \begin{bmatrix} \frac{1}{2}m_7 \\ m_8 \end{bmatrix}_q,$$

where $\boldsymbol{m} = (m_1, m_2, \ldots, m_8)$. The summation here extends over an infinite number of terms because the value of m_1 is unbounded.

To obtain a finitization $\chi^{53,75}_{a,b,c}(L)$ of $\chi^{53,75}_{17,72}$, we set $a = 72$, and can set $(b, c) = (24, 23), (24, 25), (25, 24)$. The most interesting case is the latter, so we choose that. The Takahashi trees for $a = 72$ and $b = 25$ are given in Fig. A.3.

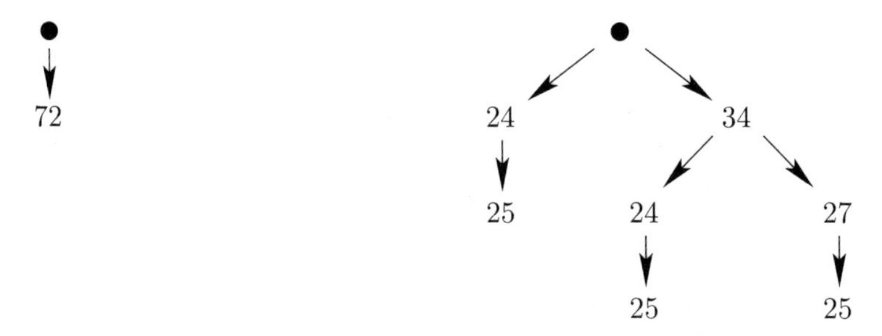

FIGURE A.3. Formations for $a = 72$ and $b = 25$ when $p' = 75$ and $p = 53$.

The single leaf-node of the Takahashi tree for $a = 72$ gives rise to the run $\mathcal{X}^L = \{\{10\}, \{2\}, \{1\}\}$. The three leaf-nodes of the Takahashi tree for $b = 25$ gives rise to the runs:

$$\mathcal{X}_1^R = \{\{10, 5\}, \{8, 0\}, \{1, -1\}\},$$
$$\mathcal{X}_2^R = \{\{10, 5, 2\}, \{7, 4, 0\}, \{-1, 1, -1\}\},$$
$$\mathcal{X}_3^R = \{\{10, 5, 2\}, \{7, 5, 1\}, \{-1, 1, 1\}\}.$$

Setting $\boldsymbol{u}^L = \boldsymbol{u}(\mathcal{X}^L)$, $\boldsymbol{\Delta}^L = \boldsymbol{\Delta}(\mathcal{X}^L)$, and $\boldsymbol{u}_j^R = \boldsymbol{u}(\mathcal{X}_j^R)$ and $\boldsymbol{\Delta}_j^R = \boldsymbol{\Delta}(\mathcal{X}_j^R)$ for $j = 1, 2, 3$, we have:

$$\boldsymbol{u}_1^L = (0, 0, 0, -1, 0, -1, 0, 0, 1), \qquad \boldsymbol{\Delta}_1^L = (0, 0, 0, -1, 0, -1, 0, 0, 1),$$
$$\boldsymbol{u}_1^R = (0, -1, 0, -1, -1, 0, 0, 1, 1), \qquad \boldsymbol{\Delta}_1^R = (0, 1, 0, 1, 1, 0, 0, 1, -1),$$
$$\boldsymbol{u}_2^R = (0, -1, 0, 0, -1, 0, 1, 0, 0), \qquad \boldsymbol{\Delta}_2^R = (0, 1, 0, 0, -1, 0, -1, 0, 0),$$
$$\boldsymbol{u}_3^R = (1, -1, 0, 0, 0, 0, 1, 0, 0), \qquad \boldsymbol{\Delta}_3^R = (1, -1, 0, 0, 0, 0, -1, 0, 0).$$

Now note that $\Delta(\boldsymbol{u}_1^R) = \Delta(\boldsymbol{u}_1^R) = -1 = c - b$, and that $\Delta(\boldsymbol{u}_1^R) = 1 = b - c$. Therefore, since b does not satisfy (1.36), expression (1.43) applies in this case to give:

$$\begin{aligned}\chi^{53,75}_{72,25,24}(L) &= F(\boldsymbol{u}^L, \boldsymbol{u}_1^R, L) + F(\boldsymbol{u}^L, \boldsymbol{u}_2^R, L) + \tilde{F}(\boldsymbol{u}^L, \boldsymbol{u}_3^R, L) \\ &= F(\boldsymbol{u}^L, \boldsymbol{u}_1^R, L) + F(\boldsymbol{u}^L, \boldsymbol{u}_2^R, L) \\ &\qquad + q^{\frac{1}{2}(L-47)} F(\boldsymbol{u}^L, \boldsymbol{u}_3^R, L) + (1 - q^L) F(\boldsymbol{u}^L, \boldsymbol{u}_3^{R+}, L-1),\end{aligned}$$

the second equality being via the third case of (1.46).

Using the definitions of Section 1.15, $\mathcal{X}_3^{R+}$ is given by $\{\{10\}, \{7\}, \{-1\}\} = \mathcal{X}^R$, as defined above. So $\boldsymbol{u}_3^{R+} = \boldsymbol{u}^R$. For convenience, we set $\mathcal{X}_4^R = \mathcal{X}^R$ and $\boldsymbol{u}_4^R = \boldsymbol{u}^R$, so that:

$$\boldsymbol{u}_4^R = (0,0,0,0,0,0,1,0,0); \qquad \boldsymbol{\Delta}_4^R = (0,0,0,0,0,0,-1,0,0).$$

For $j = 1,2,3,4$, define $\boldsymbol{m}_{1j}^{(1)} = (\boldsymbol{u}_\flat^L + \boldsymbol{u}_{j\sharp}^R) \cdot \boldsymbol{m}$ for $\boldsymbol{m} = (m_1, m_2, \ldots, m_8)$, and $\gamma'_{1j} = \gamma'(\mathcal{X}^L, \mathcal{X}_j^R)$. Then:

$$\begin{aligned}
\boldsymbol{m}_{11}^{(1)} &= -m_4 - m_6 + m_8, & \gamma'_{11} &= -1624 + 2L; \\
\boldsymbol{m}_{12}^{(1)} &= -m_6 + m_7, & \gamma'_{12} &= -1624 + 2L; \\
\boldsymbol{m}_{13}^{(1)} &= -m_6 + m_7, & \gamma'_{13} &= -1530; \\
\boldsymbol{m}_{14}^{(1)} &= -m_6 + m_7, & \gamma'_{14} &= -1624.
\end{aligned}$$

The quadratic term $\boldsymbol{m}^{(2)} = \hat{\boldsymbol{m}}^T \boldsymbol{C} \hat{\boldsymbol{m}} - L^2$, is given by $\boldsymbol{m}^{(2)} = 4\boldsymbol{n}^{(2)}$ as specified above.

With $\boldsymbol{u}_{1j} = \boldsymbol{u}^L + \boldsymbol{u}_j^R$, and $\boldsymbol{Q}_{1j} = \boldsymbol{Q}(\boldsymbol{u}_{1j})$, we have:

$$\begin{aligned}
\boldsymbol{u}_{11} &= (0,-1,0,-2,-1,-1,0,1,2), & \boldsymbol{Q}_{11} &= (1,1,1,1,0,0,1,0), \\
\boldsymbol{u}_{12} &= (0,-1,0,-1,-1,-1,1,0,1), & \boldsymbol{Q}_{12} &= (1,1,1,1,1,0,0,1), \\
\boldsymbol{u}_{13} &= (1,-1,0,-1,0,-1,1,0,1), & \boldsymbol{Q}_{13} &= (1,0,0,0,1,0,0,1), \\
\boldsymbol{u}_{14} &= (0,0,0,-1,0,-1,1,0,1), & \boldsymbol{Q}_{14} &= (0,0,0,0,1,0,0,1).
\end{aligned}$$

Then, using (1.40), the above expression for $\chi^{53,75}_{72,25,24}(L)$ produces:

$$\chi^{53,75}_{72,25,24}(L) = F_1(L) + F_2(L) + q^{\frac{1}{2}(L-47)} F_3(L) + (1 - q^L) F_4(L-1),$$

where for $j = 1,2,3,4$, we set:

$$\begin{aligned}
F_j(L) = \sum_{\boldsymbol{m} \equiv \boldsymbol{Q}_{1j}} & q^{\frac{1}{4}\boldsymbol{m}^{(2)} - \frac{1}{2}\boldsymbol{m}_{1j}^{(1)} + \frac{1}{4}\gamma'_{1j}} \begin{bmatrix} \frac{1}{2}(L + m_2 + (\boldsymbol{u}_{1j})_1) \\ m_1 \end{bmatrix}_q \\
&\times \begin{bmatrix} \frac{1}{2}(m_1 + m_2 - m_3 + (\boldsymbol{u}_{1j})_2) \\ m_2 \end{bmatrix}_q \begin{bmatrix} \frac{1}{2}(m_2 + m_4 + (\boldsymbol{u}_{1j})_3) \\ m_3 \end{bmatrix}_q \\
&\times \begin{bmatrix} \frac{1}{2}(m_3 + m_4 - m_5 + (\boldsymbol{u}_{1j})_4) \\ m_4 \end{bmatrix}_q \begin{bmatrix} \frac{1}{2}(m_4 + m_6 + (\boldsymbol{u}_{1j})_5) \\ m_5 \end{bmatrix}_q \\
&\times \begin{bmatrix} \frac{1}{2}(m_5 + m_6 - m_7 + (\boldsymbol{u}_{1j})_6) \\ m_6 \end{bmatrix}_q \begin{bmatrix} \frac{1}{2}(m_6 + m_8 + (\boldsymbol{u}_{1j})_7) \\ m_7 \end{bmatrix}_q \\
&\times \begin{bmatrix} \frac{1}{2}(m_7 + (\boldsymbol{u}_{1j})_8) \\ m_8 \end{bmatrix}_q,
\end{aligned}$$

where $\boldsymbol{m} = (m_1, m_2, \dots, m_8)$.

Note that for $j = 1, 2$, the presence of $2L$ in γ'_{1j} implies that $\lim_{L\to\infty} F_j(L) = 0$. Thus, as $L \to \infty$, only the term $F_4(L-1)$ remains. This yields precisely the fermionic expression given above for the character $\chi^{53,75}_{17,72}$.

A.3. Example 3

In this example, we provide a case when the term on the right of (1.16) appears. We will not give every detail because the expression involves many terms (thirty-one!), but will concentrate on how the extra term(s) are dealt with.

Let $p' = 118$, $p = 51$, $r = 27$ and $s = 61$. Then p'/p has continued fraction $[2, 3, 5, 3]$, so that $n = 3$, $\{t_k\}_{k=0}^3 = \{-1, 1, 4, 9, 12\}$, $t = 11$, $\{y_k\}_{k=-1}^3 = \{0, 1, 2, 7, 37, 118\}$ and $\{z_k\}_{k=-1}^3 = \{1, 0, 1, 3, 16, 51\}$. Then:

$$\mathcal{T} = \{1, 2, 3, 5, 7, 9, 16, 23, 30, 37, 44\};$$

$$\mathcal{T}' = \{117, 116, 115, 113, 111, 109, 102, 95, 88, 81, 74\};$$

$$\tilde{\mathcal{T}} = \{1, 2, 3, 4, 7, 10, 13, 16, 19\};$$

$$\tilde{\mathcal{T}}' = \{50, 49, 48, 47, 44, 41, 38, 35, 32\}.$$

Using these, we obtain the Takahashi tree for $s = 61$ and the truncated Takahashi tree for $r = 27$ shown in Fig. A.4.

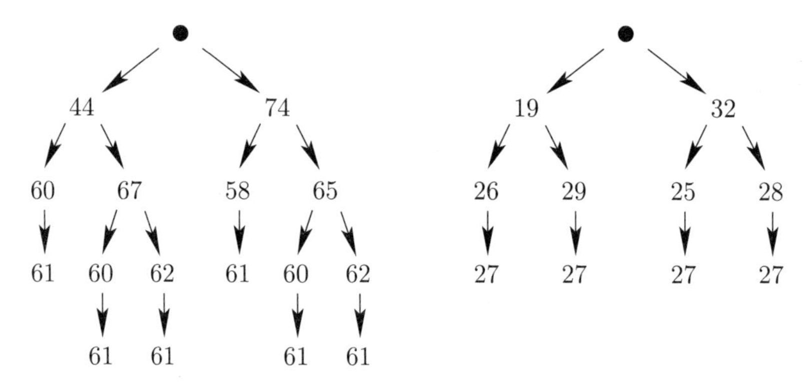

FIGURE A.4. Takahashi tree for $s = 61$ and truncated Takahashi tree for $r = 27$ when $p' = 118$ and $p = 51$.

We see that the Takahashi tree for $s = 61$ and the truncated Takahashi tree for $r = 27$ have six and four leaf-nodes respectively. Thus $|\mathcal{U}(s)| = 6$ and $|\widetilde{\mathcal{U}}(r)| = 4$. Therefore, for $\chi^{51,118}_{27,61}$, the sum in expression (1.16) runs over 24 terms $F(\boldsymbol{u}^L, \boldsymbol{u}^R)$. We will not compute these terms here, but just denote this sum by $\sum^{(11)}$, where the superscript indicates that each element of $\mathcal{U}(s)$ and $\widetilde{\mathcal{U}}(r)$ is an 11-dimensional vector.

From the prescription of Section 1.13, we obtain $\{\xi_\ell\}_{\ell=0}^5 = \{0, 37, 44, 74, 81, 118\}$ and $\{\tilde{\xi}_\ell\}_{\ell=0}^5 = \{0, 16, 19, 32, 35, 51\}$. Then $\eta(61) = \tilde{\eta}(27) = 2$ and $\hat{s} = 61 - 44 = 17$, $\hat{r} = 27 - 19 = 8$. The additional term in (1.16) is therefore present in this case. We obtain $\hat{p}' = 74 - 44 = 30$ and $\hat{p} = 32 - 19 = 13$. Thereupon:

$$\chi^{51,118}_{27,61} = \chi^{13,30}_{8,17} + \sum{}^{(11)}.$$

We now use (1.16) to determine a fermionic expression for $\chi^{13,30}_{8,17}$. So now set $p' = 30$ and $p = 13$. The continued fraction of p'/p is $[2,3,4]$. Note that this may be obtained directly from (1.35). From this, we obtain $n = 2$, $\{t_k\}_{k=0}^{3} = \{-1,1,4,8\}$, $t = 7$, $\{y_k\}_{k=-1}^{3} = \{0,1,2,7,30\}$, $\{z_k\}_{k=-1}^{3} = \{1,0,1,3,13\}$, $\mathcal{T} = \{1,2,3,5,7,9,16\}$, $\mathcal{T}' = \{29,28,27,25,23,21,14\}$, $\tilde{\mathcal{T}} = \{1,2,3,4,7\}$ and $\tilde{\mathcal{T}}' = \{12,11,10,9,6\}$. Resetting $s = \hat{s} = 17$ and $r = \hat{r} = 8$, we use these values to obtain the Takahashi tree for $s = 17$ and the truncated Takahashi tree for $r = 8$ that are shown in Fig. A.5.

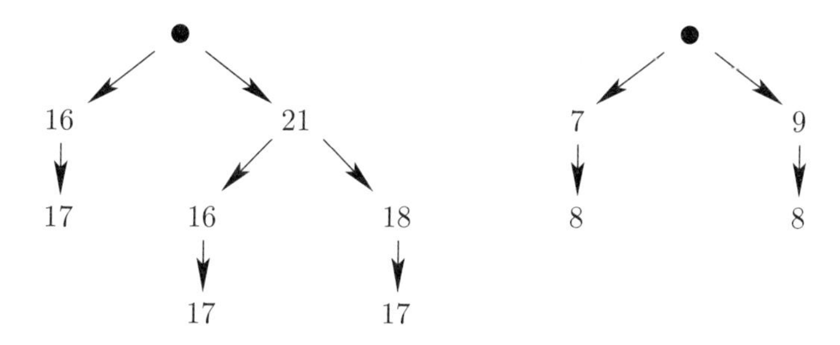

FIGURE A.5. Takahashi tree for $s = 17$ and truncated Takahashi tree for $r = 8$ when $p' = 30$ and $p = 13$.

Since these trees have three and two nodes respectively, we have $|\mathcal{U}(s)| = 3$ and $|\widetilde{\mathcal{U}}(r)| = 2$, whereupon the sum in expression (1.16) for $\chi^{13,30}_{8,17}$ runs over six terms $F(\boldsymbol{u}^L, \boldsymbol{u}^R)$. Again, we will not compute these terms here, but just denote this sum by $\sum^{(7)}$, where the superscript is used as above.

In this $p' = 30$, $p = 13$ case, the prescription of Section 1.13 yields $\{\xi_\ell\}_{\ell=0}^{7} = \{0,7,9,14,16,21,23,30\}$ and $\{\tilde{\xi}_\ell\}_{\ell=0}^{7} = \{0,3,4,6,7,9,10,13\}$. Then here, $\eta(s) = \eta(r) = 4$ and $\hat{s} = s - 16 = 1$ and $\hat{r} = r - 6 = 1$. Therefore, we again have to include the extra term in (1.16). After calculating $\hat{p}' = 21 - 16 = 5$ and $\hat{p} = 9 - 7 = 2$, we therefore have:

$$\chi^{51,118}_{27,61} = \chi^{2,5}_{1,1} + \sum{}^{(7)} + \sum{}^{(11)}.$$

A further use of (1.16) expresses $\chi^{2,5}_{1,1}$ in the fermionic form given in (1.3). We have thus expressed the Virasoro character $\chi^{51,118}_{27,61}$ in fermionic form, as a sum of thirty-one fundamental fermionic forms.

A finitization of this character is provided by $\chi^{51,118}_{a,b,c}$ with $a = 61$ and either $(b,c) = (62,63)$, $(62,61)$, $(63,64)$ or $(63,62)$. We choose the latter and give a very brief outline of how the prescription of Section 1 produces a fermionic expression for $\chi^{51,118}_{61,63,62}(L)$.

With $p = 51$ and $p' = 118$, the values pertaining to the continued fraction of p'/p calculated above yield Takahashi trees for $a = 61$ and $b = 63$ that contain six and five leaf-nodes respectively. Thus $|\mathcal{U}(a)| = 6$ and $|\mathcal{U}(b)| = 5$. Since b satisfies (1.36), expression (1.37) is the appropriate expression to use for $\chi^{51,118}_{61,63,62}(L)$. The sum in this expression thus runs over thirty terms $F(\boldsymbol{u}^L, \boldsymbol{u}^R, L)$. We denote this sum by $\sum^{(11)}(L)$.

We also find that $\eta(a) = \eta(b) = 2$ and $\hat{a} = a - 44 = 17$, $\hat{b} = b - 44 = 19$ and $\hat{c} = c - 44 = 18$. The additional term in (1.37) is therefore present in this case.

Again we have $\hat{p}' = 74 - 44 = 30$ and $\hat{p} = 32 - 19 = 13$, whereupon:

$$\chi^{51,118}_{61,63,62}(L) = \chi^{13,30}_{17,19,18}(L) + \sum{}^{(11)}(L).$$

With $p = 13$ and $p' = 30$, $b = 19$ satisfies (1.36) and therefore we use (1.37) once more to express $\chi^{13,30}_{17,19,18}(L)$ in fermionic form. The values pertaining to the continued fraction of 30/13 calculated above yield Takahashi trees for 17 and 19 that contain three and two leaf-nodes respectively. Therefore, for $\chi^{13,30}_{17,19,18}(L)$, the sum in expression (1.37) runs over six terms $F(\boldsymbol{u}^L, \boldsymbol{u}^R, L)$. We denote their sum by $\sum^{(7)}(L)$.

In this $p' = 30$, $p = 13$ case, we find that $\eta(17) = \eta(19) = 4$ and $\hat{a} = 17-16 = 1$, $\hat{b} = 19 - 16 = 3$ and $\hat{c} = 18 - 16 = 2$. Therefore, we again require the extra term from (1.37). With $\hat{p}' = 21 - 16 = 5$ and $\hat{p} = 9 - 7 = 2$, we thus have:

$$\chi^{51,118}_{61,63,62}(L) = \chi^{2,5}_{1,3,2}(L) + \sum{}^{(7)}(L) + \sum{}^{(11)}(L).$$

We may now use (1.37) yet again to express $\chi^{2,5}_{1,3,2}(L)$ in fermionic form. The result is equivalent to that given in (1.14). In this way, $\chi^{51,118}_{61,63,62}(L)$ is expressed in fermionic form, as a sum of thirty-seven fundamental fermionic forms.

In the $L \to \infty$ limit, six of the fundamental fermionic forms that comprise $\sum^{(11)}(L)$ tend to zero. The remaining thirty-one terms reproduce the fermionic expression for $\chi^{51,118}_{27,61}$ calculated above.

APPENDIX B

Obtaining the bosonic generating function

In Section 2.3, the path generating function $\chi^{p,p'}_{a,b,c}(L)$ was defined by:

$$\chi^{p,p'}_{a,b,c}(L) = \sum_{p \in \mathcal{P}^{p,p'}_{a,b,c}(L)} q^{wt(h)}$$

where $wt(h)$ was defined in (2.1). Here we prove the bosonic expression for $\chi^{p,p'}_{a,b,c}(L)$ stated in (1.9). Our proof is similar to that given in [**26**] for what, via the bijection given in [**21**], turns out to be an equivalent result.

Let $1 \leq p < p'$, $1 \leq a \leq p'-1$ and $L > 0$. If $2 \leq b \leq p'-2$, then the definition (2.1) implies the following recurrence relations for $\chi^{p,p'}_{a,b,c}(L)$: if $\lfloor bp/p' \rfloor = \lfloor (b+1)p/p' \rfloor$ then

$$\chi^{p,p'}_{a,b,b+1}(L) = q^{\frac{1}{2}(L-a+b)} \chi^{p,p'}_{a,b+1,b}(L-1) + \chi^{p,p'}_{a,b-1,b}(L-1); \tag{B.1}$$

if $\lfloor bp/p' \rfloor \neq \lfloor (b+1)p/p' \rfloor$ then

$$\chi^{p,p'}_{a,b,b+1}(L) = \chi^{p,p'}_{a,b+1,b}(L-1) + q^{\frac{1}{2}(L-a+b)} \chi^{p,p'}_{a,b-1,b}(L-1); \tag{B.2}$$

if $\lfloor bp/p' \rfloor = \lfloor (b-1)p/p' \rfloor$ then

$$\chi^{p,p'}_{a,b,b-1}(L) = q^{\frac{1}{2}(L+a-b)} \chi^{p,p'}_{a,b-1,b}(L-1) + \chi^{p,p'}_{a,b+1,b}(L-1); \tag{B.3}$$

if $\lfloor bp/p' \rfloor \neq \lfloor (b-1)p/p' \rfloor$ then

$$\chi^{p,p'}_{a,b,b-1}(L) = \chi^{p,p'}_{a,b-1,b}(L-1) + q^{\frac{1}{2}(L+a-b)} \chi^{p,p'}_{a,b+1,b}(L-1). \tag{B.4}$$

In addition, (2.1) implies that: if $\lfloor p/p' \rfloor = \lfloor 2p/p' \rfloor$ then

$$\chi^{p,p'}_{a,1,2}(L) = q^{\frac{1}{2}(L-a+1)} \chi^{p,p'}_{a,2,1}(L-1); \tag{B.5}$$

if $\lfloor p/p' \rfloor \neq \lfloor 2p/p' \rfloor$ then

$$\chi^{p,p'}_{a,1,2}(L) = \chi^{p,p'}_{a,2,1}(L-1); \tag{B.6}$$

if $\lfloor (p'-1)p/p' \rfloor = \lfloor (p'-2)p/p' \rfloor$ then

$$\chi^{p,p'}_{a,p'-1,p'-2}(L) = q^{\frac{1}{2}(L+a-p'+1)} \chi^{p,p'}_{a,p'-2,p'-1}(L-1); \tag{B.7}$$

if $\lfloor (p'-1)p/p' \rfloor \neq \lfloor (p'-2)p/p' \rfloor$ then

$$\chi^{p,p'}_{a,p'-1,p'-2}(L) = \chi^{p,p'}_{a,p'-2,p'-1}(L-1). \tag{B.8}$$

Finally, we obtain:

$$\chi^{p,p'}_{a,b,b\pm1}(0) = \delta_{a,b}. \tag{B.9}$$

Since expressions (B.1)-(B.8) enable $\chi^{p,p'}_{a,b,c}(L)$, for $1 \leq a,b,c < p'$ with $b-c = \pm1$ and $L > 0$, to be expressed in terms of various $\chi^{p,p'}_{a,b',c'}(L-1)$, with $1 \leq b', c' < p'$

and $b' - c' = \pm 1$, the above expressions together with (B.9) determine $\chi^{p,p'}_{a,b,c}(L)$ uniquely.

For $1 \le p < p'$ and $1 \le a, c < p'$ and $0 \le b \le p'$, with $L \equiv a - b \pmod 2$ and $L \ge 0$, define:

$$\ddot{\chi}^{p,p'}_{a,b,c}(L) = \sum_{\lambda=-\infty}^{\infty} q^{\lambda^2 pp' + \lambda(p' r(b,c) - pa)} \begin{bmatrix} L \\ \frac{L+a-b}{2} - p'\lambda \end{bmatrix}_q - \sum_{\lambda=-\infty}^{\infty} q^{(\lambda p + r(b,c))(\lambda p' + a)} \begin{bmatrix} L \\ \frac{L-a-b}{2} - p'\lambda \end{bmatrix}_q,$$

where

$$r(b,c) = \lfloor pc/p' \rfloor + (b - c + 1)/2. \tag{B.10}$$

Once we establish that expressions (B.1)-(B.9) are satisfied with $\ddot{\chi}^{p,p'}_{a,b,c}(L)$ in place of $\chi^{p,p'}_{a,b,c}(L)$, we can then conclude that $\chi^{p,p'}_{a,b,c}(L) = \ddot{\chi}^{p,p'}_{a,b,c}(L)$.

First consider the case $c = b+1$ with $\lfloor bp/p' \rfloor = \lfloor cp/p' \rfloor$. Set $r = r(b, b+1) = \lfloor bp/p' \rfloor$, whence $r(b+1, b) = r+1$ and $r(b-1, b) = r$. Then, on using the recurrence $\begin{bmatrix} A \\ B \end{bmatrix}_q = q^{A-B} \begin{bmatrix} A-1 \\ B-1 \end{bmatrix}_q + \begin{bmatrix} A-1 \\ B \end{bmatrix}_q$ for Gaussian polynomials, we obtain:

$$\begin{aligned}
\ddot{\chi}^{p,p'}_{a,b,b+1}(L) = q^{\frac{1}{2}(L-a+b)} \Bigg(& \sum_{\lambda=-\infty}^{\infty} q^{\lambda^2 pp' + \lambda(p'(r+1) - pa)} \begin{bmatrix} L-1 \\ \frac{(L-1)+a-(b+1)}{2} - p'\lambda \end{bmatrix}_q \\
& - \sum_{\lambda=-\infty}^{\infty} q^{(\lambda p + r + 1)(\lambda p' + a)} \begin{bmatrix} L-1 \\ \frac{(L-1)-a-(b+1)}{2} - p'\lambda \end{bmatrix}_q \Bigg) \\
+ \Bigg(& \sum_{\lambda=-\infty}^{\infty} q^{\lambda^2 pp' + \lambda(p' r - pa)} \begin{bmatrix} L-1 \\ \frac{(L-1)+a-(b-1)}{2} - p'\lambda \end{bmatrix}_q \\
& - \sum_{\lambda=-\infty}^{\infty} q^{(\lambda p + r)(\lambda p' + a)} \begin{bmatrix} L-1 \\ \frac{(L-1)-a-(b-1)}{2} - p'\lambda \end{bmatrix}_q \Bigg) \\
= q^{\frac{1}{2}(L-a+b)} & \ddot{\chi}^{p,p'}_{a,b+1,b}(L-1) + \ddot{\chi}^{p,p'}_{a,b-1,b}(L-1).
\end{aligned}$$

Thus, (B.1) holds with $\ddot{\chi}^{p,p'}_{a,b,c}(L)$ in place of $\chi^{p,p'}_{a,b,c}(L)$.

Using $r(0,1) = 0$, we calculate:

$$\begin{aligned}
\ddot{\chi}^{p,p'}_{a,0,1}(L) &= \sum_{\lambda=-\infty}^{\infty} q^{\lambda^2 pp' - \lambda pa} \begin{bmatrix} L \\ \frac{L+a}{2} - p'\lambda \end{bmatrix}_q - \sum_{\lambda=-\infty}^{\infty} q^{\lambda p(\lambda p' + a)} \begin{bmatrix} L \\ \frac{L-a}{2} - p'\lambda \end{bmatrix}_q \\
&= 0,
\end{aligned}$$

after changing the sign of the second summation parameter and using $\begin{bmatrix} A \\ B \end{bmatrix}_q = \begin{bmatrix} A \\ A-B \end{bmatrix}_q$. On substituting this into the previous expression when $b = 1$, we obtain $\ddot{\chi}^{p,p'}_{a,1,2}(L) = q^{\frac{1}{2}(L-a+1)} \ddot{\chi}^{p,p'}_{a,2,1}(L-1)$, and thus (B.5) holds with $\ddot{\chi}^{p,p'}_{a,b,c}(L)$ in place of $\chi^{p,p'}_{a,b,c}(L)$.

In a similar way, perhaps making use of the recurrence $\begin{bmatrix} A \\ B \end{bmatrix}_q = \begin{bmatrix} A-1 \\ B-1 \end{bmatrix}_q + q^B \begin{bmatrix} A-1 \\ B \end{bmatrix}_q$ for Gaussian polynomials, we obtain each of (B.2)-(B.4) and (B.6)-(B.8) with $\ddot{\chi}^{p,p'}_{a,b,c}(L)$ in place of $\chi^{p,p'}_{a,b,c}(L)$.

Finally, with $1 \le a, b, c < p$, we obtain:

$$\begin{aligned}\ddot{\chi}^{p,p'}_{a,b,c}(0) = & \sum_{\lambda=-\infty}^{\infty} q^{\lambda^2 pp' + \lambda(p' r(b,c) - pa)} \begin{bmatrix} 0 \\ \frac{a-b}{2} - p'\lambda \end{bmatrix}_q \\ & - \sum_{\lambda=-\infty}^{\infty} q^{(\lambda p + r(b,c))(\lambda p' + a)} \begin{bmatrix} 0 \\ \frac{a+b}{2} + p'\lambda \end{bmatrix}_q \\ = & \, \delta_{a,b},\end{aligned}$$

since $\begin{bmatrix} 0 \\ B \end{bmatrix}_q = \delta_{B,0}$ and $|\frac{1}{2}(a \pm b)| < p'$.

We thus conclude that if $1 \le a, b, c < p'$ with $c = b \pm 1$, then $\chi^{p,p'}_{a,b,c}(L) = \ddot{\chi}^{p,p'}_{a,b,c}(L)$, thus verifying (1.9).

In Section 2.3, we require an extension of this result to where $c = 0$ or $c = p'$. In accordance with (2.3), extend the definition (B.10) to:

$$\text{(B.11)} \qquad r(b,c) = \begin{cases} \lfloor pc/p' \rfloor + (b - c + 1)/2 & \text{if } 1 \le c < p'; \\ 1 & \text{if } c = 0 \text{ and } p' > 2p; \\ 0 & \text{if } c = 0 \text{ and } p' < 2p; \\ p - 1 & \text{if } c = p' \text{ and } p' > 2p; \\ p & \text{if } c = p' \text{ and } p' < 2p. \end{cases}$$

In the case $p' > 2p$, that the 0th and $(p'-1)$th bands are even implies that:

$$\chi^{p,p'}_{a,1,0}(L) = q^{-\frac{1}{2}(L+1-a)} \chi^{p,p'}_{a,1,2}(L); \quad \chi^{p,p'}_{a,p'-1,p'}(L) = q^{-\frac{1}{2}(L-p'+1+a)} \chi^{p,p'}_{a,p'-1,p'-2}(L).$$

Since $r(1,0) = 1$, $r(1,2) = 0$, $r(p'-1,p') = p-1$ and $r(p'-1,p'-2) = p$, it follows in this $p' > 2p$ case that $\chi^{p,p'}_{a,b,c}(L) = \ddot{\chi}^{p,p'}_{a,b,c}(L)$ for $1 \le a, b < p'$ and $0 \le c \le p'$ with $b - c = \pm 1$ and $L \ge 0$.

In the case $p' < 2p$, that the 0th and $(p'-1)$th bands are odd implies that:

$$\chi^{p,p'}_{a,1,0}(L) = q^{\frac{1}{2}(L+1-a)} \chi^{p,p'}_{a,1,2}(L); \quad \chi^{p,p'}_{a,p'-1,p'}(L) = q^{\frac{1}{2}(L-p'+1+a)} \chi^{p,p'}_{a,p'-1,p'-2}(L).$$

Since $r(1,0) = 0$, $r(1,2) = 1$, $r(p'-1,p') = p$ and $r(p'-1,p'-2) = p-1$, we conclude that $\chi^{p,p'}_{a,b,c}(L) = \ddot{\chi}^{p,p'}_{a,b,c}(L)$ for $1 \le a, b < p'$ and $0 \le c \le p'$ with $b - c = \pm 1$ and $L \ge 0$, as required.

APPENDIX C

Bands and the floor function

In this section, we derive some basic results concerning the floor function $\lfloor\cdot\rfloor$ and how it relates to the (p,p')-model and other closely related models. $\rho^{p,p'}(a)$ is defined in Section 2.2. $\delta^{p,p'}_{a,e}$ is defined in Section 2.6. $\omega^{p,p'}(a)$ is defined at the beginning of Section 5.

Lemma C.1. *Let $1 \le p < p'$. If $1 \le a < p'$, $e \in \{0,1\}$ and $a' = a + \lfloor ap/p' \rfloor + e$ then $\lfloor a'p/(p'+p) \rfloor = \lfloor ap/p' \rfloor$ and $\delta^{p,p'+p}_{a',e} = 0$. On the other hand, if $1 \le a' < p'+p$, $e \in \{0,1\}$, $\delta^{p,p'+p}_{a',e} = 0$ and $a = a' - \lfloor a'p/(p'+p) \rfloor - e$ then $\lfloor ap/p' \rfloor = \lfloor a'p/(p'+p) \rfloor$.*

In addition, if a is interfacial in the (p,p')-model and $\delta^{p,p'}_{a,e} = 0$, or if a is multifacial in the (p,p')-model, then a' is interfacial in the $(p,p'+p)$-model.

Proof: Let $r = \lfloor ap/p' \rfloor$ whence $p'r \le pa < p'(r+1)$. Then, for $x \in \{0,1\}$, we have $(p'+p)r \le p(a+r+x) < (p'+p)r + p' + xp$, so that $\lfloor (a+r+x)p/(p'+p) \rfloor = r$. In particular, $\lfloor a'p/(p'+p) \rfloor = r$, and $\lfloor (a+r+e+(-1)^e)p/(p'+p) \rfloor = r$. Thus $r = \lfloor a'p/(p'+p) \rfloor = \lfloor (a'+(-1)^e)p/(p'+p) \rfloor$ which gives the first results.

For the second statement, let $r = \lfloor a'p/(p'+p) \rfloor$. With $r = \lfloor (a'+(-1)^e)p/(p'+p) \rfloor$, this implies that $(p'+p)r \le p(a+r+x) < (p'+p)(r+1)$ for both $x \in \{0,1\}$. That $\lfloor ap/p' \rfloor = \lfloor a'p/(p'+p) \rfloor$ then follows.

Now, if a is interfacial in the (p,p')-model and $\delta^{p,p'}_{a,e} = 0$ then $\lfloor (a+(-1)^e)p/p' \rfloor = \lfloor ap/p' \rfloor \ne \lfloor (a-(-1)^e)p/p' \rfloor$. Thus $r \ne \lfloor (a-(-1)^e)p/p' \rfloor$. Clearly, this inequality also holds if a is multifacial. When $e=0$, we have $(a-1)p < rp'$ and thus $(a+r-1)p < r(p'+p)$ so that $\lfloor (a'-1)p/(p'+p) \rfloor < r$, which when compared to the above result implies that a' is interfacial in the $(p,p'+p)$-model. Similarly $e=1$ gives $(a+1)p \ge (r+1)p'$ whence $\lfloor (a'+1)p/(p'+p) \rfloor \ge r+1$, which when compared to the above result also implies that a' is interfacial in the $(p,p'+p)$-model. $\square$

Lemma C.2. *Let $1 \le p < p'$ with p coprime to p' and $1 \le a < p'$. Then $\lfloor a(p'-p)/p' \rfloor = a - 1 - \lfloor ap/p' \rfloor$.*

If, in addition, a is interfacial in the (p,p')-model and $\delta^{p,p'}_{a,e} = 0$ then a is interfacial in the $(p'-p,p')$-model and $\delta^{p'-p,p'}_{a,1-e} = 0$.

Proof: Since p and p' are coprime, $\lfloor ap/p' \rfloor < ap/p'$. Hence $\lfloor ap/p' \rfloor + \lfloor a(p'-p)/p' \rfloor = a-1$.

Since the (p,p')-model differs from the $(p'-p,p')$-model only in that corresponding bands are of the opposite parity, a being interfacial in one model implies that it also is in the other. The final part then follows immediately. $\square$

Lemma C.3. *Let $1 \le p < p' < 2p$ with p coprime to p', $1 \le a < p'$ and $e \in \{0,1\}$ and set $a' = 2a - e - \lfloor a(p'-p)/p' \rfloor$. Then $\delta^{p,p'+p}_{a',1-e} = 0$ and $\lfloor a'p/(p'+p) \rfloor = a - 1 - \lfloor a(p'-p)/p' \rfloor$.*

Furthermore, if $\delta^{p'-p,p'}_{a,e} = 0$ then a' is interfacial in the $(p, p'+p)$-model.

Proof: Lemma C.2 implies that $a' = a + 1 - e + \lfloor a'p/p' \rfloor$. Lemma C.1 immediately implies that $\delta^{p,p'+p}_{a',1-e} = 0$. Using first Lemma C.1 and then Lemma C.2 yields $\lfloor a'p/(p'+p) \rfloor = \lfloor ap/p' \rfloor = a - 1 - \lfloor a(p'-p)/p' \rfloor$.

If $\delta^{p'-p,p'}_{a,e} = 0$ then a either is interfacial in the $(p'-p,p')$-model or lies between two even bands. In the former case, Lemma C.2 implies that a is interfacial in the (p,p')-model and $\delta^{p,p'}_{a,1-e} = 0$. In the latter case, a is multifacial in the (p,p')-model. In either case, Lemma C.1 then shows that a' is interfacial in the $(p,p'+p)$-model.
□

Lemma C.4. *1) Let $0 \le a \le p'$ with a interfacial in the (p,p')-model. Let $r = \rho^{p,p'}(a)$. Then $a + r$ is interfacial in the $(p, p'+p)$-model and $\rho^{p,p'+p}(a+r) = r$. Moreover, $\omega^{p,p'+p}(a+r) = \omega^{p,p'}(a)$.*

2) Let $p' < 2p$ and $0 \le a \le p'$ with a interfacial in the $(p'-p,p')$-model. Let $r = \rho^{p'-p,p'}(a)$. Then a is interfacial in the (p,p')-model and $\rho^{p,p'}(a) = a - r$. Moreover, if $R = \lfloor rp/(p'-p) \rfloor$, then $\omega^{p'-p,p'}(a) = r^- \implies \omega^{p,p'}(a) = R^+$, and $\omega^{p'-p,p'}(a) = r^+ \implies \omega^{p,p'}(a) = (R+1)^-$.

Proof: 1) Since a is interfacial in the (p,p')-model, we have $r - 1 \le (a-1)p/p' < r \le (a+1)p/p' < r+1$ whence $(r-1)p' \le (a-1)p < rp' \le (a+1)p < (r+1)p'$ and $r(p'+p) - p' \le (a+r-1)p < r(p'+p) \le (a+r+1)p < r(p'+p) + p'$ so that $r - p'/(p'+p) \le (a+r-1)p/(p'+p) < r \le (a+r+1)p/(p'+p) < r + p'/(p'+p)$ and therefore $a+r$ is interfacial in the $(p,p'+p)$-model and $\rho^{p,p'+p}(a+r) = r$. Now, $\lfloor r(p'+p)/p \rfloor = \lfloor rp'/p \rfloor + r$, whence by definition, $\omega^{p,p'+p}(a+r) = \omega^{p,p'}(a)$.

2) Since corresponding bands in the (p,p')- and the $(p'-p,p')$-models are of opposite parity, a being interfacial in the $(p'-p,p')$-model implies that a is interfacial in the (p,p')-model. Then $\omega^{p'-p,p'}(a) = r^-$ implies that $r = \lfloor (p'-p)a/p' \rfloor + 1 = a - 1 - \lfloor pa/p' \rfloor + 1$. Thus $\lfloor pa/p' \rfloor = a - r = \lfloor rp'/(p'-p) \rfloor - r = R$, and since a has an odd band immediately below it in the (p,p')-model, $\rho^{p,p'}(a) = a - r$ and $\omega^{p,p'}(a) = R^+$. In a similar way, $\omega^{p'-p,p'}(a) = r^+$ implies that $\rho^{p,p'}(a) = a - r = R + 1$ and $\omega^{p,p'}(a) = (R+1)^-$. □

Lemma C.5. *Let $0 < a' < p' + p - 1$ with both a' and $a'+1$ interfacial in the $(p,p'+p)$-model and set $a = a' - \rho^{p,p'+p}(a')$. If $\rho^{p,p'+p}(a'+1) = \rho^{p,p'+p}(a') + 1$ then a is multifacial in the (p,p')-model.*

Proof: Let $r = \rho^{p,p'+p}(a')$. Then the rth band in the $(p,p'+p)$-model lies between heights $a'-1$ and a' so that $a' - 1 < r(p'+p)/p < a'$ whereupon $a' - r - 1 < rp'/p < a' - r$ so that the rth odd band in the (p,p')-model lies between heights $a-1$ and a. In a similar way, we see that the $(r+1)$th odd band in the (p,p')-model lies between heights a and $a+1$. The lemma follows. □

APPENDIX D

Bands on the move

In the main proof that is presented in Section 7.3, whenever Lemma 6.4 (or similarly Lemma 6.2) is invoked to obtain a generating function by extending paths, it is sometimes necessary to prove that (in the notation of Lemma 6.4) a and $a+2\Delta$ are both interfacial in the (p,p')-model. In the case of $a+2\Delta$, this is somewhat tricky. In this section, we present a number of quite technical auxiliary results which facilitate this proof. Each result examines how the relative displacement of two particular startpoints (or similarly endpoints) in a certain model changes under certain combinations of the $\mathcal{B}$- and $\mathcal{D}$-transforms.

The following result examines the differing effects of a $\mathcal{BD}$-transform on two adjacent points which are separated by an odd band in the $(p'-p,p')$-model and for which the associated pre-segments are in even bands.

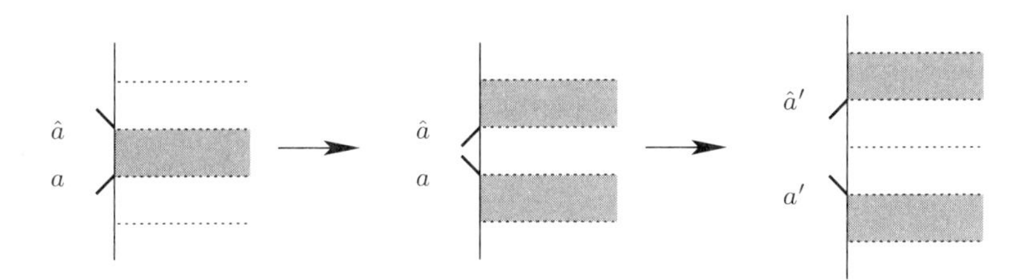

FIGURE D.1. Situation in Lemma D.1.

Lemma D.1. *Let $p'<2p$ and $1\le a\le p'-2$, with $\lfloor a(p'-p)/p'\rfloor \ne \lfloor (a+1)(p'-p)/p'\rfloor$, so that a and $\hat{a}=a+1$ are both interfacial in the $(p'-p,p')$-model. Let $e,\hat{e}\in\{0,1\}$ be such that $\delta^{p'-p,p'}_{a,e}=\delta^{p'-p,p'}_{\hat{a},\hat{e}}=0$. Let $a'=a+1-e+\lfloor ap/p'\rfloor$ and $\hat{a}'=\hat{a}+1-\hat{e}+\lfloor \hat{a}p/p'\rfloor$. Then a' and $\hat{a}'$ are both interfacial in the $(p,p'+p)$-model and $\hat{a}'=a'+2$. In addition, $\lfloor a'p/(p'+p)\rfloor=\lfloor \hat{a}'p/(p'+p)\rfloor$.*

Proof: Since a and $\hat{a}$ are separated by an even band in the (p,p')-model, $\lfloor ap/p'\rfloor = \lfloor \hat{a}p/p'\rfloor$, whence $\hat{a}'=a'+2$ immediately. In addition, $\delta^{p'-p,p'}_{a,e}=\delta^{p'-p,p'}_{\hat{a},\hat{e}}=0$ implies that $e=1$ and $\hat{e}=0$. Then $\delta^{p,p'}_{a,0}=0$ which, by Lemma C.1, implies that a' is interfacial in the $(p,p'+p)$-model. Similarly, $\delta^{p,p'}_{\hat{a},1}=0$ implies that $\hat{a}'$ is interfacial in the $(p,p'+p)$-model. Lemma C.1 also implies that $\delta^{p,p'+p}_{a',0}=\delta^{p,p'}_{\hat{a}',1}=0$, so that $\lfloor a'p/(p'+p)\rfloor = \lfloor (a'+1)p/(p'+p)\rfloor = \lfloor \hat{a}'p/(p'+p)\rfloor$. □

The conditions in the premise of Lemma D.1 of course imply that $e=1$ and $\hat{e}=0$. However, it will be more readily applied in the stated format.

The following result examines the differing effects of a $\mathcal{BD}$-transform on two adjacent points which are separated by an even band in the $(p'-p,p')$-model and for which the associated pre-segments are in even bands and in the same direction.

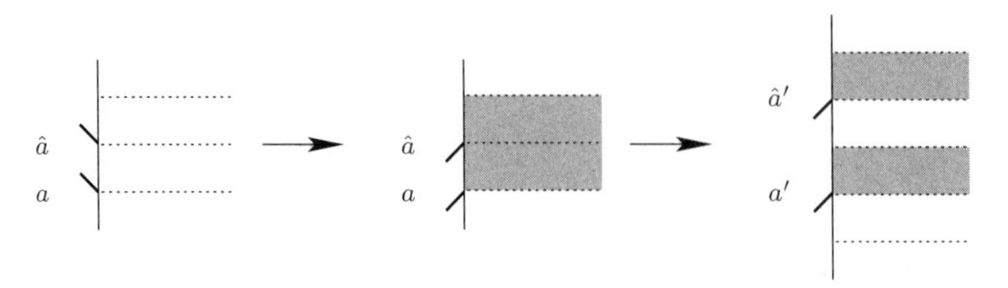

FIGURE D.2. Situation in Lemma D.2.

Lemma D.2. *Let $p' < 2p$ and $1 \le a \le p'-2$, with $\lfloor a(p'-p)/p' \rfloor = \lfloor \hat{a}(p'-p)/p' \rfloor$ and $\hat{a} = a+1$, and let $e \in \{0,1\}$ be such that $\delta^{p'-p,p'}_{a,e} = \delta^{p'-p,p'}_{\hat{a},e} = 0$. Let $a' = 2a - e - \lfloor a(p'-p)/p' \rfloor$ and $\hat{a}' = 2\hat{a} - e - \lfloor \hat{a}(p'-p)/p' \rfloor$. Then a' and $\hat{a}'$ are both interfacial in the $(p,p'+p)$-model, and $\hat{a}' = a'+2$.*

Proof: Lemma C.3 implies that both a' and $\hat{a}'$ are interfacial in the $(p,p'+p)$-model. That $\hat{a}' = a'+2$ is immediate. □

The following result examines the differing effects of a $\mathcal{BD}$-transform followed by a $\mathcal{B}$-transform on a single non-interfacial point in the $(p'-p,p')$-model with the two possible directions of the pre-segment.

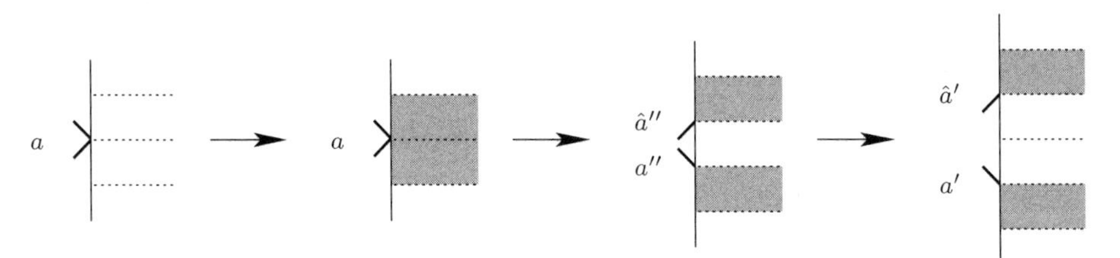

FIGURE D.3. Situation in Lemma D.3.

Lemma D.3. *Let $p' < 2p$ and $2 \le a \le p'-2$, with $\lfloor (a-1)(p'-p)/p' \rfloor = \lfloor (a+1)(p'-p)/p' \rfloor$, so that a is not interfacial in the $(p'-p,p')$-model. Let $a' = a + 2\lfloor ap/p' \rfloor$ and $\hat{a}' = a+2+2\lfloor ap/p' \rfloor$. Then a' and $\hat{a}'$ are both interfacial in the $(p,p'+2p)$-model.*

Proof: a is multifacial in the (p,p')-model, whence by Lemma C.1, both $a'' = a + \lfloor ap/p' \rfloor$ and $\hat{a}'' = a+1+\lfloor ap/p' \rfloor$ are interfacial in the $(p,p'+p)$-model, $\delta^{p,p'+p}_{a'',0} = \delta^{p,p'+p}_{\hat{a}'',1} = 0$, and $\lfloor a''p/(p'+p) \rfloor = \lfloor \hat{a}''p/(p'+p) \rfloor = \lfloor ap/p' \rfloor$. A further application of Lemma C.1 then proves the current lemma. □

Note D.4. *Once it is determined that a and $a+2$ are both interfacial in the (p,p')-model, it is readily seen that $\rho^{p,p'}(a+2) = \rho^{p,p'}(a)+1$. This holds even if $a=0$ or $a+2=p'$.*

APPENDIX E

Combinatorics of the Takahashi lengths

In this appendix, we obtain some basic but important results concerning the continued fraction of p'/p described in Section 1.3, and the Takahashi lengths $\{\kappa_i\}_{i=0}^{t}$ and truncated Takahashi lengths $\{\tilde{\kappa}_i\}_{i=0}^{t}$ that are derived from it as in Section 1.5.

For fixed p and p', let $\{\xi_\ell\}_{\ell=0}^{2c_n-1}$ be as defined in Section 1.13. In Section E.2, we show that these values partition the (p,p')-model into pieces that resemble specific other models.

E.1. Model comparisons

Here, we relate the parameters associated with the (p,p')-model for which the continued fraction is $[c_0, c_1, \dots, c_n]$ to those associated with certain 'simpler' models. In particular, if $c_0 > 1$, we compare them with those associated with the $(p, p'-p)$-model and, if $c_0 = 1$, we compare them with those associated with the $(p'-p, p')$-model.

In the following two lemmas, the parameters associated with those simpler models will be primed to distinguish them from those associated with the (p,p')-model. In particular if $c_0 > 1$, $(p'-p)/p$ has continued fraction $[c_0 - 1, c_1, \dots, c_n]$, so that in this case, $t' = t - 1$, $n' = n$ and $t'_k = t_k - 1$ for $1 \le k \le n$. If $c_0 = 1$, $p'/(p'-p)$ has continued fraction $[c_1 + 1, c_2, \dots, c_n]$, so that in this case, $t' = t$, $n' = n - 1$ and $t'_k = t_{k+1}$ for $1 \le k \le n'$.

Lemma E.1. *Let $c_0 > 1$. For $1 \le k \le n$ and $0 \le j \le t$, let y_k, z_k, κ_j and $\tilde{\kappa}_j$ be the parameters associated with the (p,p')-model as defined in Section 1.5. For $1 \le k \le n$ and $0 \le j \le t'$, let y'_k, z'_k, κ'_j and $\tilde{\kappa}'_j$ be the corresponding parameters for the $(p, p'-p)$-model. Then:*

- $y_k = y'_k + z'_k \quad (0 \le k \le n)$;
- $z_k = z'_k \quad (0 \le k \le n)$;
- $\kappa_j = \kappa'_{j-1} + \tilde{\kappa}'_{j-1} \quad (1 \le j \le t)$;
- $\tilde{\kappa}_j = \tilde{\kappa}'_{j-1} \quad (1 \le j \le t)$.

Proof: Straightforward. □

Lemma E.2. *Let $c_0 = 1$. For $1 \le k \le n$ and $0 \le j \le t$, let y_k, z_k, κ_j and $\tilde{\kappa}_j$ be the parameters associated with the (p,p')-model as defined in Section 1.5. For $1 \le k \le n'$ and $0 \le j \le t$, let y'_k, z'_k, κ'_j and $\tilde{\kappa}'_j$ be the corresponding parameters for the $(p'-p, p')$-model. Then:*

- $y_k = y'_{k-1} \quad (1 \le k \le n)$;
- $z_k = y'_{k-1} - z'_{k-1} \quad (1 \le k \le n)$;
- $\kappa_j = \kappa'_j \quad (1 \le j \le t)$;

- $\tilde{\kappa}_j = \kappa'_j - \tilde{\kappa}'_j \quad (1 \leq j \leq t)$.

Proof: Straightforward. □

E.2. Segmenting the model

Recall that the zone $\zeta(j) = k$ of an index j satisfying $0 \leq j \leq t$ is such that $t_k < j \leq t_{k+1}$. We now express an arbitrary integer as a sum of Takahashi lengths.

Lemma E.3. *Let $1 \leq s < p'$. Then, there is an expression:*

$$s = \sum_{i=1}^{g} \kappa_{\mu_i}, \tag{E.1}$$

with each $\mu_i < t_{n+1}$, such that $0 \leq \zeta(\mu_1) < \zeta(\mu_2) < \cdots < \zeta(\mu_g) \leq n$, and, in addition, for each $i < g$ such that $\mu_i = t_{\zeta(\mu_i)}$, we have $\mu_{i+1} > t_{\zeta(\mu_i)+1}$. Moreover, the expression is unique.

Proof: Use the 'greedy algorithm': let $\mu \leq t$ be the largest value such that $\kappa_\mu \leq s$. Then repeat with $s - \kappa_\mu$, if not zero. □

Informally, this expresses s as a sum of Takahashi lengths with at most one from each zone, and if one of these Takahashi lengths is the last element of a zone, the Takahashi lengths from the following zone are also excluded. Often, having formed the expression (E.1), we set $x = 0$ if $\zeta(\mu_1) > 0$ and $x = \kappa_{\mu_1}$ if $\zeta(\mu_1) = 0$ to obtain an expression of the form:

$$s = x + \sum_{i=1}^{g} \kappa_{\mu_i}, \tag{E.2}$$

with $0 \leq x \leq c_0$ and $1 \leq \zeta(\mu_1) < \zeta(\mu_2) < \cdots < \zeta(\mu_g) \leq n$, and in addition, for each $i < g$ such that $\mu_i = t_{\zeta(\mu_i)}$, we have $\zeta(\mu_{i+1}) \geq \zeta(\mu_i) + 2$, and if $x = c_0$ then $\zeta(\mu_1) \geq 2$.

By examining the definitions in Sections 1.3 and 1.5, we see that $\{\tilde{\kappa}_j : t_1 < j \leq t\}$ is the set of Takahashi lengths for $p/(p' - c_0 p)$. Therefore, we have:

Corollary E.4. *Let $1 \leq r < p$. Then, there is a unique expression:*

$$r = \sum_{i=1}^{g} \tilde{\kappa}_{\mu_i}, \tag{E.3}$$

with $t_1 < \mu_i < t_{n+1}$ for $1 \leq i \leq g$, such that $1 \leq \zeta(\mu_1) < \zeta(\mu_2) < \cdots < \zeta(\mu_g) \leq n$, and, in addition, for each $i < g$ such that $\mu_i = t_{\zeta(\mu_i)}$, we have $\mu_{i+1} > t_{\zeta(\mu_i)+1}$.

As we saw earlier, if $1 \leq r < p$, the rth odd band in the (p, p')-model lies between heights $\lfloor rp'/p \rfloor$ and $\lfloor rp'/p \rfloor + 1$. In the following result, we use Lemma E.3 and Corollary E.4 to provide a different prescription of this. This result was stated in [**12**, p333].

Lemma E.5. *1) Let $1 \leq r < p$ and use Corollary E.4 to express r in the form (E.3). Then:*

$$\sum_{i=1}^{g} \kappa_{\mu_i} = \left\lfloor \frac{rp'}{p} \right\rfloor + \delta^{(2)}_{\zeta(\mu_1),1}.$$

2) Let $1 \leq s < p'$ and use Lemma E.3 to express s in the form (E.2). Then:

$$\sum_{i=1}^{g} \tilde{\kappa}_{\mu_i} = \left\lfloor \frac{sp}{p'} \right\rfloor + \delta_{x,0}\delta^{(2)}_{\zeta(\mu_1),0}.$$

Proof: We first prove that the two statements are equivalent. First assume statement 1, with $r = \sum_{i=1}^{g} \tilde{\kappa}_{\mu_i}$. Now $\lfloor ps/p' \rfloor = r$ if and only if $\lfloor rp'/p \rfloor < s \leq \lfloor (r+1)p'/p \rfloor$. Then, on using Corollary E.4 to write $r+1 = \sum_{i=1}^{g'} \tilde{\kappa}_{\nu_i}$, we have $\lfloor ps/p' \rfloor = r$ if and only if $\sum_{i=1}^{g} \kappa_{\mu_i} - \delta^{(2)}_{\zeta(\mu_1),1} < s \leq \sum_{i=1}^{g'} \kappa_{\nu_i} - \delta^{(2)}_{\zeta(\nu_1),1}$. If $\delta^{(2)}_{\zeta(\nu_1),1} = 0$ then $s = \sum_{i=1}^{g'} \kappa_{\nu_i}$ is in this range. In this case, we obtain $\lfloor ps/p' \rfloor = r = (r+1) - 1 = \sum_{i=1}^{g'} \kappa_{\nu_i} - 1$ via statement 1, thus giving statement 2 here. Otherwise, $\sum_{i=1}^{g} \kappa_{\mu_i} - \delta^{(2)}_{\zeta(\mu_1),1} < s < \sum_{i=1}^{g'} \kappa_{\nu_i}$. We then claim that $s = x + \sum_{i=1}^{g} \kappa_{\mu_i}$ for $0 \leq x < c_1$, since if this is not the case then there exists $x' + \sum_{i=1}^{g''} \kappa_{\lambda_i}$ in this range, implying that $s' = \sum_{i=1}^{g''} \kappa_{\lambda_i}$ also is in the range. Thence if $r'' = \sum_{i=1}^{g''} \tilde{\kappa}_{\lambda_i}$ then, using statement 1, $\lfloor p'r''/p \rfloor = s' - \delta^{(2)}_{\lambda_1,1}$. This implies that $r < r'' < r+1$, which is absurd. So $a = x + \sum_{i=1}^{g} \kappa_{\mu_i}$ for $0 \leq x < c_1$, and $x = 0$ only if $\delta^{(2)}_{\mu_1,1} = 1$. Statement 2 then follows. By reversing the reasoning, it may be shown that statement 2 implies statement 1.

We now prove the two statements by using induction on the height plus the rank of p'/p. Let p'/p have continued fraction $[c_0, c_1, \ldots, c_n]$ so that the height is n, and the rank is $t = c_0 + \cdots + c_n - 2$. Assume that the two results hold in the case where rank plus height is $n + t - 1$.

First consider the case where $c_0 > 1$. For $1 \leq r < p$, write $r = \sum_{i=1}^{g} \tilde{\kappa}_{\mu_i}$. Then, in terms of the parameters of the $(p, p'-p)$-model, $r = \sum_{i=1}^{g} \tilde{\kappa}'_{\mu_i - 1}$. Thereupon, by the induction assumption, $\sum_{i=1}^{g} \kappa'_{\mu_i - 1} + \delta^{(2)}_{\zeta'(\mu_1 - 1),1} = \lfloor r(p'-p)/p \rfloor = \lfloor rp'/p \rfloor - r$. Thus, since $\zeta'(\mu_1 - 1) = \zeta(\mu_1)$, we obtain $\lfloor rp'/p \rfloor = \sum_{i=1}^{g} \kappa'_{\mu_i - 1} + \delta^{(2)}_{\zeta(\mu_1),1} + \sum_{i=1}^{g} \tilde{\kappa}'_{\mu_i - 1} = \sum_{i=1}^{g} \kappa_{\mu_i} + \delta^{(2)}_{\zeta(\mu_1),1}$, as required to prove statement 1 for $c_0 > 1$. Statement 2 then follows in this case.

Now consider the case where $c_0 = 1$ and let $1 \leq s < p'$. In terms of the parameters of the $(p'-p, p')$-model, we obtain $s = x' + \sum_{i=1}^{g} \kappa'_{\mu_i}$. Then, by the induction hypothesis, $\lfloor (p'-p)s/p' \rfloor + \delta_{x',0}\delta^{(2)}_{\zeta'(\mu_1),0} = \sum_{i=1}^{g} \tilde{\kappa}'_{\mu_i}$. Thereupon, $s - 1 - \lfloor ps/p' \rfloor + \delta_{x',0}\delta^{(2)}_{\zeta(\mu_1),1} = \sum_{i=1}^{g} \tilde{\kappa}'_{\mu_i}$, where we have made use of $\zeta(\mu_1) = \zeta'(\mu_1) + 1$. Thence, $\lfloor ps/p' \rfloor = \sum_{i=1}^{g} (\kappa'_{\mu_i} - \tilde{\kappa}'_{\mu_i}) + x' - 1 + \delta_{x',0}\delta^{(2)}_{\zeta(\mu_1),1} = \sum_{i=1}^{g} \tilde{\kappa}_{\mu_i} + x' - 1 + \delta_{x',0}\delta^{(2)}_{\zeta(\mu_1),1}$. If $x' = 0$ (so that $s = \sum_{i=1}^{g} \kappa_{\mu_i}$) then $\lfloor ps/p' \rfloor = \sum_{i=1}^{g} \tilde{\kappa}_{\mu_i} - \delta^{(2)}_{\zeta(\mu_1),0}$ as required. If $x' = 1$ (so that $s = 1 + \sum_{i=1}^{g} \kappa_{\mu_i}$) then $\lfloor ps/p' \rfloor = \sum_{i=1}^{g} \tilde{\kappa}_{\mu_i}$, as required. If $x' > 1$ (so that $s = \kappa_{x-1} + \sum_{i=1}^{g} \kappa_{\mu_i}$) then $\lfloor ps/p' \rfloor = \tilde{\kappa}_{x-1} + \sum_{i=1}^{g} \tilde{\kappa}_{\mu_i}$. We thus see that statement 2 holds for the $c_0 = 1$ case. Statement 1 then follows in this case, and the lemma is thus proved. □

Since $\lfloor rp'/p \rfloor$ is the height of the lowermost edge of the rth odd band, this theorem states that with $s = \sum_{i=1}^{g} \kappa_{\mu_i}$, then if $\zeta(\mu_1) \equiv 0 \pmod 2$, the rth odd band lies between heights s and $s+1$; and if $\zeta(\mu_1) \equiv 1 \pmod 2$, the rth odd band lies between heights $s-1$ and s.

Lemma E.6. *Let $1 < k \le n$ and $1 \le r < z_k$. If $t_k < j \le t$ then:*

$$\left\lfloor \frac{p'(r+\tilde{\kappa}_j)}{p} \right\rfloor = \left\lfloor \frac{p'r}{p} \right\rfloor + \kappa_j.$$

Proof: Write r in the form (E.3). From $1 \le r < z_k$, it follows that $g \ge 1$, $\mu_g < t_k$ and $r + \tilde{\kappa}_j < p$. Set $\mu_{g+1} = j$, whereupon $\sum_{i=1}^{g+1} \tilde{\kappa}_{\mu_i}$ is an expression of the form (E.3) for $r + \tilde{\kappa}_j$. Then:

$$\left\lfloor \frac{p'(r+\tilde{\kappa}_j)}{p} \right\rfloor = \sum_{i=1}^{g+1} \kappa_{\mu_i} - \delta^{(2)}_{\zeta(\mu_1),1} = \sum_{i=1}^{g} \kappa_{\mu_i} - \delta^{(2)}_{\zeta(\mu_1),1} + \kappa_j = \left\lfloor \frac{p'r}{p} \right\rfloor + \kappa_j,$$

where the first and third equalities follow from Lemma E.5(1). □

Lemma E.7. *Let $1 \le k \le n$ and $1 \le r < z_{k+1}$. If $r + z_k < p$ then:*

$$\left\lfloor \frac{p'(r+z_k)}{p} \right\rfloor = \left\lfloor \frac{p'r}{p} \right\rfloor + y_k.$$

Proof: Let r be expressed in the form (E.3). Note that $g \ge 1$, $\mu_1 > t_1$ and $\mu_g < t_{k+1}$. Depending on the value of μ_g, we reexpress $\kappa_{\mu_g} + y_k$ and $\tilde{\kappa}_{\mu_g} + z_k$ according to one of the following five cases: i) $t_k < \mu_g < t_{k+1}$. Here $\kappa_{\mu_g} + y_k = y_{k-1} + (\mu_g - t_k)y_k + y_k = \kappa_{\mu_g+1}$. Similarly, $\tilde{\kappa}_{\mu_g} + z_k = \tilde{\kappa}_{\mu_g+1}$. ii) $t_{k-1}+1 < \mu_g \le t_k$. Here $\kappa_{\mu_g} + y_k = y_{k-2} + (\mu_g - t_{k-1})y_{k-1} + y_k = y_{k-2} + (\mu_g - t_{k-1} - 1)y_{k-1} + (y_{k-1} + y_k) = \kappa_{\mu_g-1} + \kappa_{t_k+1}$. Similarly, $\tilde{\kappa}_{\mu_g} + z_k = \tilde{\kappa}_{\mu_g-1} + \tilde{\kappa}_{t_k+1}$. iii) $\mu_g = t_{k-1}+1$. Here $\kappa_{\mu_g} + y_k = y_{k-2} + (\mu_g - t_{k-1})y_{k-1} + y_k = y_{k-2} + y_{k-1} + y_k = y_{k-2} + \kappa_{t_k+1}$. Similarly, $\tilde{\kappa}_{\mu_g} + z_k = z_{k-2} + \tilde{\kappa}_{t_k+1}$. iv) $\mu_g = t_{k-1}$. Here $\kappa_{\mu_g} + y_k = y_{k-1} + y_k = \kappa_{t_k+1}$. Similarly, $\tilde{\kappa}_{\mu_g} + z_k = \tilde{\kappa}_{t_k+1}$. v) $\mu_g < t_{k-1}$. Here, no reexpressing is required.

We now use these to examine $\lfloor p'(r+z_k)/p \rfloor$, tackling each of the five cases in turn (but leaving case iii) until last). For convenience, we set $r' = \sum_{i=1}^{g-1} \tilde{\kappa}_{\mu_i}$ so that $r + z_k = r' + \tilde{\kappa}_{\mu_g} + z_k$ in which the last two terms will be reexpressed as above.

i) $t_k < \mu_g < t_{k+1}$. Here,

$$\begin{aligned}\left\lfloor \frac{p'(r+z_k)}{p} \right\rfloor &= \left\lfloor \frac{p'(r'+\tilde{\kappa}_{\mu_g+1})}{p} \right\rfloor = \sum_{i=1}^{g-1} \kappa_{\mu_i} + \kappa_{\mu_g+1} - \delta^{(2)}_{\zeta(\mu_1),1} \\ &= \sum_{i=1}^{g} \kappa_{\mu_i} + y_k - \delta^{(2)}_{\zeta(\mu_1),1} = \left\lfloor \frac{p'r}{p} \right\rfloor + y_k,\end{aligned}$$

where the first and third equalities follow from the case i) reexpressings above, and the second and fourth follow from Lemma E.5(1), noting when $g = 1$ that $\zeta(\mu_1 + 1) = k = \zeta(\mu_1)$.

ii) $t_{k-1} + 1 < \mu_g \le t_k$. Here,

$$\begin{aligned}\left\lfloor \frac{p'(r+z_k)}{p} \right\rfloor &= \left\lfloor \frac{p'(r'+\tilde{\kappa}_{\mu_g-1}+\tilde{\kappa}_{t_k+1})}{p} \right\rfloor = \sum_{i=1}^{g-1} \kappa_{\mu_i} + \kappa_{\mu_g-1} + \kappa_{t_k+1} - \delta^{(2)}_{\zeta(\mu_1),1} \\ &= \sum_{i=1}^{g} \kappa_{\mu_i} + y_k - \delta^{(2)}_{\zeta(\mu_1),1} = \left\lfloor \frac{p'r}{p} \right\rfloor + y_k,\end{aligned}$$

where the first and third equalities follow from the case ii) reexpressings above, and the second and fourth follow from Lemma E.5(1), noting when $g=1$ that $\zeta(\mu_1-1)=k-1=\zeta(\mu_1)$.

iv) $\mu_g=t_{k-1}$. If $g>1$ then:

$$\left\lfloor\frac{p'(r+z_k)}{p}\right\rfloor=\left\lfloor\frac{p'(r'+\tilde\kappa_{t_k+1})}{p}\right\rfloor=\sum_{i=1}^{g-1}\kappa_{\mu_i}+\kappa_{t_k+1}-\delta^{(2)}_{\zeta(\mu_1),1}$$
$$=\sum_{i=1}^{g}\kappa_{\mu_i}+y_k-\delta^{(2)}_{\zeta(\mu_1),1}=\left\lfloor\frac{p'r}{p}\right\rfloor+y_k,$$

where the first and third equalities follow from the case iv) reexpressings above, and the second and fourth follow from Lemma E.5(1). If $g=1$ then the argument is similar (with $r'=0$), noting that $\delta^{(2)}_{\zeta(t_k+1),1}=\delta^{(2)}_{\zeta(t_{k-1}),1}$ because $\zeta(t_k+1)=k$ and $\zeta(t_{k-1})=k-2$.

v) $\mu_g<t_{k-1}$. Since $z_k=\tilde\kappa_{t_k}$ and $\zeta(t_k)=k-1$, we may apply Lemma E.5(1) directly. Then:

$$\left\lfloor\frac{p'(r+z_k)}{p}\right\rfloor=\left\lfloor\frac{p'(r+\tilde\kappa_{t_k})}{p}\right\rfloor=\sum_{i=1}^{g}\kappa_{\mu_i}+\kappa_{t_k}-\delta^{(2)}_{\zeta(\mu_1),1}$$
$$=\sum_{i=1}^{g}\kappa_{\mu_i}+y_k-\delta^{(2)}_{\zeta(\mu_1),1}=\left\lfloor\frac{p'r}{p}\right\rfloor+y_k,$$

as required.

Case iii) requires the whole proof to be entombed in an induction argument. We first prove the lemma for $k=1$ and $k=2$. If $k=1$ then $g=1$ and necessarily $t_1<\mu_1<t_2$. This $k=1$ case then follows from i) as above. If $k=2$ then necessarily $t_1<\mu_g<t_3$. Except for $\mu_g=t_1+1$ this is dealt with by i) and ii) above. For $\mu_g=t_1+1$, we necessarily have $g=1$ so that $r=1$. By definition, $z_2=c_1$, $y_1=c_0$ and $y_2=c_0c_1+1$. Since $p'/p=c_0+1/(c_1+\epsilon)$ where $0<\epsilon<1$, it follows that $\lfloor p'(r+z_2)/p\rfloor=c_0c_1+c_0+1=\lfloor p'r/p\rfloor+y_2$, as required.

Having established the lemma for $k\le 2$, we now consider $k>2$. For the purposes of induction, assume that the lemma holds for k replaced by $k-2$. For all cases except where $\mu_g=t_{k-1}+1$, the lemma is immediately proved by i), ii), iv) or v) above. In the remaining case $\mu_g=t_{k-1}+1$, we immediately obtain that $\tilde\kappa_{\mu_g+1}=\tilde\kappa_{\mu_g}+z_{k-1}$. Then, because $r<\tilde\kappa_{\mu_g+1}$, we obtain $r'<z_{k-1}$ and furthermore $r'+z_{k-2}<z_{k-2}+z_{k-1}\le z_k$. Then, if $r'>0$:

$$\left\lfloor\frac{p'(r+z_k)}{p}\right\rfloor=\left\lfloor\frac{p'(r'+z_{k-2}+\tilde\kappa_{t_k+1})}{p}\right\rfloor=\left\lfloor\frac{p'(r'+z_{k-2})}{p}\right\rfloor+\kappa_{t_k+1}$$
$$=\left\lfloor\frac{p'r'}{p}\right\rfloor+y_{k-2}+\kappa_{t_k+1}=\left\lfloor\frac{p'r'}{p}\right\rfloor+\kappa_{\mu_g}+y_k$$
$$=\left\lfloor\frac{p'(r'+\tilde\kappa_{\mu_g})}{p}\right\rfloor+y_k=\left\lfloor\frac{p'r}{p}\right\rfloor+y_k,$$

where the first and fourth equalities follow from the case iii) reexpressings above, the second and fifth equalities follow from Lemma E.6, and the third follows from the induction hypothesis. The same string of equalities may be used in the case $r'=0$ provided that we subtract $\delta^{(2)}_{k-3,1}$ $(=\delta^{(2)}_{\mu_g,1})$ from the fourth and fifth expressions:

the third and fifth equalities then follow from Lemma E.5(1). This completes the induction step and the lemma follows. □

In some of the following results, we state that certain quantities have continued fractions of a certain form $[d_0, d_1, \dots, d_{m-1}, d_m]$. The case $d_m = 1$ may arise for $m > 0$. This continued fraction would then not be of the standard form stated in Section 1.3. In such a case, the continued fraction should be reinterpreted as $[d_0, d_1, \dots, d_{m-1}+1]$.

Lemma E.8. *Let p'/p have continued fraction $[c_0, c_1, \dots, c_n]$. For $0 \le j \le t$, set k such that $t_k < j \le t_{k+1}$. Then:*
1. $y_k\tilde{\kappa}_j - z_k\kappa_j = (-1)^k$;
2. κ_j and $\tilde{\kappa}_j$ are coprime;
3. $\kappa_j/\tilde{\kappa}_j$ has continued fraction $[c_0, c_1, \dots, c_{k-1}, j-t_k]$.

Proof: If $k=0$ then $y_0 = 1$, $z_0 = 0$, whereupon $\kappa_j = j+1$ and $\tilde{\kappa}_j = 1$. Each part is then seen to hold when $k=0$. For the purposes of induction, let $k>0$ and assume that each part holds for k replaced by $k-1$. Note that $\kappa_{t_k} = y_k$ and $\tilde{\kappa}_{t_k} = z_k$. The induction hypothesis then implies that $y_{k-1}z_k - z_{k-1}y_k = (-1)^{k-1}$. Statement 1. is now verified by using $\kappa_j = y_{k-1} + (j-t_k)y_k$ and $\tilde{\kappa}_j = z_{k-1} + (j-t_k)z_k$. Statement 1. then follows for general k by induction. Statement 2. follows immediately from statement 1. In the case of statement 3., the induction hypothesis states that:

$$\frac{c_{k-1}y_{k-1} + y_{k-2}}{c_{k-1}z_{k-1} + z_{k-2}} = \frac{y_k}{z_k} = c_0 + \cfrac{1}{\ddots \cfrac{}{c_{k-2} + \cfrac{1}{c_{k-1}}}}.$$

Both sides here are rational functions of c_{k-1} which are positive when c_{k-1} is positive. Therefore, we may replace c_{k-1} with $c_{k-1} + 1/(j-t_k)$. The right side yields the continued fraction $[c_0, c_1, \dots, c_{k-1}, j-t_k]$, while the left side yields:

$$\frac{(c_{k-1} + 1/(j-t_k))y_{k-1} + y_{k-2}}{(c_{k-1} + 1/(j-t_k))z_{k-1} + z_{k-2}} = \frac{y_k + y_{k-1}/(j-t_k)}{z_k + z_{k-1}/(j-t_k)} = \frac{(j-t_k)y_k + y_{k-1}}{(j-t_k)z_k + z_{k-1}} = \frac{\kappa_j}{\tilde{\kappa}_j}.$$

This completes the induction step, whereupon the lemma follows. □

Lemma E.9. *Let p'/p have continued fraction $[c_0, c_1, \dots, c_n]$. Let $0 < \ell < 2c_n - 1$, and set $\hat{p}' = \xi_{\ell+1} - \xi_\ell$ and $\hat{p} = \tilde{\xi}_{\ell+1} - \tilde{\xi}_\ell$.*
1) $\hat{p}' \ge 2$ if and only if one of the following cases applies:

i) ℓ is odd, $n \ge 2$ and if $n = 2$ then $c_0 > 1$. In this case $\hat{p}'/\hat{p}$ has continued fraction $[c_0, c_1, \dots, c_{n-2}]$;

ii) $\ell \in \{0, 2c_n - 2\}$, $n \ge 1$ and if $n = 1$ then $c_0 > 1$. In this case $\hat{p}'/\hat{p}$ has continued fraction $[c_0, c_1, \dots, c_{n-1}]$;

iii) ℓ is even, $0 < \ell < 2c_n - 2$, $c_{n-1} > 1$, $n \ge 1$ and if $n = 1$ then $c_0 > 2$. In this case $\hat{p}'/\hat{p}$ has continued fraction $[c_0, c_1, \dots, c_{n-1} - 1]$;

iv) ℓ is even, $0 < \ell < 2c_n - 2$, $c_{n-1} = 1$, $n \ge 3$ and if $n = 3$ then $c_0 > 1$. In this case $\hat{p}'/\hat{p}$ has continued fraction $[c_0, c_1, \dots, c_{n-3}]$.

2) Assume that $\hat{p}' \ge 2$, and let $\{\kappa_i\}_{i=0}^t$ and $\{\tilde{\kappa}_i\}_{i=0}^t$ be respectively the Takahashi lengths and the truncated Takahashi lengths for p'/p. Let $\{\kappa'_i\}_{i=0}^{t'}$ and $\{\tilde{\kappa}'_i\}_{i=0}^{t'}$ be the corresponding values for $\hat{p}'/\hat{p}$. Then $\kappa'_i = \kappa_i$ and $\tilde{\kappa}'_i = \tilde{\kappa}_i$ for $0 \le i \le t'$.

Proof: First note, using the definitions of Section 1.5, that (when defined) $y_{-1} = 0$, $y_0 = 1$, $y_1 = c_0$, $y_2 = c_1c_0 + 1$, $z_{-1} = 1$, $z_0 = 0$, $z_1 = 1$, and $z_2 = c_1$. We now split into the following four cases to calculate $\hat{p}'$ and $\hat{p}$ using the definitions of Section 1.13.

i). ℓ is odd: here $\hat{p}' = y_{n-1}$ and $\hat{p} = z_{n-1}$. So we certainly require $n \geq 2$. Since $y_{n-1} = \kappa_{t_{n-1}}$ and $z_{n-1} = \tilde{\kappa}_{t_{n-1}}$, Lemma E.8(3) implies that $\hat{p}'/\hat{p}$ has continued fraction $[c_0, c_1, \ldots, c_{n-2}]$. Then to ensure that $\hat{p}' \geq 2$, we must exclude the case $n = 2$ and $c_0 = 1$.

ii). $\ell \in \{0, 2c_n - 2\}$: here $\hat{p}' = y_n$ and $\hat{p} = z_n$ whereupon this case follows similarly to the first.

iii). ℓ is even, $0 < \ell < 2c_n - 2$ and $c_{n-1} > 1$: here $\hat{p}' = y_n - y_{n-1}$ and $\hat{p} = z_n - z_{n-1}$. So here we certainly require $n \geq 1$. Then $\hat{p}' = y_{n-2} + (c_{n-1} - 1)y_{n-1} = \kappa_{t_n-1}$ and similarly $\hat{p} = \tilde{\kappa}_{t_n-1}$. Lemma E.8(3) now implies that $\hat{p}'/\hat{p}$ has continued fraction $[c_0, c_1, \ldots, c_{n-1} - 1]$, whereupon we see that to ensure $\hat{p}' \geq 2$, we must exclude the case $n = 1$ and $c_0 = 2$.

iv). ℓ is even, $0 < \ell < 2c_n - 2$ and $c_{n-1} = 1$: here, $\hat{p}' = y_n - y_{n-1} = y_{n-2} + (c_{n-1} - 1)y_{n-1} = y_{n-2}$ and $\hat{p} = z_{n-2}$, whereupon this case follows similarly to the first.

Statement 2 now follows immediately via the definitions of Section 1.5. □

Theorem E.10. *Let p'/p have continued fraction $[c_0, c_1, \ldots, c_n]$ with $n \geq 1$, and if $n = 1$ then $c_0 > 1$. If $0 < \ell < 2c_n - 1$, then ξ_ℓ is interfacial in the (p, p')-model and neighbours the $\tilde{\xi}_\ell$th odd band.*

Proof: For $n = 1$, the kth odd band lies between heights kc_0 and $kc_0 + 1$ for $1 \leq k < c_0$. In this case, $\xi_{2k-1} = kc_0$, $\xi_{2k} = kc_0 + 1$, and $\tilde{\xi}_{2k-1} = \tilde{\xi}_{2k} = k$ for $1 \leq k < c_0$, whereupon this case follows immediately.

Hereafter, assume that $n \geq 2$. Note that it is sufficient to prove the $p' < 2p$ case: the $p' < 2p$ case then follows because on passing from the (p, p')-model to the $(p' - p, p')$ model, each band changes parity, and via Lemma E.2, the values ξ_ℓ are unchanged and the values of $\tilde{\xi}_\ell$ change to $\xi_\ell - \tilde{\xi}_\ell$.

If ℓ is odd, $\ell = 2k - 1$ for $1 \leq k \leq c_n - 1$, and $\tilde{\xi}_\ell = kz_n$ and $\xi_\ell = ky_n$. Then:

$$\left\lfloor \frac{p'\tilde{\xi}_\ell}{p} \right\rfloor = \left\lfloor \frac{p'(kz_n)}{p} \right\rfloor = \left\lfloor \frac{p'z_n}{p} \right\rfloor + (k-1)y_n = y_n - \delta^{(2)}_{n,0} + (k-1)y_n = \xi_\ell - \delta^{(2)}_{n,0},$$

where the second equality follows by repeated use of Lemma E.7 on noting that $kz_n < z_{n-1} + c_nz_n = z_{n+1} = p$, and the third follows from Lemma E.5(1). This implies that either the upper or lower edge of the $\tilde{\xi}_\ell$th odd band is at height ξ_ℓ. For $p' > 2p$ this immediately implies that ξ_ℓ is interfacial.

If ℓ is even, $\ell = 2k$ for $1 \leq k \leq c_n - 1$, and $\tilde{\xi}_\ell = kz_n + z_{n-1}$ and $\xi_\ell = ky_n + y_{n-1}$. Then if $n > 2$, we obtain in a similar way to the odd ℓ case above,

$$\left\lfloor \frac{p'\tilde{\xi}_\ell}{p} \right\rfloor = \left\lfloor \frac{p'(z_{n-1} + kz_n)}{p} \right\rfloor = \left\lfloor \frac{p'z_{n-1}}{p} \right\rfloor + ky_n = y_{n-1} - \delta^{(2)}_{n,1} + ky_n = \xi_\ell - \delta^{(2)}_{n,1}.$$

This calculation is also valid for $n = 2$, when the validity of the third equality follows from $z_1 = 1$ and $\lfloor p'/p \rfloor = c_0 = y_1$. As above, for $p' > 2p$ this immediately implies that ξ_ℓ is interfacial. The lemma then follows. □

Theorem E.11. *Let p'/p have continued fraction $[c_0, c_1, \ldots, c_n]$ with $n \geq 1$. Let $0 \leq \ell \leq 2c_n - 2$, let $\hat{p}' = \xi_{\ell+1} - \xi_\ell$, and let $\hat{p} = \tilde{\xi}_{\ell+1} - \tilde{\xi}_\ell$. Then for $1 \leq s \leq \hat{p}' - 2$, the parity of the sth band of the $(\hat{p}, \hat{p}')$-model is equal to the parity of the $(\xi_\ell + s)$th band of the (p, p')-model.*

Proof: Let $1 \leq r < \hat{p}$. We claim that

$$\left\lfloor \frac{p'(r + \tilde{\xi}_\ell)}{p} \right\rfloor = \left\lfloor \frac{p'r}{p} \right\rfloor + \xi_\ell,$$

In the $\ell = 0$ case this holds trivially because $\xi_\ell = \tilde{\xi}_\ell = 0$. There are two other subcases: ℓ odd and ℓ even.

For $\ell = 2k - 1$ with $1 \leq k \leq c_n - 1$, we have $\xi_\ell = ky_n$, $\tilde{\xi}_\ell = kz_n$ and $\hat{p} = \tilde{\xi}_{\ell+1} - \tilde{\xi}_\ell = z_{n-1}$. Then:

$$\left\lfloor \frac{p'(r + \tilde{\xi}_\ell)}{p} \right\rfloor = \left\lfloor \frac{p'(r + kz_n)}{p} \right\rfloor = \left\lfloor \frac{p'r}{p} \right\rfloor + ky_n, = \left\lfloor \frac{p'r}{p} \right\rfloor + \xi_\ell,$$

where the second equality is established by repeated use of Lemma E.7, the validity of which is guaranteed because $1 \leq r + kz_n < z_{n-1} + c_n z_n = z_{n+1} = p$.

For $\ell = 2k$ with $1 \leq k \leq c_n - 1$, we have $\xi_\ell = y_{n-1} + ky_n$ and $\tilde{\xi}_\ell = z_{n-1} + kz_n$. For $k < c_n - 1$, we have $\hat{p} = \tilde{\xi}_{\ell+1} - \tilde{\xi}_\ell = z_n - z_{n-1}$, whereas for $k = c_n - 1$, we have $\hat{p} = \tilde{\xi}_{\ell+1} - \tilde{\xi}_\ell = z_n$. Since $z_1 = 1$, we may restrict consideration here to $n > 1$. Then:

$$\begin{aligned}\left\lfloor \frac{p'(r + \tilde{\xi}_\ell)}{p} \right\rfloor &= \left\lfloor \frac{p'(r + z_{n-1} + kz_n)}{p} \right\rfloor \\ &= \left\lfloor \frac{p'(r + z_{n-1})}{p} \right\rfloor + ky_n = \left\lfloor \frac{p'r}{p} \right\rfloor + y_{n-1} + ky_n = \left\lfloor \frac{p'r}{p} \right\rfloor + \xi_\ell,\end{aligned}$$

where the second equality is established by repeated use of Lemma E.7, the validity of which is guaranteed because $1 \leq r + z_{n-1} + kz_n < z_{n-1} + c_n z_n = z_{n+1} = p$; and the third equality follows from a single use of Lemma E.7, and is valid because $1 \leq r < z_n$.

Lemma E.9(2) states that the sequence of truncated Takahashi lengths for $\hat{p}'/\hat{p}$ begins the sequence of truncated Takahashi lengths for p'/p. This implies that the expression (E.3) for r in terms of the truncated Takahashi lengths of p'/p is identical to that in terms of the truncated Takahashi lengths of $\hat{p}'/\hat{p}$. It then follows from Lemma E.5(1) that $\lfloor \hat{p}'r/\hat{p} \rfloor = \lfloor p'r/p \rfloor$ and therefore

$$\left\lfloor \frac{p'(r + \tilde{\xi}_\ell)}{p} \right\rfloor = \left\lfloor \frac{\hat{p}'r}{\hat{p}} \right\rfloor + \xi_\ell.$$

Thus, for $1 \leq r < \hat{p}$, the height of the rth odd band in the $(\hat{p}, \hat{p}')$-model is precisely ξ_ℓ less that the height of $(r + \tilde{\xi}_\ell)$th odd band in the (p, p')-model. The theorem is then proved if it can be shown that there are no other odd bands between heights $\xi_\ell + 1$ and $\xi_{\ell+1} - 1$ in the (p, p')-model. This is so because, by Lemma E.10, the lowermost edge of the $\tilde{\xi}_\ell$th odd band in the (p, p')-model (if there is one) is at height ξ_ℓ or $\xi_\ell - 1$, and the lowermost edge of the $\tilde{\xi}_{\ell+1}$th odd band in the (p, p')-model (if there is one) is at height $\xi_{\ell+1}$ or $\xi_{\ell+1} - 1$. □

The above result shows that for $0 \leq \ell \leq 2c_n - 2$, the band structure between heights $\xi_\ell + 1$ and $\xi_{\ell+1} - 1$ in the (p, p')-model is identical to the band structure of a certain smaller model (which is specified in Lemma E.9). This result proves useful in that a certain subset of the set of paths in the (p, p')-model will be shown to be the set of paths that lie between heights $\xi_\ell + 1$ and $\xi_{\ell+1} - 1$ for some ℓ. Their generating function will thus be given by that of the smaller model.

We demonstrate Lemma E.11 using the $(7, 24)$-model which is depicted in Fig. E.1.

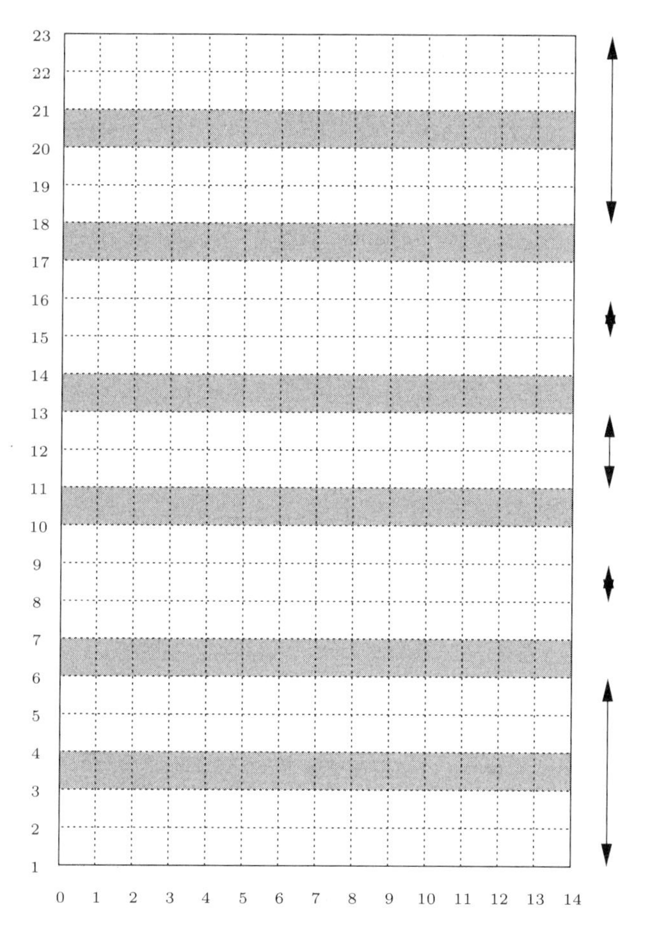

FIGURE E.1. $(7, 24)$-model.

Since $24/7$ has continued fraction $[3, 2, 3]$, we obtain $y_{n-1} = 3$ and $y_n = 7$. Thereupon $(\xi_0, \xi_1, \xi_2, \xi_3, \xi_4, \xi_5) = (0, 7, 10, 14, 17, 24)$. Lemma E.11 states that for $\ell = 0, 1, 2, 3, 4$, the band structure of the $(7, 24)$-model between heights $\xi_\ell+1$ and $\xi_{\ell+1}-1$ is identical to the band structure of a specific smaller model. The arrowed lines to the right of the grid in Fig. 8 show the extent of these smaller models. For $\ell = 0, 1, 2, 3, 4,$, they are seen (by inspection, or by using Lemma E.11) to be the $(2, 7)$-, $(1, 3)$-, $(1, 4)$-, $(1, 3)$-, and $(2, 7)$-models respectively.

Finally, we make use of Lemma E.7 to provide a result required in the proof of Theorem 9.15.

Lemma E.12. *For* $1 \le p < p'$, *let* p'/p *have continued fraction* $[c_0, c_1, \ldots, c_n]$ *with either* $n \ge 3$ *and* $c_0 = c_1 = c_2 = 1$, *or* $n = 3$ *and* $c_0 = c_1 = 1$ *and* $c_2 = 2$. *Then, for* $\Delta = \pm 1$, *there exists no value* a *with* $1 \le a, a + 5\Delta < p'$ *such that each element of* $\{a + \Delta, a + 2\Delta, a + 4\Delta\}$ *is interfacial and* $\rho^{p,p'}(a + \Delta) = \rho^{p,p'}(a + 2\Delta)$.

Proof: First note that since odd bands are separated by either c_0 or $c_0 - 1$ even bands, then necessarily $a + 3\Delta$ is interfacial. Then, without loss of generality, we may assume that $\Delta = +1$, whereupon the specified configuration arises only if $\lfloor (a+5)p/p' \rfloor = \lfloor ap/p' \rfloor + 2$.

In the case where $c_0 = c_1 = c_2 = 1$, we have $p'/p = 1 + 1/(1 + 1/(1 + \epsilon))$ with $0 < \epsilon < 1$. Then $5p/p' = \frac{5}{2}(1 + 1/(3 + 2\epsilon)) > 3$ implying that $\lfloor (a + 5)p/p' \rfloor \ge \lfloor ap/p' \rfloor + 3$ so that this case is excluded.

Now consider the case where p'/p has continued fraction $[1, 1, 2, c_3]$. Here, $y_3 = 5$, $p' = y_4 = 5c_3 + 2$, $z_3 = 3$, and $p = z_4 = 3c_3 + 1$. Use of Lemma E.7 for $k = 3$ yields $\lfloor (r + 3)p'/p \rfloor = \lfloor p'r/p \rfloor + 5$ for $1 \le r < p - 3$. This shows that the $(r + 1)$th, $(r + 2)$th and $(r + 3)$th odd bands lie within five consecutive bands. This also holds for $r = 0$ because $\lfloor 3p'/p \rfloor = \lfloor (15c_3 + 6)/(3c_3 + 1) \rfloor = 5$. Thus, for every three consecutive odd bands, there are necessarily two that are adjacent. The required result then follows. $\square$

Bibliography

[1] A.K. AGARWAL and D.M. BRESSOUD, *Lattice paths and multiple basic hypergeometric series*, Pac. J. Math. **136** (1989), 209–228.

[2] G.E. ANDREWS, *An analytic generalization of the Rogers-Ramanujan identities for odd moduli*, Proc. Nat. Acad. Sci. USA **71** (1974), 4082–4085.

[3] ______ *Multiple series Rogers-Ramanujan type identities*, Pac. J. Math. **114** (1984), 267–283.

[4] ______ *The Theory of Partitions*, Encyclopedia of Mathematics and its Applications, **2**, Addison-Wesley, Reading, 1976.

[5] G.E. ANDREWS, R.J. BAXTER, D.M. BRESSOUD, W.H. BURGE, P.J. FORRESTER and G.X. VIENNOT, *Partitions with prescribed hook differences*, Europ. J. Comb. **8** (1987), 341–350.

[6] G.E. ANDREWS, R.J. BAXTER and P.J. FORRESTER, *Eight-vertex SOS model and generalized Rogers-Ramanujan-type identities*, J. Stat. Phys. **35** (1984), 193–266.

[7] G.E. ANDREWS, A. SCHILLING and S.O. WARNAAR, *An A_2 Bailey lemma and Rogers-Ramanujan-type identities*, J. Amer. Math. Soc. **12** (1999), 677–702.

[8] R.J. BAXTER, *Exactly Solved Models in Statistical Mechanics*, Academic Press, London, 1982.

[9] A.A. BELAVIN, A.M. POLYAKOV and A.B. ZAMOLODCHIKOV, *Infinite conformal symmetry in two-dimensional quantum field theory*, Nucl. Phys. **B241** (1984), 333–380.

[10] A. BERKOVICH, *Fermionic counting of RSOS-states and Virasoro character formulas for the unitary minimal series $M(\nu, \nu+1)$. Exact results*, Nucl. Phys. **B431** (1994), 315–348.

[11] A. BERKOVICH and B.M. MCCOY, *Continued fractions and fermionic representations for characters of $M(p,p')$ minimal models*, Lett. Math. Phys. **37** (1996), 49–66.

[12] A. BERKOVICH, B.M. MCCOY and A. SCHILLING, *Rogers-Schur-Ramanujan type identities for the $M(p,p')$ minimal models of conformal field theory*, Commun. Math. Phys. **191** (1998), 325–395.

[13] D.M. BRESSOUD, *Lattice paths and the Rogers-Ramanujan identities*, in Proceedings of the international Ramanujan centenary conference, Madras, 1987, ed. K. Alladi. Lecture Notes in Mathematics **1395**, Springer, 1989.

[14] A.G. BYTSKO and A. FRING, *Factorized combinations of Virasoro characters*, Commun. Math. Phys. **209** (2000), 179–205.

[15] P. CHRISTE, *Factorized characters and form factors of descendant operators in perturbed conformal systems*, Int. J. Mod. Phys. **A 6** (1991), 5271–5286.

[16] S. DASMAHAPATRA, *On the combinatorics of row and corner transfer matrices of the $A^{(1)}_{n-1}$ restricted face models*, Int. J. Mod. Phys. **A 12** (1997), 3551–3586.

[17] S. DASMAHAPATRA and O. FODA, *Strings, paths and standard tableaux*, Int. J. Mod. Phys. **A 13** (1998), 501–522.

[18] PH. DI FRANCESCO, P. MATHIEU and D. SENECHAL, *Conformal Field Theory*, Springer, New York, 1996.

[19] B.L. FEIGIN and D.B. FUCHS, Verma modules over the Virasoro algebra, Funct. Anal. Appl. **17** (1983), 241–242.

[20] G. FELDER, *BRST approach to minimal models*, Nucl. Phys. **B317** (1989), 215–236.

[21] O. FODA, K.S.M. LEE, Y. PUGAI and T.A. WELSH, *Path generating transforms*, Contemp. Math. **254** (2000), 157–186.

[22] O. FODA, M. OKADO and S.O. WARNAAR, *A proof of polynomial identities of type $\widehat{sl(n)}_1 \otimes \widehat{sl(n)}_1/\widehat{sl(n)}_2$*, J. Math. Phys. **37** (1996), 965–986.

[23] O. FODA and Y-H. QUANO, Virasoro character identities from the Andrews-Bailey construction, Int. J. Mod. Phys. **A 12** (1997), 1651–1675.

[24] O. FODA and T.A. WELSH, *Melzer's identities revisited*, Contemp. Math. **248** (1999), 207–234.

[25] ——— *On the combinatorics of Forrester-Baxter models*, in "Physical Combinatorics", Kyoto 1999, eds. M. Kashiwara and T. Miwa. Progr. Math. **191** (2000), 49–103.

[26] P.J. FORRESTER and R.J. BAXTER, *Further exact solutions of the eight-vertex SOS model and generalizations of the Rogers-Ramanujan identities*, J. Stat. Phys. **38** (1985), 435–472.

[27] G. GASPER and M. RAHMAN, *Basic Hypergeometric Series*, Encyclopedia of Mathematics and its Applications **35**, Cambridge University Press, 1990.

[28] B. GORDON, *A combinatorial generalization of the Rogers-Ramanujan identities*, Amer. J. Math. **83** (1961), 393–399.

[29] M. JIMBO, T. MIWA and M. OKADO, *Local state probabilities of solvable lattice models: an $A_{n-1}^{(1)}$ family*, Nucl. Phys. **B300** (1988), 74–108.

[30] R. KEDEM, T.R. KLASSEN, B.M. MCCOY and E. MELZER, *Fermionic sum representations for conformal field theory characters*, Phys. Lett. **B307** (1993), 68–76.

[31] R. KEDEM, B.M. MCCOY and E. MELZER, *The sums of Rogers, Schur and Ramanujan and the Bose-Fermi correspondence in 1+1-dimensional quantum field theory*, in "Recent Progress in Statistical Mechanics and Quantum Field Theory", Los Angeles 1994, ed. P. Bouwknegt et al., World Scientific, Singapore (1995), 195-219.

[32] A. KUNIBA, T. NAKANISHI and J. SUZUKI, *Ferro- and antiferro-magnetizations in RSOS models*, Nucl. Phys. **B356** (1991), 750–774.

[33] P.A. MACMAHON, *Combinatory Analysis*, Volume 2, Cambridge University Press, 1916.

[34] E. MELZER, *Fermionic character sums and the corner transfer matrix*, Int. J. Mod. Phys. **A 9** (1994), 1115–1136.

[35] T. NAKANISHI, *Non-unitary minimal models and RSOS models*, Nucl. Phys. **B334** (1990), 745–766.

[36] A. ROCHA-CARIDI, *Vacuum vector representations of the Virasoro algebra*, in "Vertex Operators in Mathematics and Physics", eds. J. Lepowsky, S. Mandelstam and I.M. Singer, Springer, 1985.

[37] L.J. ROGERS, *Second memoir on the expansion of certain infinite products*, Proc. London Math. Soc. **25** (1894), 318–343.

[38] L.J. ROGERS and S. RAMANUJAN, *Proof of certain identities in combinatory analysis*, Proc. Cambridge Philos. Soc. **19** (1919), 211–216.

[39] A. SCHILLING, *Polynomial fermionic forms for the branching functions of the rational coset conformal field theories $\widehat{su}(2)_M \times \widehat{su}(2)_N/\widehat{su}(2)_{M+N}$*, Nucl. Phys. **B459** (1996), 393–436.

[40] A. SCHILLING and M. SHIMOZONO, *Fermionic formulas for level-restricted generalized Kostka polynomials and coset branching functions*, Commun. Math. Phys. **220** (2001), 105–164.

[41] L.J. SLATER, *Further identities of the Rogers-Ramanujan type*, Proc. London Math. Soc. (2) **54** (1952), 147–167.

[42] M. TAKAHASHI and M. SUZUKI, *One-dimensional anisotropic Heisenberg model at finite temperatures*, Prog. Theor. Phys. **48** (1972), 2187–2209.

[43] S.O. WARNAAR, *Fermionic solution of the Andrews-Baxter-Forrester model I. Unification of TBA and CTM methods*, J. Stat. Phys. **82** (1996), 657–685.

[44] ——— *Fermionic solution of the Andrews-Baxter-Forrester model II. Proof of Melzer's polynomial identities*, J. Stat. Phys. **84** (1997), 49–83.

[45] ——— *The Bailey lemma and Kostka polynomials*, J. Alg. Combin. **20** (2004), 131–171.

[46] S.O. WARNAAR and P.A. PEARCE, *A-D-E polynomial identities and Rogers-Ramanujan identities*, Int. J. Mod. Phys. **A 11** (1996), 291–311.

Editorial Information

To be published in the *Memoirs*, a paper must be correct, new, nontrivial, and significant. Further, it must be well written and of interest to a substantial number of mathematicians. Piecemeal results, such as an inconclusive step toward an unproved major theorem or a minor variation on a known result, are in general not acceptable for publication. Papers appearing in *Memoirs* are generally longer than those appearing in *Transactions*, which shares the same editorial committee.

As of January 31, 2005, the backlog for this journal was approximately 5 volumes. This estimate is the result of dividing the number of manuscripts for this journal in the Providence office that have not yet gone to the printer on the above date by the average number of monographs per volume over the previous twelve months, reduced by the number of volumes published in four months (the time necessary for preparing a volume for the printer). (There are 6 volumes per year, each containing at least 4 numbers.)

A Consent to Publish and Copyright Agreement is required before a paper will be published in the *Memoirs*. After a paper is accepted for publication, the Providence office will send a Consent to Publish and Copyright Agreement to all authors of the paper. By submitting a paper to the *Memoirs*, authors certify that the results have not been submitted to nor are they under consideration for publication by another journal, conference proceedings, or similar publication.

Information for Authors

Memoirs are printed from camera copy fully prepared by the author. This means that the finished book will look exactly like the copy submitted.

The paper must contain a *descriptive title* and an *abstract* that summarizes the article in language suitable for workers in the general field (algebra, analysis, etc.). The *descriptive title* should be short, but informative; useless or vague phrases such as "some remarks about" or "concerning" should be avoided. The *abstract* should be at least one complete sentence, and at most 300 words. Included with the footnotes to the paper should be the 2000 *Mathematics Subject Classification* representing the primary and secondary subjects of the article. The classifications are accessible from `www.ams.org/msc/`. The list of classifications is also available in print starting with the 1999 annual index of *Mathematical Reviews*. The Mathematics Subject Classification footnote may be followed by a list of *key words and phrases* describing the subject matter of the article and taken from it. Journal abbreviations used in bibliographies are listed in the latest *Mathematical Reviews* annual index. The series abbreviations are also accessible from `www.ams.org/publications/`. To help in preparing and verifying references, the AMS offers MR Lookup, a Reference Tool for Linking, at `www.ams.org/mrlookup/`. When the manuscript is submitted, authors should supply the editor with electronic addresses if available. These will be printed after the postal address at the end of the article.

Electronically prepared manuscripts. The AMS encourages electronically prepared manuscripts, with a strong preference for $\mathcal{A}\mathcal{M}\mathcal{S}$-LaTeX. To this end, the Society has prepared $\mathcal{A}\mathcal{M}\mathcal{S}$-LaTeX author packages for each AMS publication. Author packages include instructions for preparing electronic manuscripts, the *AMS Author Handbook*, samples, and a style file that generates the particular design specifications of that publication series. Though $\mathcal{A}\mathcal{M}\mathcal{S}$-LaTeX is the highly preferred format of TeX, author packages are also available in $\mathcal{A}\mathcal{M}\mathcal{S}$-TeX.

Authors may retrieve an author package from e-MATH starting from www.ams.org/tex/ or via FTP to ftp.ams.org (login as anonymous, enter username as password, and type cd pub/author-info). The *AMS Author Handbook* and the *Instruction Manual* are available in PDF format following the author packages link from www.ams.org/tex/. The author package can be obtained free of charge by sending email to pub@ams.org (Internet) or from the Publication Division, American Mathematical Society, 201 Charles St., Providence, RI 02904, USA. When requesting an author package, please specify $\mathcal{A}\mathcal{M}\mathcal{S}$-LaTeX or $\mathcal{A}\mathcal{M}\mathcal{S}$-TeX, Macintosh or IBM (3.5) format, and the publication in which your paper will appear. Please be sure to include your complete mailing address.

Sending electronic files. After acceptance, the source file(s) should be sent to the Providence office (this includes any TeX source file, any graphics files, and the DVI or PostScript file).

Before sending the source file, be sure you have proofread your paper carefully. The files you send must be the EXACT files used to generate the proof copy that was accepted for publication. For all publications, authors are required to send a printed copy of their paper, which exactly matches the copy approved for publication, along with any graphics that will appear in the paper.

TeX files may be submitted by email, FTP, or on diskette. The DVI file(s) and PostScript files should be submitted only by FTP or on diskette unless they are encoded properly to submit through email. (DVI files are binary and PostScript files tend to be very large.)

Electronically prepared manuscripts can be sent via email to pub-submit@ams.org (Internet). The subject line of the message should include the publication code to identify it as a Memoir. TeX source files, DVI files, and PostScript files can be transferred over the Internet by FTP to the Internet node e-math.ams.org (130.44.1.100).

Electronic graphics. Comprehensive instructions on preparing graphics are available at www.ams.org/jourhtml/graphics.html. A few of the major requirements are given here.

Submit files for graphics as EPS (Encapsulated PostScript) files. This includes graphics originated via a graphics application as well as scanned photographs or other computer-generated images. If this is not possible, TIFF files are acceptable as long as they can be opened in Adobe Photoshop or Illustrator. No matter what method was used to produce the graphic, it is necessary to provide a paper copy to the AMS.

Authors using graphics packages for the creation of electronic art should also avoid the use of any lines thinner than 0.5 points in width. Many graphics packages allow the user to specify a "hairline" for a very thin line. Hairlines often look acceptable when proofed on a typical laser printer. However, when produced on a high-resolution laser imagesetter, hairlines become nearly invisible and will be lost entirely in the final printing process.

Screens should be set to values between 15% and 85%. Screens which fall outside of this range are too light or too dark to print correctly. Variations of screens within a graphic should be no less than 10%.

Inquiries. Any inquiries concerning a paper that has been accepted for publication should be sent directly to the Electronic Prepress Department, American Mathematical Society, 201 Charles St., Providence, RI 02904, USA.

Editors

This journal is designed particularly for long research papers, normally at least 80 pages in length, and groups of cognate papers in pure and applied mathematics. Papers intended for publication in the *Memoirs* should be addressed to one of the following editors. In principle the Memoirs welcomes electronic submissions, and some of the editors, those whose names appear below with an asterisk (*), have indicated that they prefer them. However, editors reserve the right to request hard copies after papers have been submitted electronically. Authors are advised to make preliminary email inquiries to editors about whether they are likely to be able to handle submissions in a particular electronic form.

Titles in This Series

828 **Joel Berman and Paweł M. Idziak,** Generative complexity in algebra, 2005

827 **Trevor A. Welsh,** Fermionic expressions for minimal model Virasoro characters, 2005

826 **Guy Métivier and Kevin Zumbrun,** Large viscous boundary layers for noncharacteristic nonlinear hyperbolic problems, 2005

825 **Yaozhong Hu,** Integral transformations and anticipative calculus for fractional Brownian motions, 2005

824 **Luen-Chau Li and Serge Parmentier,** On dynamical Poisson groupoids I, 2005

823 **Claus Mokler,** An analogue of a reductive algebraic monoid whose unit group is a Kac-Moody group, 2005

822 **Stefano Pigola, Marco Rigoli, and Alberto G. Setti,** Maximum principles on Riemannian manifolds and applications, 2005

821 **Nicole Bopp and Hubert Rubenthaler,** Local zeta functions attached to the minimal spherical series for a class of symmetric spaces, 2005

820 **Vadim A. Kaimanovich and Mikhail Lyubich,** Conformal and harmonic measures on laminations associated with rational maps, 2005

819 **F. Andreatta and E. Z. Goren,** Hilbert modular forms: Mod p and p-adic aspects, 2005

818 **Tom De Medts,** An algebraic structure for Moufang quadrangles, 2005

817 **Javier Fernández de Bobadilla,** Moduli spaces of polynomials in two variables, 2005

816 **Francis Clarke,** Necessary conditions in dynamic optimization, 2005

815 **Martin Bendersky and Donald M. Davis,** V_1-periodic homotopy groups of $SO(n)$, 2004

814 **Johannes Huebschmann,** Kähler spaces, nilpotent orbits, and singular reduction, 2004

813 **Jeff Groah and Blake Temple,** Shock-wave solutions of the Einstein equations with perfect fluid sources: Existence and consistency by a locally inertial Glimm scheme, 2004

812 **Richard D. Canary and Darryl McCullough,** Homotopy equivalences of 3-manifolds and deformation theory of Kleinian groups, 2004

811 **Ottmar Loos and Erhard Neher,** Locally finite root systems, 2004

810 **W. N. Everitt and L. Markus,** Infinite dimensional complex symplectic spaces, 2004

809 **J. T. Cox, D. A. Dawson, and A. Greven,** Mutually catalytic super branching random walks: Large finite systems and renormalization analysis, 2004

808 **Hagen Meltzer,** Exceptional vector bundles, tilting sheaves and tilting complexes for weighted projective lines, 2004

807 **Carlos A. Cabrelli, Christopher Heil, and Ursula M. Molter,** Self-similarity and multiwavelets in higher dimensions, 2004

806 **Spiros A. Argyros and Andreas Tolias,** Methods in the theory of hereditarily indecomposable Banach spaces, 2004

805 **Philip L. Bowers and Kenneth Stephenson,** Uniformizing dessins and Belyĭ maps via circle packing, 2004

804 **A. Yu Ol'shanskii and M. V. Sapir,** The conjugacy problem and Higman embeddings, 2004

803 **Michael Field and Matthew Nicol,** Ergodic theory of equivariant diffeomorphisms: Markov partitions and stable ergodicity, 2004

802 **Martin W. Liebeck and Gary M. Seitz,** The maximal subgroups of positive dimension in exceptional algebraic groups, 2004

801 **Fabio Ancona and Andrea Marson,** Well-posedness for general 2×2 systems of conservation law, 2004

800 **V. Poénaru and C. Tanas,** Equivariant, almost-arborescent representation of open simply-connected 3-manifolds; A finiteness result, 2004

799 **Barry Mazur and Karl Rubin,** Kolyvagin systems, 2004

For a complete list of titles in this series, visit the AMS Bookstore at **www.ams.org/bookstore/**.